Vibration Mitigation Systems in Structural Engineering

Vibration Mitigation Systems in Structural Engineering

Okyay Altay

CRC Press
Taylor & Francis Group
Boca Raton London New York

CRC Press is an imprint of the
Taylor & Francis Group, an **informa** business

First edition published 2021
by CRC Press
6000 Broken Sound Parkway NW, Suite 300, Boca Raton, FL 33487-2742

and by CRC Press
2 Park Square, Milton Park, Abingdon, Oxon, OX14 4RN

First edition published by CRC Press 2021

CRC Press is an imprint of Taylor & Francis Group, LLC

ISBN: 9781138564169 (hbk)
ISBN: 9781032038025 (pbk)
ISBN: 9781315122243 (ebk)

Typeset in Times
by KnowledgeWorks Global Ltd.

To my family

Contents

viii

Preface

Vibrations jeopardize both the integrity and the serviceability of structures and can be induced by natural events, such as wind and earthquake, and anthropogenic activities, such as traffic. Particularly modern structures built by lightweight materials and with slender architecture exhibit low damping and respond highly sensitive to vibrations. Traditional structural design has relied so far on the strength and ductility of structures. However, this approach seems to reach its limits and cannot provide sufficient protection to prevent vibrations and their aftereffects. Developments of the modern era both in economics and architectural design and rapidly changing environmental conditions require a more efficient approach. Based on recent progresses in computer science and cybernetics, modern structural control is initiating an increasing number of alternative methods, such as active and semi-active damping systems, which enable a new level of safety for vibration-engineering structures. This book provides the necessary theoretical background and a comprehensive overview of conventional vibration control methods followed by detailed insights into some of the newest developments.

The first part of the book covers the theoretical background, which is required for the implementation of general structural control methods in the context of structural engineering. In Chapter 1, the most important aspects of structural dynamics are summarized. Governing mathematical context is presented with examples. The analytical solutions are provided as MATLAB® codes. In Chapter 2, a historical development of structural control is introduced together with a general definition and classification of the so far applied devices, materials and strategies. In Chapter 3, the most important principles of structural control are highlighted including state-space representation of systems with a general overview of structural control algorithms. For the application of state-space representation, examples are provided.

The second part of the book presents numerous examples for conventional vibration damping systems, which are grouped as dissipators and tuned mass dampers. In Chapter 4, examples of dissipators, metallic, friction, viscoelastic and viscous fluid dampers are introduced. In Chapter 5, examples of tuned mass dampers, classical tuned mass dampers, pendulum tuned mass dampers, tuned liquid dampers and tuned liquid column dampers are introduced. In both chapters, the mathematical background is described and design examples are provided.

The third part of the book is concerned with some advanced newly developed damping systems. In this regard, in Chapter 6, active and semi-active damping systems are described and application examples are given. Chapter 7 presents two semi-active tuned liquid column dampers. Mathematical modeling approaches are proposed and validated by experiments. Numerical studies are conducted to explore the control capability. Chapter 8 treats shape memory alloy-based damping systems and focuses on challenges regarding the constitutive modeling of their dynamic behavior. Experimental and numerical studies are conducted including real-time hybrid simulation. Chapter 9 connects the control approaches with monitoring by providing a KALMAN filter-based system identification algorithm, which detects structural parameters via sensors for the applied damping systems.

MATLAB® is registered trademark of The MathWorks, Inc. For product information, please contact:

The MathWorks, Inc.
3 Apple Hill Drive
Natick, MA 01760-2098 USA
Tel: 508 647 7000
Fax: 508-647-7000
E-mail: info@mathworks.com
Web: www.mathworks.com

Acknowledgments

First I would like to thank my research collaborators, who contributed to the presented research content particularly in the third part of the book. They gave me the opportunity to develop the presented systems in their lab and provided me with the necessary resources. In this regard, I would like to express my sincere thanks to RWTH Aachen University, particularly to my dear professor Sven Klinkel (Chair of Structural Analysis and Dynamics) for his friendly supervision, Prof. em. Konstantin Meskouris for his mentoring as well as Prof. Dirk Abel (Chair of Automatic Control) for his support. I would like to thank also Tsinghua University in Beijing, especially to my dear research partner Prof. Jinting Wang. In particular, Dr. Sebastian Stemmler and Dr. Fei Zhu as well as the PhD students Andreas Kaup, Behnam Mehrkian, Pavle Milicevic, Simon Schleiter, Markus Zimmer and Hao Ding are researchers from both universities, who have the most important impact on this book. Your hard work, time and effort are sincerely appreciated. Furthermore, I would like to thank Dr. Michael Reiterer for his pioneering works and inspiration as well as my previous supervisors Prof. Christoph Butenweg and Dr. Philipp Renault for their motivation. My research has been financially supported primarily by the German Research Foundation (Deutsche Forschungsgemeinschaft, DFG), the German Federal Ministry of Education and Research (Bundesministerium fr Bildung und Forschung, BMBF) as well as the National Natural Science Foundation (NSFC) of China. I would like to express my sincerest gratitude.

I extend my special appreciation to Dr. Gagandeep Singh, project editor, CRC Press/Taylor & Francis, and to his team, for their trust and kind assistance.

Finally, I would like to dedicate this book to my family, particularly to my parents and to my dearest wife Dr. Margarita Chasapi. I am grateful for your sacrifices, patience and help.

List of Figures

List of Tables

List of Abbreviations

AMD: Active mass damper

ARMA: Autoregressive moving average

A-TMD: Active tuned mass damper

A-UKF: Adaptive unscented Kalman filter

AVS: Active variable stiffness system

DAQ: Data acquisition

DoF: Degree of freedom

EKF: Extended Kalman filter

EoM: Equation of motion

FDD: frequency-domain decomposition

FE: Finite element

FFT: Fast Fourier transformation

FLC: Fuzzy logic control

HMD: Hybrid mass damper

KF: Kalman filter

LCD-PA: Liquid column damper with period adjustment

LCVA: Liquid column vibration absorber

LQG: Linear quadratic Gaussian controller

LQR: Linear quadratic regulator

LSE: Least squares estimation

MAX: Maximum value

MDoF: Multi degree of freedom

MEMS: Microelectromechanical system

MF: Membership function

MPC: Model predictive control

MR: Magnetorheological

PGA: Peak ground acceleration

RAT: Response analysis task

RMS: Root mean square

SDoF: Single degree of freedom

SGT: Signal generation task

SMA: Shape memory alloy

SMC: Sequential Monte Carlo

SSI: Soil structure interaction

O-TLCD: Omnidirection tuned liquid column damper

PF: Particle filter

P-TMD: Pendulum tuned mass damper

RMS: Root mean square

RTHS: Real time hybrid simulation

SCRAMNet: Shared common RAM network

TCP/IP: Transmission control protocol / internet protocol

TLCD: Tuned liquid column damper

TLCGD: Tuned liquid column gas damper

TLD: Tuned liquid damper

TMD: Tuned mass damper

TSD: Tuned sloshing damper

UKF: Unscented Kalman filter

UT: Unscented transformation

ZOH: Zero-order hold

Part I

Fundamentals

1

Theory of Structural Vibration

1.1 Introduction

The design of structural control systems requires a deep knowledge of structural dynamics, particularly regarding the theory of structural vibrations. This chapter summarizes the most important mathematical fundamentals required for the analysis of structural vibrations.

Sections 1.2 and 1.3 are concerned with the dynamic responses of single-degree-of-freedom (SDoF) systems and classify responses considering the damping characteristics of systems. Furthermore, the initiation process of the vibrations is distinguished as free and forced. Sections 1.4 and 1.5 introduce frequency- and time-domain-based methods for the calculation of such vibrations. Section 1.6 enhances the fundamentals, which are introduced for SDoF systems, to multi-degree-of-freedom (MDoF) systems. Section 1.7 describes the basic theory of modal analysis for the computation of these systems. Section 1.8 gives a general overview of damping models, which are relevant for the mathematical description of the material behavior of vibration control devices. Finally, Section 1.9 covers the topic of nonlinear vibrations, which can occur particularly in controlled structures.

The chapter aims to give a general overview of the theory. For more information, particularly in the general framework of structural dynamics and earthquake engineering, the interested readers are referred to reference books, such as Chopra [2], Clough and Penzien [3], Humar [4], Inman and Singh [5], Meskouris [6] as well as de Silva [7].

1.2 Vibrations without Damping

This section is the response of SDoF systems without damping. The *degree-of-freedom* (DoF) of a system is defined as the number of independent vectors required to define the motion characteristics of a system. The general dynamic response of most of engineering structures can be approximated by using SDoF systems. Typical examples are high-rise structures, such as buildings, towers and wind turbines, as well as single-span bridges as shown in Figure 1.1 (a)

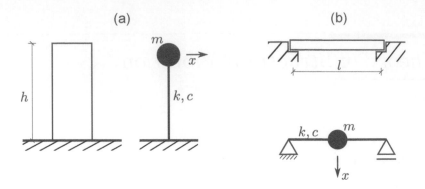

FIGURE 1.1
Examples to SDoF systems: High-rise structure, such as a building or a tower (a) and single-span structure, such as a bridge (b).

and (b), respectively. The single DoF of the systems is represented in figure by x. The dynamic behavior of these systems can be described by their mass m, stiffness k and damping coefficient c.

1.2.1   Free Vibrations without Damping

Structures respond to an initial displacement and/or an initial velocity by free vibrations without requiring any further external dynamic force. For the SDoF system shown in Figure 1.2 (a), the initial displacement and velocity are represented by x_o and $\dot{x}_0$, respectively. These correspond to the vibration direction and, accordingly, the only possible DoF x of the system.

The characteristics of the vibration response depend beside the amplitudes of the initial perturbation also on the structural parameters: mass m and stiffness k. A further parameter is the damping, which is in the context of structural vibration control particularly important. In the shown SDoF example, a linear damping behavior is assumed, which is represented by the damping coefficient c.

Figure 1.2 (b) depicts an alternative representation of the SDoF system by a mass, which is attached to spring and damping elements.

This section is concerned with the free vibration of SDoF systems without damping. The effects of damping on the free vibration response will be introduced in Section 1.3.1. An SDoF system without damping, i.e., $c = 0$, is shown in Figure 1.3. The equation of motion (EoM) of the system can be determined from the NEWTON's second law of motion

$$F = m\ddot{x}, \tag{1.1}$$

where F represents external forces applied to the system and $\ddot{x}$ the acceleration

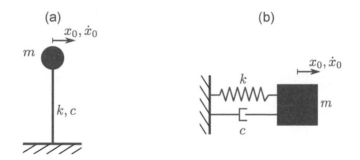

FIGURE 1.2
SDoF system disturbed from its static equilibrium position by the initial displacement x_0 and/or the initial velocity $\dot{x}_0$ (a). Alternative representation of the SDoF system using a mass-spring-damping element (b).

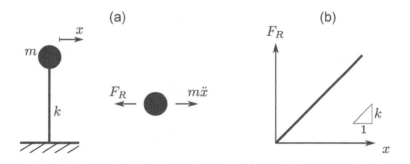

FIGURE 1.3
Forces acting on an SDoF system without damping (a). F_R: Internal restoring force. Linear elastic deformation curve of the system (b).

response. The forces acting on the system are shown in Figure 1.3 (a). In case of a free vibration, the external force is zero, $F = 0$. The only force acting on the mass is the restoring force F_R caused by the stiffness of the system. The direction of this force is opposite to the motion direction. Therefore, we consider this force in NEWTON's second law of motion with a negative sign, which yields

$$-F_R = m\ddot{x}. \tag{1.2}$$

As shown in Figure 1.3 (a), if we assume that the structure is performing a linear elastic behavior, we can introduce the stiffness k in Equation 1.2 by $F_R = kx$ and obtain the general EoM of a free vibrating SDoF system as

$$m\ddot{x} + kx = 0 \tag{1.3}$$

$$\Leftrightarrow \ddot{x} + \frac{k}{m}x = 0, \tag{1.4}$$

which is a linear and homogeneous differential equation with constant coefficients. To solve this equation, we substitute an exponential function as

$$x = e^{\lambda t}, \quad \dot{x} = \lambda e^{\lambda t}, \quad \ddot{x} = \lambda^2 e^{\lambda t}. \tag{1.5}$$

The EoM can then be simplified by introducing $\omega_1 = \sqrt{k/m}$ as

$$\lambda^2 + \omega_1^2 = 0. \tag{1.6}$$

The solution is

$$\lambda_{1,2} = \pm\sqrt{-\omega_1^2} = \pm i\omega_1 \tag{1.7}$$

$$\Leftrightarrow x_1 = e^{i\omega_1 t}, \quad x_2 = e^{-i\omega_1 t}, \tag{1.8}$$

where $i = \sqrt{-1}$ is the imaginary number. Accordingly, the general solution can be written using two yet undetermined constants A and B as

$$x = A e^{i\omega_1 t} + B e^{-i\omega_1 t}. \tag{1.9}$$

The real solution is obtained by using DE MOIVRE's theorem with

$$\cos(\omega_1 t) = \frac{e^{i\omega_1 t} + e^{-i\omega_1 t}}{2}, \quad \sin(\omega_1 t) = \frac{e^{i\omega_1 t} - e^{-i\omega_1 t}}{2i}, \tag{1.10}$$

which yields

$$x = C_1 \cos(\omega_1 t) + C_2 \sin(\omega_1 t). \tag{1.11}$$

The constants C_1 and C_2 can be determined from the initial conditions x_0 and $\dot{x}_0$ as

$$C_1 = x_0, \quad C_2 = \frac{\dot{x}_0}{\omega_1}. \tag{1.12}$$

The time histories of an SDoF system without damping are shown in Figure 1.4. Here, the free vibration is caused by the initial conditions x_0 and/or $\dot{x}_0$. The maximum ordinate values are referred to as *amplitude* and shown in the plot as $\pm a_i$ with $i = a, b, c$ corresponding to the presented cases a, b, c. From the responses, we observe that the amplitude remains constant for the free vibration without damping.

Furthermore, the vibration repeats itself at every $2\pi/\omega_1$ seconds, where ω_1 is the same variable we used in Equation 1.4 to substitute $\sqrt{k/m}$. This variable is characteristic for the dynamic behavior of systems and referred to as the *natural (angular) frequency* or the *natural (circular) frequency* of vibration. It has the dimension [rad s^{-1}]. The duration of one oscillation is referred to as *natural period* of vibration, which has the dimension [s] and calculated as

$$T_1 = \frac{2\pi}{\omega_1}. \tag{1.13}$$

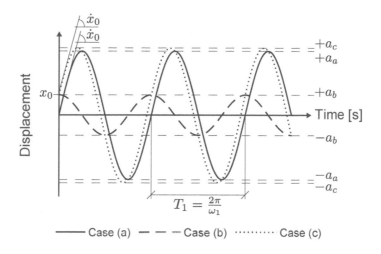

FIGURE 1.4
Time histories of an SDoF system without damping. Free vibrations are caused by the initial displacement x_o (Case a), the initial velocity $\dot{x}_0$ (Case b) and simultaneous x_0 and $\dot{x}_0$ (Case c).

The number of oscillations per second is calculated from the reciprocal value of T_1 as

$$f_1 = \frac{1}{T_1} = \frac{\omega_1}{2\pi} \tag{1.14}$$

and referred to as the *natural (cyclic) frequency* of vibration. f_1 has the dimension $[s^{-1}]$ or a further SI unit, [Hz], which is named after the physicist Heinrich Rudolf Hertz.

It can be seen from the plotted curves that all free vibration responses of the SDoF system are sinusoidal. Accordingly, these time histories can be mathematically described by using the harmonic functions sin or cos. Therefore, free vibrations are also referred to as *harmonic vibrations* .

In Equations 1.15–1.17, corresponding displacement $x(t)$, velocity $\dot{x}(t)$ and acceleration $\ddot{x}(t)$ functions are given for an undamped SDoF, which is performing a free vibration due to an initial displacement of x_0. The equations show on the right side also the corresponding maximum amplitudes of the motion.

$$x(t) = x_0 \cos(\omega_1 t) \quad \rightarrow \quad x_{max} = x_0 \tag{1.15}$$
$$\dot{x}(t) = -x_0\omega_1 \sin(\omega_1 t) \quad \rightarrow \quad \dot{x}_{max} = x_0\omega_1 \tag{1.16}$$
$$\ddot{x}(t) = -x_0\omega_1^2 \cos(\omega_1 t) \quad \rightarrow \quad \ddot{x}_{max} = x_0\omega_1^2 \tag{1.17}$$

As the system is undamped, it cannot dissipate energy. Accordingly, its kinetic and potential energies must be equal:

$$E_k = E_p. \tag{1.18}$$

For the calculation of the kinetic energy E_k, we need the velocity, which has the maximum value $x_0\omega_1$ according to Equation 1.16. Assuming that the SDoF system has a linear elastic deformation behavior, we can calculate its potential energy from the area below the force-deformation curve, as shown in Figure 1.3 (b). Accordingly, the energy equation yields

$$\frac{1}{2}m(x_0\omega_1)^2 = \frac{1}{2}x_0kx_0.$$

(1.19)

By modifying this equation, we obtain the natural frequency of the SDoF system as

$$\Leftrightarrow \omega_1 = \sqrt{\frac{k}{m}},$$

(1.20)

as already introduced in Equation 1.4. The derivation method can be conducted in a similar manner also using other free vibration responses, which are induced by an initial velocity or both by an initial displacement and a velocity.

1.2.2   Forced Vibrations without Damping

Dynamic excitation forces can be induced both by natural events, such as wind and earthquake, and by anthropogenic activities, such as traffic. These time-dependent forces cause on structures apart from static displacements also oscillations about their static equilibrium position. Time-dependent harmonic, periodic, step, pulse and other arbitrary functions can be used to model these forces. The response of an SDoF system without damping to a dynamic excitation force $F(t)$ is determined from the EoM

$$\ddot{x} + \frac{k}{m}x = F(t),$$

(1.21)

where t represents the time variable. Compared to the free vibration case without any external forces (cf. Figure 1.3), this time the time-dependent excitation force $F(t)$ force is applied to the system. Therefore, we add this force to the right side of EoM in Equation 1.4. Similar to Equation 1.20, the natural circular frequency ω_1 is again introduced and simplifies the EoM as

$$\ddot{x} + \omega_1^2 x = f(t),$$

(1.22)

where $f(t) = F(t)/m$ is the *mass normalized time-dependent force function*. Equation 1.22 is a linear and, due to the non-zero force function $f(t)$, a non-homogeneous differential equation. Accordingly, the general solution of the EoM consists a homogeneous x_h and a particular part x_p as

$$x = x_h + x_p,$$

(1.23)

where the homogeneous solution corresponds to the solution of the EoM for free vibrations and can be obtained from Equation 1.11.

The particular solution can be determined using the DUHAMEL integral. As shown in Figure 1.5, this integral interprets the applied force as a sequence of infinitesimally short impulses and sums their corresponding responses to obtain the total response of the system. In the figure, $F(t)$ is the time-dependent excitation force, τ a time instant during the excitation, $d\tau$ the infinitesimal time increment and dI the infinitesimal short impulse occurred during τ corresponding to the area below force-time curve. The infinitesimal short impulse depends on mass and velocity. It is calculated for an SDoF system as

$$dI = F(\tau)\,d\tau = m\dot{x}(\tau), \tag{1.24}$$

from which the velocity is obtained as

$$\dot{x}(\tau) = \frac{F(\tau)\,d\tau}{m}. \tag{1.25}$$

The response of SDoF systems unit impulses is equivalent to free vibrations, which occur after an initial velocity $\dot{x}(\tau) \neq 0$ without initial displacement $x(\tau) = 0$. The general solution of such a response can be calculated using the homogeneous differential equation according to Equation 1.11 with $C_1 = 0$ and $C_2 = \dot{x}(\tau)/\omega_1$ as

$$x(t) = \frac{\dot{x}(\tau)}{\omega_1} \sin\left(\omega_1(t - \tau)\right) = \frac{F(\tau)}{\omega_1 m} \sin\left(\omega_1(t - \tau)\right) \tag{1.26}$$

after substituting Equation 1.25 for the initial velocity $\dot{x}(\tau)$. After replacing $F(\tau)/m$ with $f(\tau)$ and integrating over all differential impulses dI, we obtain the particular solution as

$$x_p = \frac{1}{\omega_1} \int_0^t f(\tau) \sin\left(\omega_1(t - \tau)\right) d\tau. \tag{1.27}$$

Systems under dynamic forces exhibit two distinct vibration types. These are

a. *transient vibrations*, which depend on the initial conditions x_0 and/or $\dot{x}_0$,

b. *steady-state vibrations*, which include only vibrations caused by the applied force $F(t)$ and are independent from the initial conditions.

In case of a harmonic excitation with the excitation force

$$F(t) = F_0 \sin(\omega t), \tag{1.28}$$

and the excitation frequency $\omega \neq \omega_1$, the particular solution of the EoM reads

$$x_p = C_3 \sin(\omega t), \tag{1.29}$$

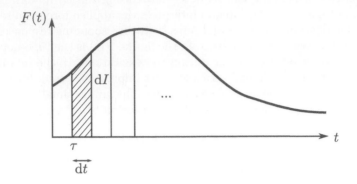

FIGURE 1.5
Time-dependent excitation force $F(t)$ represented by infinitesimal short impulses, the responses to which are calculated according to the DUHAMEL integral approach.

where C_3 is the amplitude of the response, which can be calculated by

$$C_3 = \frac{F_0}{k} \frac{1}{1 - \eta^2}. \tag{1.30}$$

where, $\eta = \omega/\omega_1$ denotes the frequency ratio. The case $\omega = \omega_1$ is referred to as the *resonance state*. At this state, the vibration of the SDoF system increases continuously and tends to infinity. Furthermore, it is noteworthy that

$$R_d = \frac{x_{dy}}{x_{st}} = \frac{1}{|1 - \eta^2|}, \tag{1.31}$$

represents the ratio of the maximum dynamic amplitude to the static displacement ($x_{st} = F_0/k$) and referred to as the *deformation response factor* of the undamped SDoF system.

    Considering both the homogeneous and the particular parts, the general solution of the EoM of a harmonically excited undamped SDoF can be determined from

$$x = C_1 \cos(\omega_1 t) + C_2 \sin(\omega_1 t) + C_3 \sin(\omega t). \tag{1.32}$$

Accordingly, we obtain the velocity of the system as

$$\dot{x} = -\omega_1 C_1 \sin(\omega_1 t) + \omega_1 C_2 \cos(\omega_1 t) + \omega C_3 \cos(\omega t), \tag{1.33}$$

where the constants C_1 and C_2 are determined from the initial conditions at $t = 0$ as

$$C_1 = x_0, \quad C_2 = \frac{\dot{x}_0}{\omega_1} - \frac{F_0}{k} \frac{\eta}{1 - \eta^2}. \tag{1.34}$$

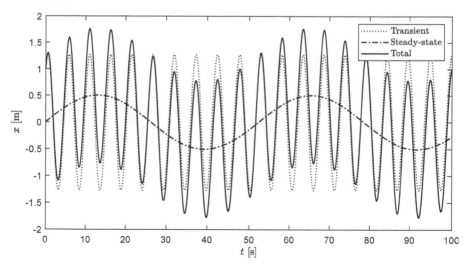

FIGURE 1.6

Transient, steady-state and total response of an SDoF system to a simultaneous excitation by the initial displacement x_0, the initial velocity $\dot{x}_0$ and the harmonic force $F = F_0 \sin(\omega t)$ with $\omega \neq \omega_1$.

Finally, the solution of the EoM reads

$$x = x_0 \cos(\omega_1 t) + \left(\frac{\dot{x}_0}{\omega_1} - \frac{F_0}{k} \frac{\eta}{1 - \eta^2} \right) \sin(\omega_1 t) \tag{1.35}$$

$$+ \frac{F_0}{k} \frac{1}{1 - \eta^2} \sin(\omega t), \tag{1.36}$$

where the first row represents the transient response and the second row the steady state response of the system.

Example 1.1 *A structure is subjected to a simultaneous excitation by the initial displacement $x_0 = 1$ m, initial velocity $\dot{x}_0 = 1$ m s^{-1} and the harmonic force $F(t) = 0.5 \sin(0.12t)$ kN. The natural frequency and the stiffness of the structure are given as omega$_1$ = 1.2 rad s^{-1} and $k = 1$ kN m^{-1} respectively. The transient, steady-state and total displacement responses are depicted in Figure 1.6. In Section A.1, a MATLAB code is provided for the calculation of the shown responses.*

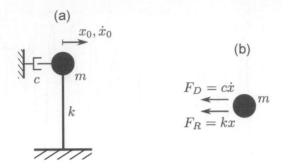

FIGURE 1.7

A free vibrating SDoF system with damping (a) and the forces acting on its mass (b). F_R: Internal restoring force, F_D: Internal damping force.

1.3 Vibrations with Damping

The term *damping* refers to the process, by which the vibration energy of a structure is converted to heat energy and diminished by various mechanisms such as

a. external friction at connections, cracks and between structural and nonstructural elements,

b. internal friction due to material deformation or phase transformation.

1.3.1 Free Vibrations with Damping

As shown in Figure 1.7, a dashpot can be used to represent damping mechanisms of a structure in an idealized manner. The internal damping force can be calculated linearly from the vibration velocity by using the *viscous damping coefficient c* . Accordingly, the EoM of a free vibrating SDoF system with damping reads

$$m\ddot{x} + c\dot{x} + kx = 0, \tag{1.37}$$

which can be rewritten by introducing the *damping ratio D* as

$$\ddot{x} + 2D\omega_1\dot{x} + \omega_1 x = 0. \tag{1.38}$$

The damping ratio D is a dimensionless measure of damping, which compares the viscous damping coefficient c of a structure with its *critical damping coefficient* c_{cr}. The critical damping coefficient is the smallest viscous damping coefficient, at which no vibration occurs. The damping ratio is calculated as

$$D = \frac{c}{c_{cr}} = \frac{c}{2m\omega_1}. \tag{1.39}$$

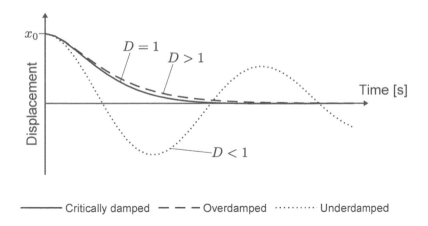

FIGURE 1.8
Vibration responses of critically damped, overdamped and underdamped
SDoF systems to an initial displacement x_0.

Depending on the value of the damping ratio D the damping behavior of
systems is classified as

   a.   *underdamped* $(c < c_{cr} \Leftrightarrow D < 1)$,

   b.   *critically damped* $(c = c_{cr} \Leftrightarrow D = 1)$,

   c.   *overdamped* $(c > c_{cr} \Leftrightarrow D > 1)$.

Figure 1.8 compares the vibration responses of underdamped, critically
damped and overdamped SDoF systems after an initial displacement. In case
of a free vibration, underdamped systems vibrate about their equilibrium position with an amplitude decay. Civil engineering structures and most of mechanical engineering systems are typically underdamped systems. Steel structures, e.g., wind turbines, exhibit usually a low inherent damping. Accordingly, their damping ratio is usually around $D = 1$ %. The damping ratio of
reinforced concrete structures is around $D = 2 - 3$ %. Due to cracking, the
damping ratio may reach 5 %. Wood structures can dissipate high amount of
energy especially at nailed or bolted joints and have a damping ratio usually
above 5 %.

As illustrated in Figure 1.8, both critically and overdamped systems return
to their equilibrium position without any vibration. Critically damped systems
return faster to the equilibrium.

The differential equation of the free vibration of an SDoF system with
damping reads

$$\ddot{x} + \frac{c}{m}\dot{x} + \frac{k}{m}x = 0 \qquad (1.40)$$

and can be solved using the exponential functions $x = e^{\lambda t}$, $\dot{x} = \lambda e^{\lambda t}$ and $\ddot{x} = \lambda^2 e^{\lambda t}$, which yields

$$\lambda^2 + \frac{c}{m}\lambda + \omega_1^2. \tag{1.41}$$

The solution of this equation is given by

$$\lambda_{1,2} = -\frac{c}{2m} \pm \sqrt{\left(\frac{c}{2m}\right)^2 - \omega_1^2}. \tag{1.42}$$

Depending on the radical of the equation, the characteristics of the solution changes. For critically damped systems with $c = c_{cr} = 2m\omega_1$, the radical becomes 0. For underdamped structures, the radical is less than 0 and the solution values $\lambda_{1,2}$ are complex numbers. For overdamped structures, the radical is more than 0 and the solution values $\lambda_{1,2}$ are real numbers.

The general solution for free oscillating underdamped SDoF systems ($c < c_r$ or $D < 1$) with the initial conditions x_0 and/or $\dot{x}_0$ equals to

$$x = e^{-D\omega_1 t}\left(x_0 \cos(\omega_D t) + \frac{\dot{x}_0 + D\omega_1 x_0}{\omega_D}\sin(\omega_D t)\right). \tag{1.43}$$

where ω_D is referred to as the *damped (angular) frequency of vibration* and reads

$$\omega_D = \omega_1\sqrt{1 - D^2}. \tag{1.44}$$

For underdamped structures, the damped angular frequency is almost equal to the natural angular frequency. For instance, even for $D = 0.10\,\%$ the damped circular frequency is very close to the natural frequency and reads

$$\omega_D = 0.995\,\omega_1. \tag{1.45}$$

The responses of SDoF systems to an initial displacement and an velocity are shown in Figure 1.9 for three cases: without damping, with damping D_1 and with a higher damping $D_2 > D_1$. The system without damping is vibrating with a constant amplitude, while the vibration amplitude of remaining systems with damping is decreasing. The amplitude decay of the system with higher damping D_2 is more significant than the system with less damping D_1.

From the amplitude decay of the response curves, the damping ratio of systems can be calculated by using the *logarithmic decrement* Λ. The time histories of structural responses, such as displacement, velocity or acceleration can be used for this purpose. First, the natural logarithm of the ratio of the amplitudes of two successive peaks p_i and p_{i+n} is computed as

$$\Lambda = \frac{1}{n}\ln\frac{p_i}{p_{i+n}}, \tag{1.46}$$

where n is the number of oscillations between both peaks. The damping ratio can then be approximated by

$$D \approx \frac{\Lambda}{2\pi}. \tag{1.47}$$

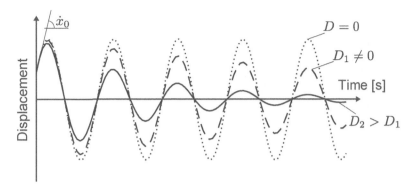

FIGURE 1.9
Response of SDoF systems with different damping characteristics to an initial displacement x_0 and an initial velocity $\dot{x}_0$.

This assumption is accurate for civil engineering and most mechanical engineering structures, which are normally underdamped with

$$\sqrt{1 - D^2} \approx 1. \tag{1.48}$$

Example 1.2 *An industrial chimney is shown in Figure 1.10. The damped structure is free oscillating due to the initial conditions x_0 and $\dot{x}_0$ applied on top of the structure. The same figure shows also the measured displacement response of the structure.*

To determine the damping ratio D of the structure, we apply the logarithmic decrement method as

$$\lambda = \frac{1}{n} \ln \frac{p_i}{p_{i+n}} = \frac{1}{3} \ln \frac{0.017}{0.007} = 0.30 \tag{1.49}$$

$$D \approx \frac{\lambda}{2\pi} = 0.05 = 5 \ \% \tag{1.50}$$

by using the peak values x_1 and x_4 with the number of oscillations $n = 3$. Accordingly, the damping ratio D of the structure corresponds to 5 %.

A further method for the identification of damping parameter from measurement signals is introduced in Section 1.4. Furthermore, for the identification of system parameters, Section 9.3 introduces the KALMAN filter method. Besides linear viscous damping, further nonlinear damping models exist and some examples are introduced in Section 1.8 and applied to dissipators in Section 4.4.

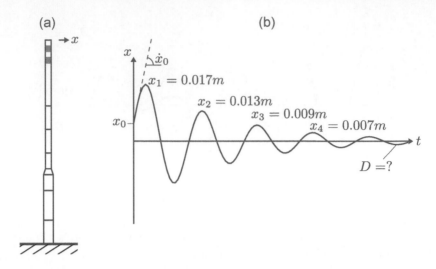

FIGURE 1.10
An industrial chimney (a) and its displacement history (b) after an initial
perturbation by the initial displacement x_0 and the initial velocity $\dot{x}_0$. The
damping ratio D can be calculated from the displacement peak values x_i using
the logarithmic decrement method.

1.3.2 Forced Vibrations with Damping

In the case of forced vibrations with damping, the general solution of the
differential EoM

$$\ddot{x} + 2D\omega_1\dot{x} + \omega_1^2 x = \frac{F(t)}{m} = f(t) \tag{1.51}$$

is again given by the sum of the homogeneous solution of the free vibration
response and the particular solution of the forced vibration response as

$$x = x_h + x_p. \tag{1.52}$$

The particular solution can be calculated using again the DUHAMEL integral:

$$x_p = \frac{1}{\omega_D} \int_0^t f(\tau) e^{-D\omega_1(-\tau)} \sin(\omega_D(t - \tau)) \, d\tau . \tag{1.53}$$

The general solution of the differential EoM shows also for systems with damping the two distinct vibration components:

a. transient vibrations,

b. steady-state vibrations.

Due to damping, after a while, the transient vibrations disappear and
only the steady-state response remains. Accordingly, the steady-state response

dominates the vibrations of the system. However, the largest deformation peak may occur before the system has reached steady state.

In case of a harmonic excitation with the excitation force

$$F(t) = F_0 \sin(\omega t), \tag{1.54}$$

and the excitation frequency ω, the particular solution reads

$$x_p = C_3 \sin(\omega t - \varphi), \tag{1.55}$$

where C_3 corresponds to the amplitude of the steady-state response x_{dy} and can be calculated by

$$C_3 = x_{dy} = \frac{F_0}{k} R_d \tag{1.56}$$

with the deformation response factor

$$R_d = \frac{1}{\sqrt{(1 - \eta^2)^2 + (2D\eta)^2}}. \tag{1.57}$$

Here, $\eta = \omega/\omega_1$ is the frequency ratio. The *phase shift* φ is initiated by the damping and can be calculated by

$$\varphi = \arctan\left(\frac{2D\eta}{1 - \eta^2}\right). \tag{1.58}$$

The homogeneous solution of the EoM is given by

$$x_h = e^{-D\omega_1 t} \left(C_1 \cos(\omega_D t) + C_2 \sin(\omega_D t)\right) \tag{1.59}$$

with the damped (angular) vibration frequency ω_D corresponding to Equation 1.43. Analogous to the approach for free vibrations, the constants C_1 and C_2 are calculated from the initial conditions $x(t = 0)$ and $\dot{x}(t = 0)$. Considering both homogeneous and particular parts, the total response at $t = 0$ is given as

$$x(t = 0) = x_0 = C_1 + C_3 \sin(-\varphi) \tag{1.60}$$

and the initial velocity at $t = 0$ reads

$$\dot{x}(t = 0) = \dot{x}_0 = -C_1 D\omega_1 + C_2 \omega_D + C_3 \omega_e \cos(-\varphi), \tag{1.61}$$

where x_0 and $\dot{x}_0$ are the initial displacement and the initial velocity of the SDoF system, respectively. By solving for C_1 and C_2, we obtain both constants as

$$C_1 = x_0 + C_3 \sin \varphi, \tag{1.62}$$

$$C_2 = \frac{1}{\sqrt{1 - D^2}} \left(\frac{\dot{x}_0}{\omega_1} + C_3 D \sin \varphi - C_3 \eta \cos \varphi\right). \tag{1.63}$$

In case of resonance state with $\omega = \omega_1$, the phase shift becomes $\varphi = \pi/2$.

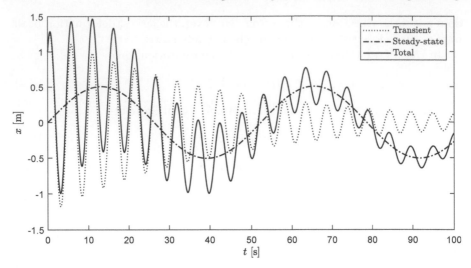

FIGURE 1.11
Transient, steady-state and total response of a damped SDoF system with to
a simultaneous excitation by the initial displacement x_0, the initial velocity
$\dot{x}_0$ and the harmonic force $F = F_0 \sin(\omega t)$ with $\omega \neq \omega_1$.

Example 1.3 *We consider again the structure from Example 1.1 with the
natural frequency and the stiffness omega$_1$ = 1.2 rad s^{-1} and $k = 1$ kN m^{-1}.
However, this time, we assume that the structure is damped with the damping ratio $D = 0.02$. The transient, steady-state and total displacement responses of the structure to the simultaneous excitation by the initial displacement $x_0 = 1$ m, initial velocity $\dot{x}_0 = 1$ m s^{-1} and the harmonic force
$F(t) = 0.5 \sin(0.12t)$ kN are depicted in Figure 1.11. As mentioned previously, the transient vibration disappears after a while due to damping and the
total vibration is dominated by the steady-state response. In Section A.2, a
MATLAB code is provided for the calculation of the shown responses.*

1.4   Frequency-Domain Methods

Figure 1.12 shows the vibration response time history of an SDoF system to
the harmonic excitation force $F = F_0 \sin(\omega t)$. Depending on both the excitation frequency ω and the damping ratio D the response character changes.
The highest amplitude is reached for the resonance state with $\omega = \omega_1$. If the
structure is undamped $D = 0$, the vibration amplitude tends to infinity. For
$\omega \neq \omega_1$, the response reaches after a transient phase the steady-state case.

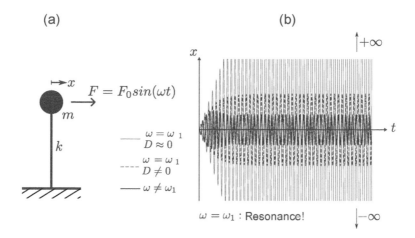

FIGURE 1.12
An SDoF system (a) and its displacement response (b) to the harmonic excitation force $F = F_0 \sin(\omega_1 e)$ for different damping ratios and excitation frequencies.

The shown time histories can be calculated using the approach introduced in Section 1.3.1.

1.4.1 Transfer Function

In the particular solution, using the DUHAMEL integral,

$$x_p = \frac{1}{\omega_D} \int_0^t f(\tau)e^{-D\omega_1(-\tau)} \sin(\omega_D(t - \tau))\,d\tau \qquad (1.64)$$

the excitation function $f(t)$ is folded by the *impulse response function* $h(t)$. Accordingly, the solution can be rewritten as

$$x_p = \int_0^t f(\tau)h(t - \tau)\,d\tau, \qquad (1.65)$$

where

$$h(t) = \frac{1}{\omega_D}e^{-D\omega_1 t} \sin(\omega_D t). \qquad (1.66)$$

This operation can be also represented as

$$x_p = f(t) * h(t), \qquad (1.67)$$

where $*$ means *folding* and not multiplication.

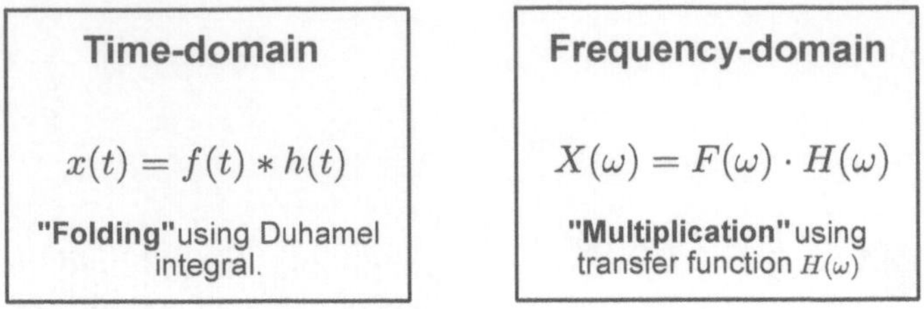

Time-domain	Frequency-domain
$x(t) = f(t) * h(t)$	$X(\omega) = F(\omega) \cdot H(\omega)$
"Folding" using Duhamel integral.	**"Multiplication"** using transfer function $H(\omega)$

FIGURE 1.13

In time domain, the dynamic response is calculated by the *folding* operation using DUHAMEL integral. In frequency domain, the corresponding operation is *multiplication*.

Both the displacement $x(t)$ and the excitation function $f(t)$ can be transferred from time-domain to frequency-domain using FOURIER transformation. FOURIER transform of the displacement response is given by

$$X(\omega) = \int_{-\infty}^{\infty} x(t) e^{-i\omega t} \, dt, \tag{1.68}$$

where $i = \sqrt{-1}$ is the imaginary number. By introducing the DUHAMEL integral for $x(t)$, the equation can be rewritten as

$$X(\omega) = \int_{-\infty}^{\infty} \left(\int_{-\infty}^{\infty} f(\tau) h(t - \tau) \, d\tau \right) e^{-i\omega t} \, dt. \tag{1.69}$$

By substituting $(t - \tau)$ with λ, the equation can be also written as

$$X(\omega) = \int_{-\infty}^{\infty} f(\tau) e^{-i\omega\tau} \, d\tau \cdot \int_{-\infty}^{\infty} h(\lambda) e^{-i\omega\tau} \, d\lambda, \tag{1.70}$$

which is equal to

$$X(\omega) = F(\omega) \cdot H(\omega). \tag{1.71}$$

This equation shows that folding of the functions $f(t)$ and $h(t)$ in time-domain is equivalent to a multiplication of their FOURIER transforms $F(\omega)$ and $H(\omega)$ in frequency-domain. Using this approach, the differential EoMs of SDoF systems can be transformed from time-domain into algebraic equations in frequency-domain. The corresponding operations of time- and frequency-domains are depicted in Figure 1.13.

The FOURIER transform of the impulse response function $h(t)$ is $H(\omega)$, which is referred to as the *transfer function* or the *frequency response function* of the system. We consider a general harmonic unit excitation of an SDoF system at the excitation frequency ω as

$$f = 1 \cdot e^{i\omega t} = 1 \cdot (\cos(\omega t) + i \sin(\omega t)) \tag{1.72}$$

and its FOURIER transform as

$$F = \int_{-\infty}^{\infty} 1 \, dt \, . \tag{1.73}$$

Now, we consider the FOURIER transform of the response as

$$X = \int_{-\infty}^{\infty} x e^{-i\omega t} \, dt, \tag{1.74}$$

and rewrite it as

$$X = H \int_{-\infty}^{\infty} 1 \, dt = \int_{-\infty}^{\infty} x e^{-i\omega t} \, dt \, . \tag{1.75}$$

The steady-state response of a system, which is excited by a harmonic function, must be in time-domain also a harmonic, such as

$$x = a e^{i\omega t}, \tag{1.76}$$

which gives us

$$\Leftrightarrow X = H \int_{-\infty}^{\infty} 1 \, dt = a \int_{-\infty}^{\infty} 1 \, dt \, . \tag{1.77}$$

Accordingly, the motion at the excitation frequency ω can be expressed as

$$x = H e^{i\omega t} \tag{1.78}$$

$$\Leftrightarrow \dot{x} = i\omega H e^{i\omega t} \tag{1.79}$$

$$\Leftrightarrow \ddot{x} = -\omega^2 H e^{i\omega t}, \tag{1.80}$$

where $H(\omega)$ scales the harmonic unit excitation $f(t)$. By introducing these equations into the EoM of an damped SDoF system, which is excited by a harmonic function with the excitation frequency ω, we determine the corresponding transfer function as

$$H(\omega_e) = \frac{1}{k(1 - \eta^2 + i2D\eta)}, \tag{1.81}$$

where k and D are the stiffness and the damping ratio of the SDoF system, respectively. $\eta = \omega/\omega_1$ is the frequency ratio with ω and ω_1 are the excitation and the natural frequencies, respectively. The complex frequency-response function H can be separated to real and imaginary parts as

$$H = H_R + iH_I, \tag{1.82}$$

where its absolute value is given by

$$|H| = \sqrt{H_R^2 + H_I^2}. \tag{1.83}$$

and calculated as

$$|H| = \frac{1}{k\left(\sqrt{(1 - \eta^2) + (2D\eta)^2}\right)}, \tag{1.84}$$

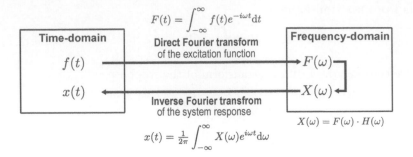

FIGURE 1.14
General approach to calculate the dynamic response to an arbitrary excitation function $f(t)$ using frequency-domain methods.

Using the absolute value, the steady-state response amplitude can be determined as

$$x_{dy} = F_0|H| = \frac{F_0}{k}R_d, \tag{1.85}$$

which matches with the expression of previous Equation 1.56. Here, F_0 is the amplitude of the time-dependent excitation function $F(t)$ and R_d is the deformation response factor.

The angle between real and imaginary parts of the transfer function $H(\omega)$ corresponds to the phase shift as also introduced previously in Equation 1.58:

$$\varphi = \arctan\left(\frac{-H_I}{H_R}\right) = \arctan\left(\frac{2D\eta}{1-\eta^2}\right). \tag{1.86}$$

The general approach for the calculation of the response of an SDoF system to an arbitrary excitation in the frequency-domain is shown in Figure 1.14. The arbitrary excitation function $f(t)$ is transferred to frequency-domain as $F(\omega)$ by the direct FOURIER transform. In frequency-domain, the force function is multiplied by the transfer function H to calculate the frequency-dependent response $X(\omega)$, which is transferred back to time-domain by the inverse FOURIER transform as the time-dependent response function $x(t)$.

1.4.2   Filtering

The frequency content of a function can be modified by transfer functions H. This operation is also referred to as *filtering*. The general filter types and their functions are

a.   *High-pass filters*: To eliminate low-frequency components,

b.   *Low-pass filters*: To eliminate high-frequency components,

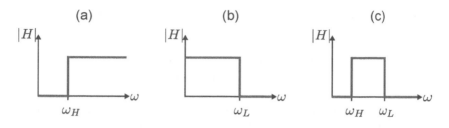

FIGURE 1.15
Absolute values of the transfer function $|H|$ for the three filter types: High-pass filter (a), low-pass filter (b) and band-pass filter (c). ω_H and ω_L are the cutoff frequencies of the filter.

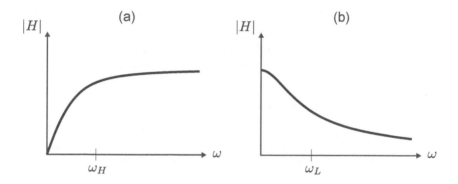

FIGURE 1.16
First-order high-pass (a) and low-pass (b) filters.

   c. *Band-pass filters*: To eliminate frequencies, which are out of a certain frequency range.

The absolute values of corresponding transfer functions are shown schematically in Figure 1.15 for the three general filter types, where ω_H and ω_L are *cutoff frequencies* of the filter.

An example high-pass filter is given by the transfer function

$$H(\omega) = \frac{\omega^2 + i\omega\omega_H}{\omega_H^2 + \omega^2}, \tag{1.87}$$

which is a first order high-pass filter with the cutoff frequency ω_H as shown in Figure 1.16 (a). An example to a low-pass filter reads

$$H(\omega) = \frac{\omega_L^2 + i\omega\omega_L}{\omega_L^2 + \omega^2}, \tag{1.88}$$

which is a first order low-pass filter defined by the transfer function $H(\omega)$ with the cutoff frequency ω_L as shown in Figure 1.16 (b). Another low-pass filter example reads

$$H(\omega) = \frac{1 + \frac{\omega^2}{\omega_g^2}(4D_g^2 - 1) - 2iD_g\frac{\omega^3}{\omega_g^3}}{\left(1 - \left(\frac{\omega^2}{\omega_g^2}\right)^2\right)^2 + 4D_g^2\left(\frac{\omega^2}{\omega_g^2}\right)^2}, \tag{1.89}$$

which is a second order low-pass filter of an SDoF system defined by the transfer function $H(\omega)$. This is the well-known KANAI-TAJIMI filter, which is often used in earthquake engineering and characterized by the natural frequency ω_g and the damping ratio D_g of the ground. A band-pass filter can be constructed as a product of the transfer function of the first order high-pass filter and the first order low-pass filter as

$$H(\omega) = \frac{\omega^2 + i\omega\omega_H}{\omega_H^2 + \omega^2} \cdot \frac{\omega_L^2 - i\omega\omega_L}{\omega_L^2 + \omega^2}, \tag{1.90}$$

which is characterized by the cutoff frequencies ω_L and ω_H and can be utilized for eliminating frequencies out of the range $\omega_H < \omega < \omega_L$, cf. Section 1.4.

1.4.3   Deformation Response Factor

The *deformation response factor* R_d is the ratio of the amplitude of the dynamic steady-state response x_{dy} of an SDoF system excited by a harmonic force to its static deformation x_{st}. For damped systems it reads

$$R_d(\omega_e) = \frac{x_{dy}}{x_{st}} = \frac{1}{\sqrt{(1 - \eta^2)^2 + (2D\eta)^2}}, \tag{1.91}$$

where ω is the excitation frequency and $\eta = \omega/\omega_1$ is the frequency ratio. The natural frequency and the damping ratio are ω_1 and D, respectively. The static deformation of the SDoF system can be calculated by

$$x_{st} = \frac{F_0}{k}, \tag{1.92}$$

where F_0 and k are the load amplitude and stiffness of the SDoF system, respectively.

Using the deformation response factor, the *frequency response curve* of SDoF systems can be drawn as shown in Figure 1.17 for different damping ratios. The $D_1 = 0$ represents the case without damping and $D_4 = 1$ depicts the response of a *critically damped* SDoF system.

For $\eta = 1$ the deformation response factor R_d equals to 1, representing the static loading case. If η is small, R_d is only slightly larger than 1 and the amplitude of the dynamic deformation x_{dy} is essentially the same as the static deformation x_{st}. If η is close to 1, R_d becomes significantly larger than 1. Accordingly, in this region, the dynamic deformation x_{dy} becomes much larger than the static deformation x_{st}.

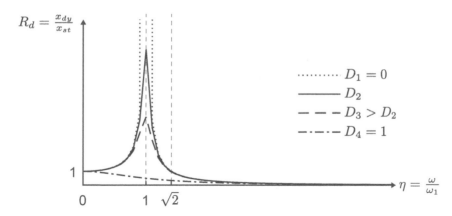

FIGURE 1.17
Frequency response curve of an SDoF system to a harmonic excitation for different damping ratios.

The *resonance state* of an undamped SDoF system occurs if the harmonic excitation frequency ω matches with the natural (angular) frequency ω_1 and accordingly if $\eta = 1$. At this state, R_d tends infinity and the excitation frequency ω is referred to as the *resonant frequency*. The resonance state of a damped SDoF system occurs, when the excitation frequency matches the damped natural (angular) frequency ω_D. As discussed previously in Section 1.3.1, the damped natural frequency is for underdamped structures usually very close to the natural frequency ω_1. For undamped structures, at resonance state, R_d does not tend infinity but reaches its maximum.

For $\eta > \sqrt{2}$, the deformation response factor is less than 1 and, accordingly, the dynamic deformation x_{dy} is less than the static deformation x_{st}. For $\eta > \sqrt{2}$, R_d approaches 0.

Example 1.4 *Figure 1.18 shows a damped SDoF system with its parameters. The system is exposed to the harmonic loading $F(t)$ with the excitation frequency ω, which corresponds to the natural frequency ω_1 of the system.*

The maximum dynamic displacement x_{dy} of the system can be calculated by using the deformation response factor R_d. For this purpose, first, we calculate the static displacement as

$$x_{st} = \frac{F_0}{k} = \frac{5}{3750} = 1.\bar{3} \cdot 10^{-3} \ m. \tag{1.93}$$

Next, we determine the natural angular frequency of the system from its stiffness and mass as

$$\omega_1 = \sqrt{\frac{k}{m}} = \sqrt{\frac{3750}{10}} = 19.36 \ rad \ s^{-1}. \tag{1.94}$$

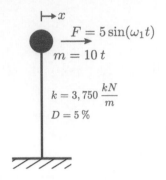

FIGURE 1.18
A damped SDoF system under the harmonic loading $F(t)$.

At resonance state, due to $\omega_1 = \omega$, the deformation response factor reaches its maximum value and corresponds

$$R_{d,max} = \frac{1}{2D} = \frac{1}{2 \cdot 0.05} = 10. \tag{1.95}$$

Finally, we calculate the maximum dynamic displacement as

$$x_{dy} = x_{st} \, R_{d,max} = 1.\bar{3} \cdot 10^{-3} \cdot 10 = 13.\bar{3} \cdot 10^{-3} \ m. \tag{1.96}$$

From the frequency response curve of an SDoF system the damping ratio can be determined by using the *bandwidth method* as

$$D = \frac{\eta_2 - \eta_1}{2}, \tag{1.97}$$

where η_1 and η_2 are frequency ratios corresponding to the deformation response factor $R_{d,max}/\sqrt{2}$.

Example 1.5 *Figure 1.19 shows the frequency response curve of a damped SDoF system. The system is excited by an actuator in vertical direction. The response is measured by an accelerometer.*

To determine the damping ratio, we determine the frequency ratios η_1 and η from the frequency response curve as shown in Figure 1.19 and substitute them as

$$D = \frac{\eta_2 - \eta_1}{2} = \frac{1.05 - 0.95}{2} = 0.05 = 5 \ \%. \tag{1.98}$$

Remark: To obtain the frequency response curve, the measured acceleration signal must be transformed to a displacement signal. Due to harmonic excitation the displacement response is also expected to be harmonic and can be

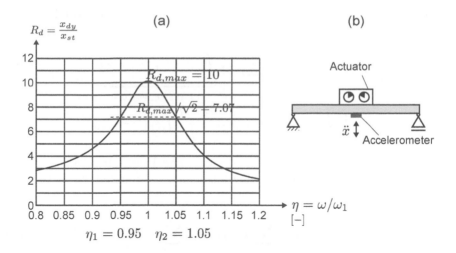

FIGURE 1.19
The frequency response curve of a damped SDoF system (a). An actuator
applies harmonic forces on the structure (b).

calculated as

$$x = x_0 sin(\omega t) \tag{1.99}$$

$$\Leftrightarrow \dot{x} = x_0 \omega cos(\omega t) \tag{1.100}$$

$$\Leftrightarrow \ddot{x} = -x_0 \omega^2 sin(\omega t) \tag{1.101}$$

$$\Leftrightarrow x_0 = \frac{\ddot{x}_0}{\omega^2}. \tag{1.102}$$

1.4.4 Fourier Transformation

For a periodic time function $f(t)$ with the period $T = 2\pi/\omega$, it holds

$$f(t + jT) = f(t), \tag{1.103}$$

where $j = \{-\infty, \ldots, -1, 0, 1, \ldots, \infty\}$. Using the definition of exponential functions $e^{\pm i\omega t} = cos(\omega t) \pm i \sin(\omega t)$, the periodic function can be rewritten as

$$f(t) = a_o + a_1 e^{i\omega t} + a_2 e^{i2\omega t} + \cdots + a_n e^{in\omega t} \tag{1.104}$$

$$+ a_{-1} e^{-i\omega t} + a_{-2} e^{-i2\omega t} + \cdots + a_{-n} e^{-in\omega t}.$$

Based on this formulation, complex FOURIER series can be introduced to define
the periodic function as

$$f(t) = \sum_{j=-\infty}^{\infty} a_j e^{i(j\omega t)}. \tag{1.105}$$

From the FOURIER series expression, the so-called FOURIER coefficients a_n can be determined as

$$\int_0^T f(t)e^{-i(n\omega t)}\,\mathrm{d}t = \sum_{j=-\infty}^{\infty} a_j \underbrace{\int_0^T e^{i(j\omega t)}e^{-i(n\omega t)}\,\mathrm{d}t}_{T \text{ if } j\neq n, \text{ else } 0} \tag{1.106}$$

$$\Leftrightarrow a_n = \frac{1}{T}\int_0^T f(t)e^{-i(n\omega t)}\,\mathrm{d}t = \frac{1}{T}\int_0^T f(t)e^{-i(\omega_n t)}\,\mathrm{d}t, \tag{1.107}$$

where $n = \{0, \pm 1, \pm 2, \ldots\}$ and $\omega_n = n\omega$. The FOURIER series expression can also be rewritten by introducing trigonometric functions in place of the complex exponential as

$$f(t) = a_0 + \sum_{j=1}^{\infty} a_j \cos \omega_j t + \sum_{j=1}^{\infty} b_j \sin \omega_j t. \tag{1.108}$$

The coefficient a_0 is given by

$$a_o = \frac{1}{T}\int_0^T f(t)\,\mathrm{d}t \tag{1.109}$$

and the coefficients a_j and b_j are determined by

$$a_j = \frac{2}{T}\int_0^T f(t)\cos(\omega_j t)\,\mathrm{d}t \quad \text{and} \quad b_j = \frac{2}{T}\int_0^T f(t)\sin(\omega_j t)\,\mathrm{d}t \tag{1.110}$$

Aperiodic time functions can be seen as periodic functions with periods approaching infinity ($T \to \infty$). In order to transform an aperiodic time function into the frequency-domain, the discrete FOURIER coefficients a_n and the discrete frequencies $\omega_n = n\omega$ are replaced by the inverse FOURIER transform

$$x(t) = \frac{1}{2\pi}\int_{-\infty}^{\infty} X(\omega)e^{i\omega t}\,\mathrm{d}\omega. \tag{1.111}$$

Accordingly, the *direct* FOURIER transform is given as

$$X(\omega) = \frac{1}{2\pi}\int_{-\infty}^{\infty} x(t)e^{-i\omega t}\,\mathrm{d}t. \tag{1.112}$$

The *discrete* FOURIER transformation (DFT) is used to transform a time series x_r consisting N discrete points with a constant time step Δt, where $r = \{0, 1, 2, \ldots, N-1\}$. The direct transform of x_r computes N complex coefficients

$$X_c = \frac{1}{N}\sum_{r=0}^{N-1} x_r \left(\cos(\frac{2\pi c r}{N}) - i\sin(\frac{2\pi c r}{N})\right), \tag{1.113}$$

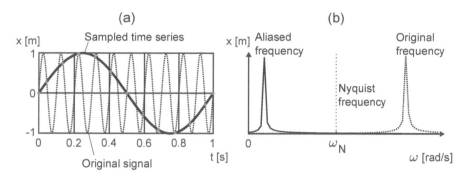

FIGURE 1.20
A displacement response in time (a) and frequency (b) domains. Low sampling of the response in time domain causes the aliasing effect. Instead of the original frequency of the response an aliased frequency occurs in the frequency domain.

where $c = \{0, 1, 2, \ldots, N-1\}$. The inverse transform is given by

$$x_r = \sum_{c=0}^{N-1} X_c e^{i\frac{2\pi c r}{N}}. \tag{1.114}$$

The maximum frequency of the harmonic components of a time series is called as the NYQUIST frequency ω_N. The transformed values for the frequencies higher are the same as those for the corresponding smaller frequency values. Therefore, it is also called as the *folding frequency*. The NYQUIST frequency is determined from

$$\omega_N = \frac{\pi}{\Delta t}, \tag{1.115}$$

where Δt is the time difference between discrete time instants and referred to as *sampling rate*. The NYQUIST frequency should be chosen to be sufficiently large so that the highest frequency of significance is represented in the time series. This implies that the sampling rate Δt of the time series should be chosen sufficiently small.

If the time series is not sampled with low sampling rate and contains harmonic components with frequencies higher than the NYQUIST frequency *aliasing* occurs. As shown in Figure 1.20, in case of aliasing, the original frequencies higher than the NYQUIST frequency are folded back causing aliased frequencies.

Fast FOURIER transformation (FFT) algorithms are much more efficient by saving around 99 % of mathematical operations compared to the DFT algorithm [6]. For the classical COOLEY-TUKEY FFT algorithm the number of data points N must be equal to 2^n with n integer. This is usually done by *padding* the time series with zeros. The basic idea of the FFT algorithm is to re-arrange the elements of the time series $x(t)$ into two sub-series $x_1(t) = x(2t)$

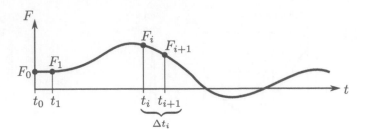

FIGURE 1.21
Time-dependent force function $F(t)$ defined at discrete time instants t_i for $i = \{0, 1, \ldots, N\}$.

containing all signal values with even index t and $x_2(t) = x(2t+1)$ containing all signal values with uneven index t. By dividing the DFT with N input values into two separate DFTs each with $N/2$, FFT reaches a higher computation efficiency.

1.5   Time-Domain Methods

Analytical solution of the differential EoMs gets difficult if the excitation function is complex or the system inhabits nonlinearities. In such cases, *time stepping methods* can be applied to solve the EoMs by integrating at discrete time instants.

For this purpose, first, we consider the time-dependent arbitrary excitation function at the discrete time instant t_i for $i = \{0, 1, \ldots, N\}$ as $F_i = F(t_i)$ with the time interval $\Delta t_i = t_{i+1} - t_i$ for each time step as shown in Figure 1.21. From the solution of the EoM

$$m\ddot{x} + c\dot{x} + kx = F(t). \tag{1.116}$$

at the time step t_i, we determine the responses x_i, $\dot{x}_i$ and $\ddot{x}_i$ of the damped SDoF system that is excited by the function $F(t)$. Accordingly, the EoM at this time step reads

$$m\ddot{x}_i + c\dot{x}_i + kx_i = F_i. \tag{1.117}$$

When we repeat the solution of the EoM successively also for other time steps with $i = \{0, 1, \ldots, N\}$, the SDoF response can be determined at all time instants as shown in Figure 1.22. The initial response at t_0 occurs due to the initial conditions x_o and $\dot{x}_0$.

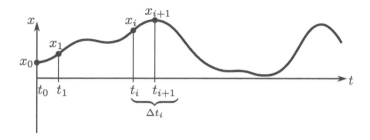

FIGURE 1.22
The response of an SDoF system calculated at discrete time instants t_i for $i = \{0, 1, \ldots, N\}$.

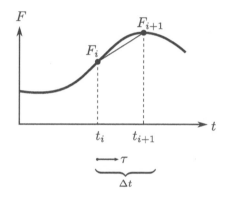

FIGURE 1.23
Linear interpolation of the time-dependent excitation function $F(t)$ in the time interval Δt.

The following sections cover three numerical methods, which can be applied to solve the EoMs at discrete time steps.

1.5.1 Interpolation of Excitation Method

In this method, the time-dependent excitation function $F(t)$ is interpolated over each time interval Δt_i. If the time intervals are short, linear interpolation is satisfactory as shown in Figure 1.23. However, with increasing time interval length and complexity of the excitation function, the methods loses its accuracy.

The excitation function between time steps t_i and t_{i+1} is approximated by

$$F(\tau) = F_i + \frac{\Delta F_i}{\Delta t}\tau, \tag{1.118}$$

where τ is computing the time starting from t_i and the change of force is defined as $\Delta F_i = F_{i+1} - F_i$. Accordingly, between the time steps, the EoM of an undamped SDoF system is given by

$$m\ddot{x} + kx = F_i + \frac{\Delta F_i}{\Delta t}\tau. \tag{1.119}$$

In this time interval, the solution of the EoM consists three parts:

a. Homogeneous solution due to the initial conditions x_i and $\dot{x}_i$ at $\tau = 0$,

b. particular solution due to step force F_i,

c. particular solution due to ramp force $\frac{\Delta F_i}{\Delta t}\tau$,

which yields the displacement at the time instant τ as

$$x(\tau) = \underbrace{x_i \cos(\omega_1\tau) + \frac{\dot{x}_i}{\omega_1}\sin(\omega_1\tau)}_{\text{a. Homogeneous sol.}} + \underbrace{\frac{F_i}{k}(1 - \cos(\omega_1\tau))}_{\text{b. Particular sol.}} \tag{1.120}$$

$$+ \underbrace{\frac{\Delta F_i}{k}\left(\frac{\tau}{\Delta t} - \frac{\sin(\omega_1\tau)}{\omega_1\Delta t}\right)}_{\text{c. Particular sol.}}.$$

To calculate the response at the time step t_{i+1}, we evaluate the previous equation for $\tau = \Delta t$ as

$$x_{i+1} = x_i \cos(\omega_1\Delta t) + \frac{\dot{x}_i}{\omega_1}\sin(\omega_1\Delta t) + \frac{F_i}{k}(1 - \cos(\omega_1\Delta t)) \tag{1.121}$$

$$+ \frac{\Delta F_i}{k}\left(1 - \frac{\sin(\omega_1\Delta t)}{\omega_1\Delta t}\right).$$

For an damped SDoF system, the general solution is given by the following recurrence equation with the coefficients A, B, C and D as

$$x_{i+1} = Ax_i + B\dot{x}_i + C\ddot{x}_i + DF_{i+1}, \tag{1.122}$$

where

$$A = e^{-D\omega_1 \Delta t} \left(\frac{D}{\sqrt{1-D^2}} \sin(\omega_D \Delta t) + \cos(\omega_D \Delta t) \right), \tag{1.123}$$

$$B = e^{-D\omega_1 \Delta t} \left(\frac{1}{\omega_D} \sin(\omega_D \Delta t) \right), \tag{1.124}$$

$$C = \frac{1}{k} \left(\frac{2D}{\omega_1 \Delta t} + e^{-D\omega_1 \Delta t} \left(\left(\frac{1 - 2D^2}{\omega_D \Delta t} - \frac{D}{\sqrt{1-D^2}} \right) \sin(\omega_D \Delta t) \right. \right. \tag{1.125}$$
$$\left. \left. - \left(1 + \frac{2D}{\omega_1 \Delta t} \right) \cos(\omega_D \Delta t) \right) \right),$$

$$D = \frac{1}{k} \left(1 - \frac{2D}{\omega_1 \Delta t} + e^{-D\omega_1 \Delta t} \left(\frac{2D^2 - 1}{\omega_D \Delta t} \sin(\omega_D \Delta t) \right. \right. \tag{1.126}$$
$$\left. \left. + \frac{2D}{\omega_1 \Delta t} \cos(\omega_D \Delta t) \right) \right).$$

The general approach of the interpolation of the excitation method is summarized below for a linear damped SDoF system.

Interpolation of Excitation Method

1. Initial calculations:
   $e^{-D\omega_1 \Delta t}$, $\omega_D = \sqrt{1-D^2}$, $\sin(\omega_D \Delta t)$, $\cos(\omega_D \Delta t)$
   A, B, C and D

2. Calculation for the time step t_i:
   $x_{i+1} = Ax_i + B\dot{x}_i + C\ddot{x}_i + DF_{i+1}$

3. Repetition for the next time step t_{i+1}:
   Replace i by $i+1$ and return to calculation step 2.

1.5.2 Newmark's Method

The EoMs can be solved by the direct integration in the time-domain using the NEWMARK's integration method [8] based on the following equations:

$$\dot{x}_{i+1} = \dot{x}_i + (1-\gamma)\Delta t \ddot{x}_i + \gamma \Delta t \ddot{x}_{i+1}, \tag{1.127}$$

$$x_{i+1} = x_i + \Delta t \dot{x}_i + (\frac{1}{2} - \beta)\Delta t^2 \ddot{x}_i + \beta \Delta t^2 \ddot{x}_{i+1}. \tag{1.128}$$

Because of the unknown parameter $\ddot{x}_{i+1}$, an iterative approach is required for nonlinear systems. For linear systems, the NEWMARK's method can be applied without iteration, which will be introduce in following. The computation of nonlinear systems is covered in Section 1.9.

The operators γ and β determine the stability and accuracy of the solution. Typical operator values are $\gamma = 1/2$ and $1/6 \leq \beta \leq 1/4$. NEWMARK's method

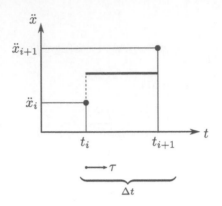

FIGURE 1.24

Assumption of a constant acceleration during the time interval Δt according to the average acceleration case of NEWMARK's method.

bases on assumed variation of acceleration during each time interval. Depending on the chosen assumption (average acceleration or linear acceleration) the operators change.

In *average acceleration* case, the acceleration over a time interval is assumed to be constant and equal to the average acceleration as shown in Figure 1.24. Accordingly the acceleration reads

$$\ddot{x}(\tau) = \frac{1}{2}(\ddot{x}_{i+1} + \ddot{x}_i). \tag{1.129}$$

The velocity and displacement of the time step t_{i+1} yield

$$\dot{x}_{i+1} = \dot{x}_i + \frac{\Delta t}{2}(\ddot{x}_{i+1} + \ddot{x}_i), \tag{1.130}$$

$$x_{i+1} = x_i + \Delta t \dot{x}_i + \frac{\Delta t^2}{4}(\ddot{x}_{i+1} + \ddot{x}_i), \tag{1.131}$$

from which we observe for the average acceleration that $\gamma = 1/2$ and $\beta = 1/4$ applies.

In *linear acceleration* case, the acceleration over a time interval is assumed to be linear as shown in Figure 1.25. Accordingly the acceleration reads

$$\ddot{x}(\tau) = \ddot{x}_i + \frac{\tau}{\Delta t}(\ddot{x}_{i+1} - \ddot{x}_i). \tag{1.132}$$

The velocity and displacement of the time step t_{i+1} yield

$$\dot{x}_{i+1} = \dot{x}_i + \frac{\Delta t}{2}(\ddot{x}_{i+1} + \ddot{x}_i), \tag{1.133}$$

$$x_{i+1} = x_i + \Delta t \dot{x}_i + \Delta t^2 \left(\frac{\ddot{x}_{i+1}}{6} + \frac{\ddot{x}_i}{3}\right), \tag{1.134}$$

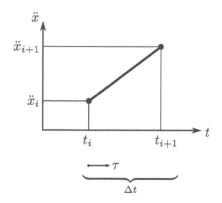

FIGURE 1.25
Assumption of a linear acceleration change during the time interval Δt according to the linear acceleration case of NEWMARK's method.

from which we observe for the linear acceleration that $\gamma = 1/2$ and $\beta = 1/6$.

For linear systems, the non-iterative formulation of NEWMARK's method can be achieved by reformulating the equations. For this purpose, we introduce the first incremental quantities as

$$\Delta x_i = x_{i+1} - x_i, \quad \Delta \dot{x}_i = \dot{x}_{i+1} - \dot{x}_i,$$
$$\Delta \ddot{x}_i = \ddot{x}_{i+1} - \ddot{x}_i, \quad \Delta F_i = F_{i+1} - F_i, \tag{1.135}$$

which define the change in displacement, velocity, acceleration and excitation force. Substituting in $\Delta \dot{x} = \dot{x}_{i+1} - \dot{x}_i$, the NEWMARK's equation for $\dot{x}_{i+1}$ yields

$$\Delta \dot{x}_i = (1 - \gamma)\Delta t \ddot{x}_i + \gamma \Delta t \ddot{x}_{i+1}. \tag{1.136}$$

With $\ddot{x}_{i+1} = \Delta \ddot{x}_i + \ddot{x}_i$, we determine the velocity change as

$$\Delta \dot{x}_i = \Delta t \ddot{x}_i + \gamma \Delta t \Delta \ddot{x}_i. \tag{1.137}$$

After substituting $\Delta x_i = x_{i+1} - x_i$, the Newmark's equation for x_{i+1} yields

$$\Delta x_i = \Delta t \dot{x}_i + (\frac{1}{2} - \beta)\Delta t^2 \ddot{x}_i + \beta \Delta t^2 \ddot{x}_{i+1}. \tag{1.138}$$

Again with $\ddot{x}_{i+1} = \Delta \ddot{x}_i + \ddot{x}_i$, we determine the displacement change as

$$\Delta x_i = \Delta t \dot{x}_i + \frac{\Delta t^2}{2}\ddot{x}_i + \beta \Delta t^2 \Delta \ddot{x}_i, \tag{1.139}$$

from which we obtain the change of acceleration as

$$\Delta \ddot{x}_i = \frac{1}{\beta \Delta t^2}\Delta x_i - \frac{1}{\beta \Delta t}\dot{x}_i - \frac{1}{2\beta}\ddot{x}_i \tag{1.140}$$

and finally after substituting the change of acceleration in Equation 1.137, we calculate the change of velocity as

$$\Delta \dot{x}_i = \frac{\gamma}{\beta \Delta t} \Delta x_i - \frac{\gamma}{\beta} \dot{x}_i + \Delta t \left(1 - \frac{\gamma}{2\beta} \right) \ddot{x}_i. \tag{1.141}$$

After substituting the incremental response quantities, the EoM of the linear SDoF system reads

$$m \Delta \ddot{x}_i + c \Delta \dot{x}_i + k \Delta x_i = \Delta F_i, \tag{1.142}$$

which can be rewritten as

$$\hat{k} \Delta x_i = \Delta \hat{F}_i, \tag{1.143}$$

where

$$\hat{k} = k + \frac{\gamma}{\beta \Delta t} c + \frac{1}{\beta \Delta t^2} m, \tag{1.144}$$

$$\Delta \hat{F}_i = \Delta F_i + \left(\frac{1}{\beta \Delta t} m + \frac{\gamma}{\beta} c \right) \dot{x}_i + \left(\frac{1}{2\beta} m + \Delta t \left(\frac{\gamma}{2\beta} - 1 \right) c \right) \ddot{x}_i. \tag{1.145}$$

Here, $\hat{k}$ and $\Delta \hat{F}_i$ are calculated from the system parameters m, k and c, operators γ and β, and the system responses $\dot{x}_i$ and $\ddot{x}_i$ at the beginning of each time step. The incremental displacement of the linear SDoF system is then computed from

$$\Delta x_i = \frac{\Delta \hat{F}_i}{\hat{k}}. \tag{1.146}$$

Once Δx_i is known, $\Delta \dot{x}_i$ and $\Delta \ddot{x}_i$ can also be obtained using Equations 1.137 and 1.140, from which the system responses x_{i+1}, $\dot{x}_{i+1}$ and $\ddot{x}_{i+1}$ of the next time step t_{i+1} can be determined using Equation 1.135 without iteration.

    The general approach for the solution of a linear damped SDoF system using the NEWMARK's method is summarized below.

Newmark's Method

1. Initial calculations:
   $\gamma = 1/2$ and $\beta = 1/4$ for average acceleration method
   $\gamma = 1/2$ and $\beta = 1/6$ for linear acceleration method
   $\ddot{x}_0 = \frac{F_0 - c\dot{x}_0 - kx_0}{m}$
   $\hat{k} = k + \frac{\gamma}{\beta \Delta t}c + \frac{1}{\beta \Delta t^2}m$
   $a = \frac{1}{\beta \Delta t}m + \frac{\gamma}{\beta}c$ and $b = \frac{1}{2\beta}m + \Delta t\left(\frac{\gamma}{2\beta} - 1\right)c$

2. Calculations at time step t_i:
   $\Delta \hat{F}_i = \Delta F_i + a\dot{x}_i + b\ddot{x}_i$
   $\Delta x_i = \frac{\Delta \hat{F}_i}{\hat{k}}$
   $\Delta \dot{x}_i = \frac{\gamma}{\beta \Delta t}\Delta x_i - \frac{\gamma}{\beta}\dot{x}_i + \Delta t\left(1 - \frac{\gamma}{2\beta}\right)\ddot{x}_i$
   $\Delta \ddot{x}_i = \frac{1}{\beta \Delta t^2}\Delta x_i - \frac{1}{\beta \Delta t}\dot{x}_i - \frac{1}{2\beta}\ddot{x}_i$
   $x_{i+1} = x_i + \Delta x_i$, $\dot{x}_{i+1} = \dot{x}_i + \Delta \dot{x}_i$ and $\ddot{x}_{i+1} = \ddot{x}_i + \Delta \ddot{x}_i$

3. Repetition for the next time step t_{i+1}:
   Replace i by $i + 1$ and return to calculation step 2.

In NEWMARK's method, the solution x_{i+1} at time instant t_{i+1} is determined from the incremental EoM

$$m\Delta \ddot{x}_i + c\Delta \dot{x}_i + k\Delta x_i = \Delta F_i \tag{1.147}$$

by using the equilibrium condition

$$m\ddot{x}_{i+1} + c\dot{x}_{i+1} + kx_{i+1} = F_{i+1}. \tag{1.148}$$

Such methods are called as *implicit methods*.

Example 1.6 *Figure 1.26 shows an SDoF system and the load function $F(t)$ applied to it. The discrete values of the load function are given with the corresponding time instants. The initial load is $F_0 = 0$. The system is at rest at time $t = 0$. Accordingly, the initial conditions read*

$$x_0 = 0, \quad \dot{x}_0 = 0. \tag{1.149}$$

The time step size is given as $\Delta t = 0.1$ s. The mass, damping coefficient and stiffness of the system are given as

$$m = 1 \ t, \quad c = 0.316 \ kNs \ m^{-1}, \quad k - 10 \ kN \ m^{-1}. \tag{1.150}$$

We apply NEWMARK's time integration method for the average acceleration case to calculate the response of the system for the first two time steps.

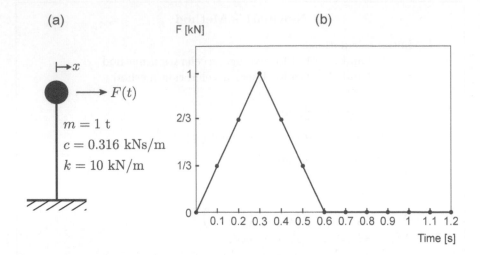

FIGURE 1.26
An SDoF system (a) under the dynamic loading $F(t)$ (b).

With the assumption of average acceleration $(\gamma = 0.5$ and $\beta = 0.25)$, we conduct the initial calculations as

$$\ddot{x}_0 = \frac{F_0 - c\dot{x}_0 - kx_0}{m} = 0, \tag{1.151}$$

$$\hat{k} = k + \frac{\gamma}{\beta\Delta t}c + \frac{\gamma}{\beta\Delta t^2}m = 416.32 \ kN \ m^{-1}, \tag{1.152}$$

$$a = \frac{1}{\beta\Delta t}m + \frac{\gamma}{\beta}c = 40.63 \ t \ s^{-1}, \tag{1.153}$$

$$b = \frac{1}{2\beta}m + \Delta t\left(\frac{\gamma}{2\beta} - 1\right)c = 2 \ t. \tag{1.154}$$

We continue with the calculations of the first time step $(i = 0)$ as

$$\Delta\hat{F}_0 = \Delta F_0 + a\dot{x}_0 + b\ddot{x}_0 = 0.\bar{3} \ kN, \tag{1.155}$$

$$\Delta x_0 = \frac{\Delta\hat{F}_0}{\hat{k}} = 0.0008 \ m, \tag{1.156}$$

$$\Delta\dot{x}_0 = \frac{\gamma}{\beta\Delta t}\Delta x_0 - \frac{\gamma}{\beta}\dot{x}_0 + \Delta t\left(1 - \frac{\gamma}{2\beta}\right)\ddot{x}_0 = 0.016 \ m \ s^{-1} \tag{1.157}$$

$$\Delta\ddot{x}_0 = \frac{1}{\beta\Delta t^2}(\Delta x_0 - \Delta t\dot{x}_0) - \frac{1}{2\beta}\ddot{x}_0 = 0.32 \ m \ s^{-2}, \tag{1.158}$$

where the initial load change is obtained from the given load function as $\Delta F_0 =$

$0.\bar{3}$ *kN. Accordingly, the system responses read*

$$x_1 = 0 + \Delta x_0 = 0.0008 \ m, \tag{1.159}$$

$$\dot{x}_1 = 0 + \Delta \dot{x}_0 = 0.016 \ m \ s^{-1}, \tag{1.160}$$

$$\ddot{x}_1 = 0 + \Delta \ddot{x}_0 = 0.32 \ m \ s^{-2}. \tag{1.161}$$

Now, we repeat the calculations for the second time step (i = 1) as

$$\Delta \hat{F}_1 = \Delta F_1 + a\dot{x}_1 + b\ddot{x}_1 = 1.62 \ kN, \tag{1.162}$$

$$\Delta x_1 = \frac{\Delta \hat{F}_1}{\hat{k}} = 0.0039 \ m, \tag{1.163}$$

$$\Delta \dot{x}_1 = \frac{\gamma}{\beta \Delta t} \Delta x_1 - \frac{\gamma}{\beta} \dot{x}_1 + \Delta t \left(1 - \frac{\gamma}{2\beta}\right) \ddot{x}_1 = 0.046 \ m \ s^{-1}, \tag{1.164}$$

$$\Delta \ddot{x}_1 = \frac{1}{\beta \Delta t^2} (\Delta x_1 - \Delta t \dot{x}_1) - \frac{1}{2\beta} \ddot{x}_1 = 0.28 \ m \ s^{-2}, \tag{1.165}$$

where the load change corresponding to the first time step is obtained from the given load function as $\Delta F_1 = 0.\bar{3}$ *kN. Accordingly, the system responses corresponding to the second time step read*

$$x_2 = x_1 + \Delta x_1 = 0.0047 \ m, \tag{1.166}$$

$$\dot{x}_2 = \dot{x}_1 + \Delta \dot{x}_1 = 0.062 \ m \ s^{-1}, \tag{1.167}$$

$$\ddot{x}_2 = \ddot{x}_1 + \Delta \ddot{x}_1 = 0.60 \ m \ s^{-2}. \tag{1.168}$$

The solution can be proceeded for further time steps in the same fashion. The calculated responses of the system are depicted in Figure 1.27. In Figure 1.28, the calculation is repeated for a higher sampling rate using $\Delta t = 0.01$ *s. From the comparison of both figures it can be seen that a further increase in time increment has only a marginal effect on the response accuracy.*

In Section A.3, a MATLAB code of the Newmark's method with the calculation of the example is provided.

1.5.3 Central Difference Method

From the finite difference approximation of the time derivatives of displacement, the *central difference* expression for velocity at time t_i is obtained as

$$\dot{x}_i = \frac{x_{i+1} - x_{i-1}}{2\Delta t} \tag{1.169}$$

and the corresponding acceleration reads

$$\ddot{x}_i = \frac{x_{i+1} - 2x_i + x_{i-1}}{\Delta t^2}, \tag{1.170}$$

where x_i and x_{i-1} are assumed to be known from the previous time steps. Substituting these approximate expressions for velocity and acceleration into

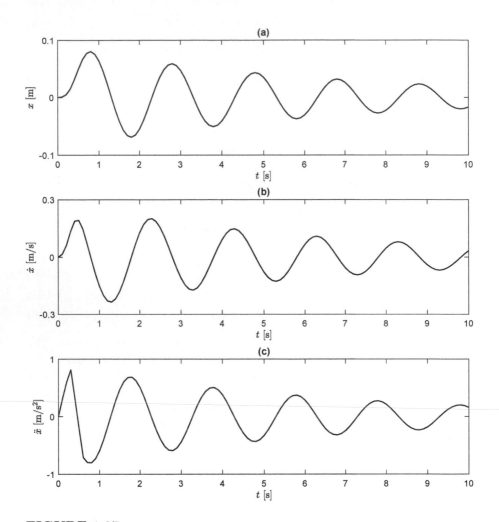

FIGURE 1.27
Displacement (a), velocity (b) and acceleration (c) responses of the SDoF system to the dynamic loading $F(t)$. Calculations are conducted using the Newmark's integration method with a time increment of $\Delta t = 0.1$ s.

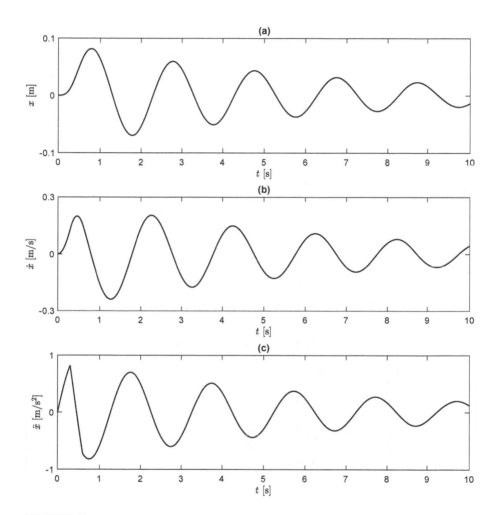

FIGURE 1.28

Displacement (a), velocity (b) and acceleration (c) responses of the SDoF system to the dynamic loading $F(t)$. Calculations are conducted using the Newmark's integration method with a time increment of $\Delta t = 0.01$ s.

the EoM of a linear SDoF system gives

$$m\frac{x_{i+1} - 2x_i + x_{i-1}}{\Delta t^2} + c\frac{x_{i-1} - x_i}{2\Delta t} + kx_i = F_i \qquad (1.171)$$

$$\Leftrightarrow \left(\frac{m}{\Delta t^2} + \frac{c}{2\Delta t}\right) x_{i+1} = F_i \qquad (1.172)$$

$$- \left(\frac{m}{\Delta t^2} + \frac{c}{2\Delta t}\right) x_{i-1} - \left(k - \frac{2m}{2\Delta t^2}\right) x_i.$$

This equation can also be rewritten to determine the displacement of the next time step as

$$\hat{k}x_{i+1} = \hat{F}_i \qquad (1.173)$$

$$\Leftrightarrow x_{i+1} = \frac{\hat{F}_i}{\hat{k}}, \qquad (1.174)$$

where

$$\hat{k} = \frac{m}{\Delta t^2} + \frac{c}{2\Delta t} \qquad (1.175)$$

and

$$\hat{F}_i = F_i - \left(\frac{m}{\Delta t^2} - \frac{c}{2\Delta t}\right) x_{i-1} - \left(k - \frac{2m}{\Delta t^2}\right) x_i. \qquad (1.176)$$

The general approach for the solution of a linear damped SDoF system using the central difference method is summarized below.

Central Difference Method

1. Initial calculations:
   $\ddot{x}_0 = \frac{F_0 - c\dot{x}_0 - kx_0}{m}$
   $x_{-1} = x_0 - \Delta t\dot{x}_0 + \frac{\Delta t^2}{2}\ddot{x}_0$
   $\hat{k} = \frac{m}{\Delta t^2} + \frac{c}{2\Delta t}$
   $a = \frac{m}{\Delta t^2} - \frac{c}{2\Delta t}$ and $b = k - \frac{2m}{\Delta t^2}$

2. Calculations for the time step t_i:
   $\hat{F}_i = F_i - ax_{i-1} - bx_i$
   $x_{i+1} = \frac{\hat{F}_i}{\hat{k}}$

3. Repetition for the next time step t_{i+1}:
   Replace i by $i+1$ and return to the calculation step 2.

The x_{-1} term is derived from the finite difference approximation of the initial acceleration as

$$\ddot{x}_0 = \frac{x_1 - 2x_0 + x_{-1}}{\Delta t^2} \qquad (1.177)$$

$$\Leftrightarrow x_{-1} = \Delta t^2 \ddot{x}_0 - x_1 + 2x_0. \qquad (1.178)$$

Accordingly, from the finite difference approximation of the initial velocity, we obtain

$$\dot{x}_0 = \frac{x_1 - x_{-1}}{2\Delta t} \tag{1.179}$$

$$\Leftrightarrow x_1 = 2\Delta t \dot{x}_0 + x_{-1}. \tag{1.180}$$

The combination of these two expressions yields

$$x_{-1} = \Delta t^2 \ddot{x}_0 - 2\Delta t \dot{x}_0 - x_{-1} + 2x_0 \tag{1.181}$$

and finally

$$x_{-1} = x_0 - \Delta t \dot{x}_0 + \frac{\Delta t^2}{2} \ddot{x}_0. \tag{1.182}$$

In the central difference method, the solution x_{i+1} at time t_{i+1} is determined from the EoM

$$m\ddot{x}_i + c\dot{x}_i + kx_i = F_i, \tag{1.183}$$

without using the equilibrium condition

$$m\ddot{x}_{i+1} + c\dot{x}_{i+1} + kx_{i+1} = F_{i+1}. \tag{1.184}$$

Such methods are referred to as *explicit methods*. Explicit algorithms calculate the solution at time t_{i+1} directly, while in implicit methods the unknowns appear on both sides of equations and must be determined by solving a linear equation system. For nonlinear systems, implicit methods require iterations, which is critical particularly for real-time applications. This shortcoming of implicit methods is offset by their better superior stability properties. If the chosen time step is not short enough, the central difference method can yield inaccurate results.

Example 1.7 *We apply the central difference method to calculate the response of the SDoF system from Example 1.6 for the same dynamic load $F(t)$. We start with the initial conditions as*

$$\ddot{x}_0 = \frac{F_0 - c\dot{x}_0 - kx_0}{m} = 0, \tag{1.185}$$

$$x_{-1} = x_0 - \Delta t \dot{x}_0 + \frac{\Delta t^2}{2}\ddot{x}_0 = 0, \tag{1.186}$$

$$\hat{k} = \frac{m}{\Delta t^2} + \frac{c}{2\Delta t} = 101.58 \ kN \ m^{-1}, \tag{1.187}$$

$$a = \frac{m}{\Delta t^2} - \frac{c}{2\Delta t} = 98.42 \ kN \ m^{-1}, \tag{1.188}$$

$$b = k - \frac{2m}{\Delta t^2} = -190 \ kN \ m^{-1}. \tag{1.189}$$

We continue with the calculations for the first time step ($i = 1$) as

$$x_1 = 2\,\Delta t\,\dot{x}_0 + x_{-1} = 0, \tag{1.190}$$

$$\hat{F}_1 = F_1 - ax_{-1} - bx_1 = 0.\bar{3}, \tag{1.191}$$

$$x_2 = \frac{\hat{F}_1}{\hat{k}} = 0.00\bar{3}m, \tag{1.192}$$

$$\dot{x}_1 = \frac{x_2 - x_0}{2\,\Delta t} = 0.016 \; m \; s^{-1}, \tag{1.193}$$

$$\ddot{x}_1 = \frac{x_2 - 2x_1 + x_0}{\Delta t^2} = 0.32 \; m \; s^{-2}. \tag{1.194}$$

Accordingly, the calculations for the second time step ($i = 2$) read

$$x_2 = 2\,\Delta t\,\dot{x}_1 + x_0 = 0.00\bar{3}m, \tag{1.195}$$

$$\hat{F}_2 = F_2 - ax_1 - bx_2 = 1.29 \; kN, \tag{1.196}$$

$$x_3 = \frac{\hat{F}_2}{\hat{k}} = 0.013 \; m, \tag{1.197}$$

$$\dot{x}_2 = \frac{x_3 - x_1}{2\,\Delta t} = 0.064 \; m \; s^{-1}, \tag{1.198}$$

$$\ddot{x}_2 = \frac{x_3 - 2x_2 + x_1}{\Delta t^2} = 0.61 \; m \; s^{-2}. \tag{1.199}$$

The calculated response deviates slightly from Example 1.6, in which the calculation is conducted by the NEWMARK method. For instance, the acceleration of the second time step $t = 0.2$ s is determined with the NEWMARK method as $\ddot{x}_2 = 0.60 \; m \; s^{-2}$. The difference shows the sensitivity of the central difference method regarding time step size.

The calculated responses of the system are depicted in Figure 1.29. In Figure 1.30, the calculation is repeated for $\Delta t = 0.01$ s. From the comparison of both figures it can be seen that a further increase in time increment slightly increases the response accuracy, e.g., at $t = 0.2$ s, the acceleration is now determined also as $\ddot{x}_2 = 0.60 \; m \; s^{-2}$, which corresponds with the result of the Example 1.6.

In Section A.4, a MATLAB code of the central difference method with the calculation of the example is provided.

1.5.4 Stability and Computational Error

Conditionally stable time-domain methods lead to accurate solutions, only if the time step Δt complies with a stability limit. On the other hand, *unconditionally stable* methods lead to accurate solutions regardless of the length of the time step Δt.

Table 1.1 shows the classification of the so far described approaches as conditionally and unconditionally stable methods, where T_1 represents the natural period of the calculated SDoF system. The given time step limits are

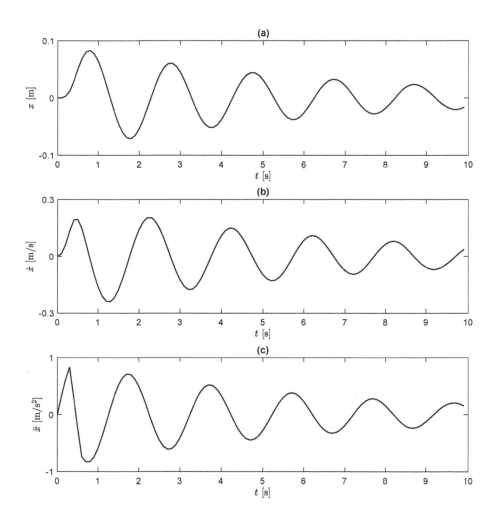

FIGURE 1.29
Displacement (a), velocity (b) and acceleration (c) responses of the SDoF system to the dynamic loading $F(t)$. Calculations are conducted using the central difference method with a time increment of $\Delta t = 0.1$ s.

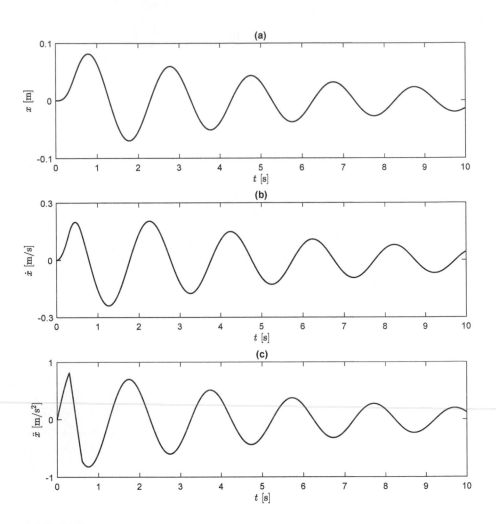

FIGURE 1.30

Displacement (a), velocity (b) and acceleration (c) responses of the SDoF system to the dynamic loading $F(t)$. Calculations are conducted using the central difference method with a time increment of $\Delta t = 0.01$ s.

TABLE 1.1

Examples to conditionally and unconditionally stable time-domain methods.

	Conditionally stable	Unconditionally stable	Stability limit
Average acceleration		•	
Linear acceleration	•		$\Delta t < T_1/1.814$
Central difference	•		$\Delta t < T_1/\pi$

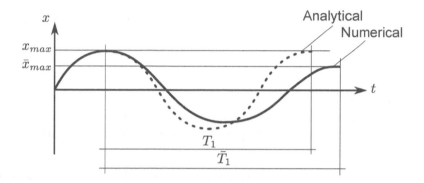

FIGURE 1.31

Comparison of the numerical solution with the theoretical solution. The computational error causes an amplitude decay $(AD = x_{max} - \bar{x}_{max})$ and a period elongation $(PE = (\bar{T}_1 - T_1)/T_1)$ in the calculated time history.

necessary to ensure the stability of conditionally stable methods. However, in structural engineering, usually calculations require shorter time steps to be able to represent the dynamics of systems by involving relevant frequencies, cf. Section 1.4.

Numerical calculations of system dynamics involve always *computational error*. The effects of this error can be observed by the *amplitude decay*, which is associated with the numerical damping. Furthermore, depending on the applied method and chosen time step size a *period elongation* can also be observed as shown in Figure 1.31.

1.6 Multi-Degree-of-Freedom Systems

For some systems it is not accurate to describe their dynamic behavior by using a single DoF. Systems, which encompass more than one DoF, are referred to as *multi-degree-of-freedom systems* (MDoF). The EoMs of these systems

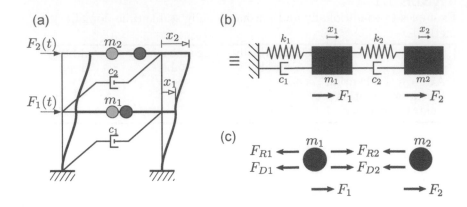

FIGURE 1.32

A 2-DoF system (a) and its alternative representation with mass-spring-damper elements (b) with corresponding internal and external forces (c).

are formulated by vectors of their acceleration, velocity and displacement responses at the DoFs. Matrices are used to describe their mass, damping and stiffness properties. For an MDoF system with n DoFs, the vectors and matrices are of size $n \times 1$ and $n \times n$, respectively.

Figure 1.32 (a) shows a 2-DoF damped system, which is subjected to external dynamic forces F_1 and F_2. Figure 1.32 (b) shows an equivalent mass-spring-damper system. Internal and external forces of the system are depicted in Figure 1.32 (c).

According to NEWTON's second law of motion, we can write the acting internal and external forces as

$$F_1 - F_{D1} - F_{R1} + F_{D2} + F_{R2} = m_1\ddot{x}_1, \tag{1.200}$$

$$F_2 - F_{D2} - F_{R2} = m_2\ddot{x}_2, \tag{1.201}$$

where the damping forces F_D are calculated by using the damping coefficients c and the restoring forces F_R are calculated by using the stiffness values k as

$$F_{D1} = c_1\dot{x}_1, \quad F_{D2} = c_2(\dot{x}_2 - \dot{x}_1),$$
$$F_{R1} = k_1 x_1, \quad F_{R2} = k_2(x_2 - x_1). \tag{1.202}$$

By substituting the internal damping and restoring force in Equations 1.200 and 1.201, we determine the EoMs, which represent the dynamic equilibrium of the 2-DoF system, as

$$m_1\ddot{x}_1 + c_1\dot{x}_1 - c_2(\dot{x}_2 - \dot{x}_1) + k_1 x_1 - k_2(x_2 - x_1) = F_1, \tag{1.203}$$

$$m_2\ddot{x}_2 + c_2(\dot{x}_2 - \dot{x}_1) + k_2(x_2 - x_1) = F_2, \tag{1.204}$$

The EoMs can be represented in matrix form as

$$\Leftrightarrow \begin{bmatrix} m_1 & 0 \\ 0 & m_2 \end{bmatrix} \begin{bmatrix} \ddot{x}_1 \\ \ddot{x}_2 \end{bmatrix} + \begin{bmatrix} c_1 + c_2 & -c_2 \\ -c_2 & c_2 \end{bmatrix} \begin{bmatrix} \dot{x}_1 \\ \dot{x}_2 \end{bmatrix}$$

$$+ \begin{bmatrix} k_1 + k_2 & -k_2 \\ -k_2 & k_2 \end{bmatrix} \begin{bmatrix} x_1 \\ x_2 \end{bmatrix} = \begin{bmatrix} F_1 \\ F_2 \end{bmatrix}, \quad (1.205)$$

or in short form as

$$\mathbf{M}\ddot{\mathbf{x}} + \mathbf{C}\dot{\mathbf{x}} + \mathbf{K}\mathbf{x} = \mathbf{f}(t). \quad (1.206)$$

1.6.1 Natural Frequencies and Modes

MDoF systems exhibit after certain initial displacement configurations a simple harmonic motion corresponding to their initial deflected shape. These characteristic deflected shapes are referred to as *natural modes* or *mode shapes*. An MDoF system with n DoFs encompasses n natural modes.

For each mode $\boldsymbol{\phi}_i$, a corresponding *natural period* T_i exists, which is equal to the time required for one cycle of the simple harmonic motion after the associated initial displacement. The corresponding *natural (angular) frequency* is ω_i and the corresponding *natural (cyclic) frequency* is f_i.

Figure 1.33 illustrates modes shapes of a 2-DoF system. The system exhibits corresponding to the number of DoF two mode shapes $\boldsymbol{\phi}_1$ and $\boldsymbol{\phi}_2$. The coordinates of the mode shape are denoted for the first mode shape as ϕ_{11} and ϕ_{21} corresponding to the first and second DoFs, respectively. Accordingly, for the second mode shape, ϕ_{12} and ϕ_{22} are used.

In mode shapes, the points of zero displacement are referred to as *nodes*. The shown 2-DoF system exhibits its first node in the second mode shape. With increasing mode number, the number of nodes increases as well. Examples of MDoF systems with three DoFs are shown in Figure 1.34. These systems exhibit two nodes in their third mode shape.

Furthermore, Figure 1.35 shows the mode shapes of a bridge structure. The structure is modeled by finite-elements (FEs). The model has several thousands of DoFs and exhibits, consequently a high number of mode shapes. The first five modes are depicted.

The natural frequencies and modes of undamped MDoF systems are determined from the nontrivial solution of the eigenvalue problem, which is given in Equation 1.212. This approach is valid also for systems with low damping, e.g., civil engineering structures. To formulate the eigenvalue problem, first, we consider the EoM of the undamped MDoF system

$$\mathbf{M}\ddot{\mathbf{x}} + \mathbf{K}\mathbf{x} = \mathbf{0}. \quad (1.207)$$

The free vibration of the system in one of its modes is

$$\mathbf{x}(t) = q_i(t)\boldsymbol{\phi}_i, \quad (1.208)$$

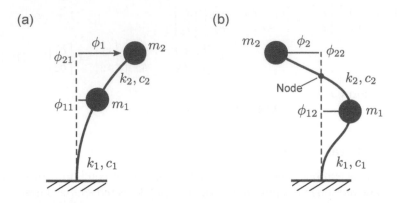

FIGURE 1.33
First (a) and second (b) mode shapes of a 2-DoF cantilever beam.

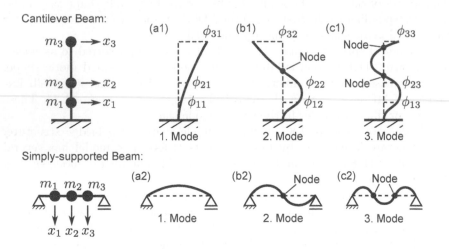

FIGURE 1.34
First (a1), second (b1) and third mode shapes (c1) of a 3-DoF cantilever beam. First (a2), second (b2) and third mode shapes (c2) of a 3-DoF simply supported beam.

$$f_1 = 0.70 Hz \qquad\qquad f_2 = 0.89 Hz$$

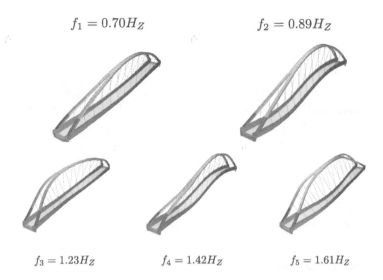

$$f_3 = 1.23 Hz \qquad\qquad f_4 = 1.42 Hz \qquad\qquad f_5 = 1.61 Hz$$

FIGURE 1.35
The first five natural modes calculated from the FE model of a bridge structure.

where $\phi_i \in \mathbb{R}^{n \times 1}$ for $i = \{1, 2, \ldots, n\}$ is the time independent vector of the characteristic deflected shape and q_i is the harmonic function, which can be written as

$$q_i = A_i \cos(\omega_i t) + B_i \sin(\omega_i t). \tag{1.209}$$

By substituting these terms in the EoM, we obtain

$$\left(-\omega_i^2 \mathbf{M}\phi_i + \mathbf{K}\phi_i\right) q_i = \mathbf{0}. \tag{1.210}$$

The nontrivial solution must satisfy the matrix *eigenvalue problem*

$$\mathbf{K}\phi_i = \omega_i^2 \mathbf{M}\phi_i, \tag{1.211}$$

The eigenvalue problem can be rewritten as

$$\left(\mathbf{K} - \omega_i^2 \mathbf{M}\right) \phi_i = \mathbf{0}. \tag{1.212}$$

The equation has a nontrivial solution, if the following *characteristic equation* (*the frequency equation*) is satisfied:

$$\det\left(\mathbf{K} - \omega_i^2 \mathbf{M}\right) = 0 \tag{1.213}$$

The scalar natural frequencies ω_i are determined from the n roots of this equation. Natural modes ϕ_i of the system are calculated by substituting each ω_i for $i = \{1, 2, \ldots, n\}$ in Equation 1.212. The application is shown in Example 1.9.

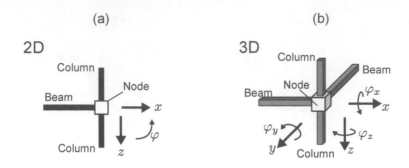

FIGURE 1.36
Discretization of the DoFs in 2D (a) and 3D (b) spaces.

1.6.2   Discretization

Structures can be discretized by elements and nodes, at which the elements are connected and the responses of the system are calculated corresponding to the DoFs. In a 2D-system, each node involves 3xDoFs, which are two translations and one rotation. In a 3D-system, each node has 6xDoFs, which are three translations and three rotations. Both cases are shown in Figure 1.36.

   The structural mass can be approximated by distributing it to the nodes as *lumped masses* corresponding to the DoFs of the system. The *lumped mass matrix* **M** of an n-DoF system reads

$$\mathbf{M} = \begin{bmatrix} m_1 & \cdots & 0 \\ \vdots & \ddots & \vdots \\ 0 & \cdots & m_n \end{bmatrix}_{n \times n}, \tag{1.214}$$

where, depending on the discretization, if the inertial forces of the DoFs are independent from each other, the off-diagonal terms vanish with $m_{ij} = 0$ for $i \neq j$.

1.6.3   Reduction of Degrees-of-Freedom

For the simplification of the computation, numerous approaches have been proposed to reduce the DoFs of systems. The following section will introduce three of these methods.

1.6.3.1   Static Condensation

By the *static condensation* method, the DoFs with zero mass can be eliminated. For this purpose, the DoFs are grouped as master DoFs $\mathbf{x}_t$ and slave

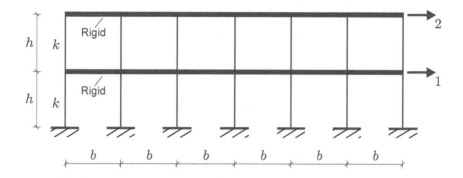

FIGURE 1.37

Reduction of DoFs of a frame structure by kinematic constraints. Floor elements are rigid in translational direction. Furthermore, the columns are rigid in vertical direction. Accordingly, rotational and vertical DoFs can be eliminated. The response of the structure can be represented by two horizontal DoFs.

DoFs $\mathbf{x}_0$. Accordingly, the EoMs are partitioned and read

$$\begin{bmatrix} \mathbf{M}_{tt} & \mathbf{0} \\ \mathbf{0} & \mathbf{0} \end{bmatrix} \begin{bmatrix} \ddot{\mathbf{x}}_t \\ \ddot{\mathbf{x}}_0 \end{bmatrix} + \begin{bmatrix} \mathbf{K}_{tt} & \mathbf{K}_{t0} \\ \mathbf{K}_{0t} & \mathbf{K}_{00} \end{bmatrix} \begin{bmatrix} \mathbf{x}_t \\ \mathbf{x}_0 \end{bmatrix} = \begin{bmatrix} \mathbf{f}_t \\ \mathbf{0} \end{bmatrix}, \qquad (1.215)$$

for an undamped MDoF system. The stiffness matrix of the system, which is condensed to the master DoFs, is determined as

$$\tilde{\mathbf{K}}_{tt} = \mathbf{K}_{tt} - \mathbf{K}^{\mathbf{T}}_{0t} \mathbf{K}^{-1}_{00} \mathbf{K}_{0t}. \qquad (1.216)$$

1.6.3.2 Kinematic Constraints

The number of DoFs of a system can be also reduced by kinematic constraints, in which the displacements of many DoFs are expressed in a condensed manner.

Figure 1.37 shows an example frame structure for this approach. The floors of the structure are stiff and can be assumed as rigid in translational direction. Accordingly, rotational DoFs can be neglected and the same horizontal DoF is applied to the nodes of each floor. Furthermore, columns of such structures are generally rigid in axial direction. Accordingly, the vertical DoFs at the nodes can be neglected as well. The DoFs of the structure are reduced to two horizontal DoFs.

1.6.3.3 Rayleigh-Ritz Method

The RAYLEIGH-RITZ method is a general approach to reduce the original set of n equations of system responses $\mathbf{x}(t)$ to a smaller set of j equations in a generalized coordinate represented by $\mathbf{z}(t)$. Accordingly, if we consider the general EoM of a damped MDoF system

$$\mathbf{M}\ddot{\mathbf{x}} + \mathbf{C}\dot{\mathbf{x}} + \mathbf{K}\mathbf{x} = \mathbf{f}(t) \tag{1.217}$$

the time-dependent system response reads

$$\mathbf{x}(t) = \sum_{r=1}^{j} \boldsymbol{\psi}_r z_r(t) = \boldsymbol{\Psi}\mathbf{z}(t). \tag{1.218}$$

The RITZ vectors $\boldsymbol{\psi}_r$ (shape vectors) are linearly independent and satisfy the geometric conditions of the MDoF system. The Ritz vectors can be determined from natural mode shapes by visually approximating. The matrix $\boldsymbol{\Psi}$ involves the RITZ vectors. Substituting the RITZ transformation $\mathbf{x} = \boldsymbol{\Psi}\mathbf{z}$ in EoM of the MDoF system gives

$$\mathbf{M}\boldsymbol{\Psi}\ddot{\mathbf{z}} + \mathbf{C}\boldsymbol{\Psi}\dot{\mathbf{z}} + \mathbf{K}\boldsymbol{\Psi}\mathbf{z} = \mathbf{s}f(t), \tag{1.219}$$

where we introduce the incidence vector $\mathbf{s}$ to distribute the time-dependent load function $f(t)$ to the DoFs of the system. By multiplying the EoM with $\boldsymbol{\Psi}^\top$ we get

$$\boldsymbol{\Psi}^\top\mathbf{M}\boldsymbol{\Psi}\ddot{\mathbf{z}} + \boldsymbol{\Psi}^\top\mathbf{C}\boldsymbol{\Psi}\dot{\mathbf{z}} + \boldsymbol{\Psi}^\top\mathbf{K}\boldsymbol{\Psi}\mathbf{z} = \boldsymbol{\Psi}^\top\mathbf{s}f(t), \tag{1.220}$$

from which the reduced mass $\tilde{\mathbf{M}}$, damping $\tilde{\mathbf{C}}$ and stiffness $\tilde{\mathbf{K}}$ matrices as well as the reduced load vector $\tilde{\mathbf{F}}$ are determined as

$$\begin{aligned} \tilde{\mathbf{M}} &= \boldsymbol{\Psi}^\top\mathbf{M}\boldsymbol{\Psi}, \quad \tilde{\mathbf{C}} = \boldsymbol{\Psi}^\top\mathbf{C}\boldsymbol{\Psi}, \\ \tilde{\mathbf{K}} &= \boldsymbol{\Psi}^\top\mathbf{K}\boldsymbol{\Psi}, \quad \tilde{\mathbf{F}} = \boldsymbol{\Psi}^\top\mathbf{s}. \end{aligned} \tag{1.221}$$

With the reduced matrices and the reduced load vector, we can compute the structural response in the generalized coordinate $\mathbf{z}(t)$, such as by using time-domain methods, cf. Section 1.5. Furthermore, we can calculate the natural frequencies and modes of the system by solving the eigenvalue problem, cf. Section 1.6.1.

Example 1.8 *A tower structure with four DoFs is given as shown in Figure 1.38 with its the first mode vector $\boldsymbol{\phi}_1$. Furthermore, mass and stiffness parameters are provided as $m = 100$ t and $k = 10,000$ kN m^{-1}. The damping ratio of the structure is $D = 1$ %. A harmonic load with the load amplitude $F_0 = 10$ kN is applied to x_4.*

We reduce the 4-DoF structure to a SDoF system by the RAYLEIGH-RITZ method to calculate its responses. The mass and stiffness matrices of the initial system read

$$\mathbf{M} = \begin{bmatrix} 2m & 0 & 0 & 0 \\ 0 & 2m & 0 & 0 \\ 0 & 0 & m & 0 \\ 0 & 0 & 0 & m \end{bmatrix}, \quad \mathbf{K} = \begin{bmatrix} 2k & -k & 0 & 0 \\ -k & 2k & -k & 0 \\ 0 & -k & 2k & -k \\ 0 & 0 & -k & k \end{bmatrix}. \tag{1.222}$$

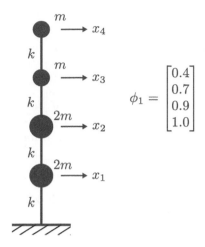

FIGURE 1.38
Tower structure and its approximated first natural mode ϕ_1. The DoFs of the structure can be reduced by the RAYLEIGH-RITZ method.

The RITZ *vector is determined from the first mode shape*

$$\psi = \phi_1 = [0.4 \; 0.7 \; 0.9 \; 1.0]^\top. \tag{1.223}$$

Accordingly, we calculate the reduced mass $\tilde{\mathbf{M}}$ *and stiffness* $\tilde{\mathbf{K}}$ *matrices as*

$$\tilde{\mathbf{M}} = \boldsymbol{\Psi}^\top \mathbf{M} \boldsymbol{\Psi} \tag{1.224}$$

$$= [0.4 \quad 0.7 \quad 0.9 \quad 1.0] \cdot m \cdot \begin{bmatrix} 2 & 0 & 0 & 0 \\ 0 & 2 & 0 & 0 \\ 0 & 0 & 1 & 0 \\ 0 & 0 & 0 & 1 \end{bmatrix} \cdot \begin{bmatrix} 0.4 \\ 0.7 \\ 0.9 \\ 1.0 \end{bmatrix} \tag{1.225}$$

$$= 311 \; t. \tag{1.226}$$

$$\tilde{\mathbf{K}} = \boldsymbol{\Psi}^\top \mathbf{K} \boldsymbol{\Psi} \tag{1.227}$$

$$= [0.4 \quad 0.7 \quad 0.9 \quad 1.0] \cdot k \cdot \begin{bmatrix} 2 & -1 & 0 & 0 \\ -1 & 2 & -1 & 0 \\ 0 & -1 & 2 & -1 \\ 0 & 0 & -1 & 1 \end{bmatrix} \cdot \begin{bmatrix} 0.4 \\ 0.7 \\ 0.9 \\ 1.0 \end{bmatrix} \tag{1.228}$$

$$= 3,000 \; kN \; m^{-1}. \tag{1.229}$$

The system is now reduced to an SDoF system with the mass and stiffness parameters $m_1 = 311 \; t$ *and* $k_1 = 3,000 \; kN \; m^{-1}$, *respectively. From the amplitude* F_0 *of the harmonic load, we calculate the static* x_{st} *and maximum* displacements x_{max} as

$$x_{st} = \frac{F_0}{k_1} = 0.003 \ m, \tag{1.230}$$

$$x_{max} = \frac{x_{st}}{2D} = 0.150 \ m. \tag{1.231}$$

Due to harmonic excitation, the response of the system is also expected to be harmonic. Accordingly, we determine the maximum velocity $\dot{x}_{max}$ and the corresponding acceleration $\ddot{x}_{max}$ of the system from its displacement amplitude x_{max} as

$$\dot{x}_{max} = x_{max}\omega_1 = 0.47 \ m \ s^{-1}, \tag{1.232}$$

$$\ddot{x}_{max} = x_{max}\omega_1^2 = 1.45 \ m \ s^{-2}, \tag{1.233}$$

where the natural frequency is calculated from the mass and stiffness parameters as

$$\omega_1 = \sqrt{k_1/m_1} = 3.11 \ rad \ s^{-1}. \tag{1.234}$$

1.7   Modal Analysis

We consider an MDoF system with n DoFs. The EoM of such a system reads

$$\mathbf{M\ddot{x} + C\dot{x} + Kx = f}(t) \tag{1.235}$$

and is coupled through the off-diagonal terms of the matrices $\mathbf{M}$, $\mathbf{C}$ and $\mathbf{K}$, which are referred to as *coupling terms*.

As an example, the matrices of a 2-DoF frame structure are given below. Here, the off-diagonal terms c_2 and k_2 are coupling the equations.

$$\mathbf{M} = \begin{bmatrix} m_1 & 0 \\ 0 & m_2 \end{bmatrix} \quad \mathbf{C} = \begin{bmatrix} c_1 + c_2 & -c_2 \\ -c_2 & c_2 \end{bmatrix} \quad \mathbf{K} = \begin{bmatrix} k_1 + k_2 & -k_2 \\ -k_2 & k_2 \end{bmatrix} \tag{1.236}$$

Due to coupling terms, the solution of the differential equations of the EoM must be solved together. Accordingly, the previously introduced frequency- and time-domain methods cannot be applied to compute the structural response without uncoupling of the equations. For this purpose, the EoM of the time-dependent structural response x_i can be rewritten in modal coordinates q_i in a similar manner to the RAYLEIGH-RITZ reduction method, cf. Section 1.6.3.3, and reads

$$\mathbf{x}(t) = \sum_{r=1}^{N} \boldsymbol{\phi}_r q_r(t) = \boldsymbol{\Phi}\mathbf{q}(t), \tag{1.237}$$

where the vectors $\boldsymbol{\phi}_r$ with $r = \{1, \ldots, N\}$ are the *mode shapes* of the MDoF system, cf. Section 1.6.1, and $N \leq n$. The matrix $\boldsymbol{\Phi} \in \mathbb{R}^{n \times N}$ is called as the *modal matrix*. Finally, $\mathbf{q}(t) \in \mathbb{R}^{n \times 1}$ represents the vector of *modal coordinates*.

1.7.1 Modal Analysis without Damping

The EoM of a linear undamped MDoF system with n-DoFs reads

$$\mathbf{M}\ddot{\mathbf{x}} + \mathbf{K}\mathbf{x} = \mathbf{f}(t). \tag{1.238}$$

After transforming the system response to modal coordinates using Equation 1.237, we rewrite the EoM as

$$\sum_{r=1}^{N} \mathbf{M}\boldsymbol{\phi}_r \ddot{q}_r + \sum_{r=1}^{N} \mathbf{K}\boldsymbol{\phi}_r q_r = \mathbf{f}(t). \tag{1.239}$$

where $N \leq n$. Furthermore, we multiply each term with the transpose vector of the i. mode shape $\boldsymbol{\phi}_i^\top$ and get

$$\sum_{r=1}^{N} \boldsymbol{\phi}_i^\top \mathbf{M}\boldsymbol{\phi}_r \ddot{q}_r + \sum_{r=1}^{N} \boldsymbol{\phi}_i^\top \mathbf{K}\boldsymbol{\phi}_r q_r = \boldsymbol{\phi}_i^\top \mathbf{f}(t). \tag{1.240}$$

Here, because of the orthogonality, all terms in each of the summations vanish, except the $r = i$ term, which yields

$$\left(\boldsymbol{\phi}_i^\top \mathbf{M}\boldsymbol{\phi}_i\right) \ddot{q}_i + \left(\boldsymbol{\phi}_i^\top \mathbf{K}\boldsymbol{\phi}_i\right) q_i = \boldsymbol{\phi}_i^\top \mathbf{f}(t). \tag{1.241}$$

We can shorten this equation as

$$\hat{m}_i \ddot{q}_i + \hat{k}_i q_i = \hat{F}_i, \tag{1.242}$$

where $\hat{m}_i$, $\hat{k}_i$ and $\hat{F}_i$ are referred to as the *generalized mass, generalized stiffness* and *generalized force* corresponding to the i. mode of the MDoF system. These terms are computed as

$$\hat{m}_i = \boldsymbol{\phi}_i^\top \mathbf{M}\boldsymbol{\phi}_i, \qquad \hat{k}_i = \boldsymbol{\phi}_i^\top \mathbf{K}\boldsymbol{\phi}_i, \qquad \hat{F}_i = \boldsymbol{\phi}_i^\top \mathbf{f}(t). \tag{1.243}$$

In Equation 1.242, the only unknown is the system response in the modal coordinate q_i. There are in total N possible equations corresponding to the number of mode shapes. The set of n coupled differential EoMs are now transformed to the set of N uncoupled equations representing generalized SDoF systems for each mode as shown in Figure 1.39. In matrix form, the set of N uncoupled equations are written as

$$\hat{\mathbf{M}}\ddot{\mathbf{q}} + \hat{\mathbf{K}}\mathbf{q} = \hat{\mathbf{f}}(t). \tag{1.244}$$

Dividing Equation 1.242 by $\hat{m}_i$ gives the *modal equation* of the generalized SDoF system without damping as

$$\ddot{q}_i + \omega_i^2 q_i = \frac{\hat{F}_i}{\hat{m}_i} = f_i, \tag{1.245}$$

where ω_i is the natural angular frequency of vibration for the i. mode.

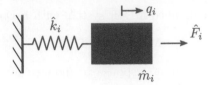

FIGURE 1.39

Generalized SDoF system without damping for the i. mode represented by the uncoupled EoM in modal coordinate q_i.

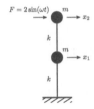

FIGURE 1.40

MDoF system without damping under harmonic excitation $F(t)$ with the excitation frequency ω.

Example 1.9 *The undamped MDoF system shown in Figure 1.40 is subjected to a harmonic load $F(t)$ with the excitation frequency $\omega = 1$ rad s^{-1}. System mass and stiffness parameters are given as $m = 1$ t, $k = 5$ kN m^{-1}, respectively.*

To calculate the dynamic response of the system, a modal analysis is necessary. For this purpose, we first write the mass and stiffness matrices

$$\mathbf{M} = \begin{bmatrix} m & 0 \\ 0 & m \end{bmatrix}, \quad \mathbf{K} = \begin{bmatrix} 2k & -k \\ -k & k \end{bmatrix} \qquad (1.246)$$

and solve the eigenvalue problem with

$$det(\mathbf{K} - \omega_n^2\mathbf{M}) = (2k - m\omega^2)(k - m\omega^2)$$
$$= m^2\omega^4 - (3km)\omega^2 + k^2 = 0 \qquad (1.247)$$

to determine the natural frequencies of the system:

$$\omega_1 = 1.38 \ rad \ s^{-1}, \quad \omega_2 = 3.62 \ rad \ s^{-1}. \qquad (1.248)$$

Furthermore, using the natural frequencies, we calculate the mode shapes. For the first natural mode, we solve

$$(\mathbf{K} - \omega_1^2\mathbf{M})\boldsymbol{\phi}_1 = \begin{bmatrix} 8.09 & -5 \\ -5 & 3.09 \end{bmatrix} \begin{bmatrix} \phi_{11} \\ \phi_{21} \end{bmatrix} = \begin{bmatrix} 0 \\ 0 \end{bmatrix}, \qquad (1.249)$$

an get for $\phi_{11} = 1$

$$\phi_1 = \begin{bmatrix} \phi_{11} \\ \phi_{21} \end{bmatrix} = \begin{bmatrix} 1 \\ 0.62 \end{bmatrix}. \tag{1.250}$$

Accordingly, for the second natural mode, we solve the eigenvalue problem

$$(\mathbf{K} - \omega_2^2 \mathbf{M})\phi_2 = 0 \tag{1.251}$$

and determine the corresponding natural frequency as $\omega_2^2 = 13.09\ rad\ s^{-1}$. Consequently, the second natural mode reads for $\phi_{22} = 1$

$$\phi_2 = \begin{bmatrix} -0.62 \\ 1 \end{bmatrix}. \tag{1.252}$$

Using the modes, we determine the generalized mass of the system as

$$\hat{m}_1 = \phi_1^\top \mathbf{M} \phi_1 = 1.38\ t, \quad \hat{m}_2 = \phi_2^\top \mathbf{M} \phi_2 = 1.38\ t \tag{1.253}$$

and the generalized stiffness as

$$\hat{k}_1 = \phi_1^\top \mathbf{M} \phi_1 = 5.73\ kN\ m^{-1}, \quad \hat{k}_2 = \phi_2^\top \mathbf{K} \phi_2 = 15\ kN\ m^{-1} \tag{1.254}$$

as well as the generalized forces as

$$\hat{F}_1(t) = \phi_1^\top \mathbf{F} = [\phi_{11}\ \phi_{21}] \begin{bmatrix} 0 \\ 2\ sin(\omega t) \end{bmatrix} = 1.24\ sin(\omega t), \tag{1.255}$$

$$\hat{F}_2(t) = \phi_2^\top \mathbf{F} = [\phi_{12}\ \phi_{22}] \begin{bmatrix} 0 \\ 2\ sin(\omega t) \end{bmatrix} = 2\ sin(\omega t). \tag{1.256}$$

Using the generalized mass, stiffness and forces, we write the modal equations as

$$\begin{aligned} \hat{m}_1 \ddot{q}_1 + \hat{k}_1 q_1 = \hat{F}_1 &\quad \Leftrightarrow 1.38 \ddot{q}_1 + 5.73 q_1 = 1.24\ sin(\omega t), \\ \hat{m}_2 \ddot{q}_2 + \hat{k}_2 q_2 = \hat{F}_2 &\quad \Leftrightarrow 1.38 \ddot{q}_2 + 15 q_2 = 2\ sin(\omega t), \end{aligned} \tag{1.257}$$

and solve for q_1 and q_2

$$q_1 = \frac{\hat{F}_1/\hat{k}_1}{1 - \left(\frac{\omega}{\omega_1}\right)^2} sin(\omega t) = 0.45 sin(\omega t), \tag{1.258}$$

$$q_2 = \frac{\hat{F}_2/\hat{k}_2}{1 - \left(\frac{\omega}{\omega_2}\right)^2} sin(\omega t) = 0.15 sin(\omega t). \tag{1.259}$$

It is noteworthy that only the steady state solution is provided here and the transient solution is neglected. We calculate the modal responses as

$$\mathbf{x}_1 = \phi_1 q_1 = \begin{bmatrix} 0.45 \\ 0.28 \end{bmatrix} sin(\omega t), \tag{1.260}$$

$$\mathbf{x}_2 = \phi_2 q_2 = \begin{bmatrix} -0.09 \\ 0.15 \end{bmatrix} sin(\omega t) \tag{1.261}$$

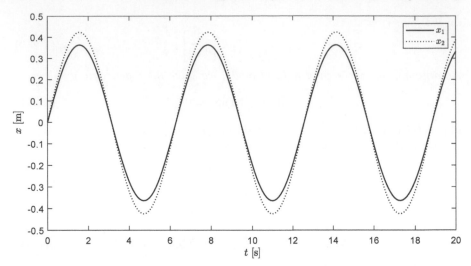

FIGURE 1.41
Responses of the 2DoF system from the modal analysis.

and combine them to find the total response of the system in x_1 and x_2 directions as

$$\mathbf{x} = \mathbf{x}_1 + \mathbf{x}_2 = \begin{bmatrix} x_1 \\ x_2 \end{bmatrix} = \begin{bmatrix} 0.36 \\ 0.42 \end{bmatrix} sin(\omega t). \tag{1.262}$$

The time histories of the system responses x_1 and x_2 are depicted in Figure 1.41. Furthermore, in Section A.5, a MATLAB code of the modal analysis method with the calculation of the example is provided.

1.7.2　Modal Analysis with Damping

The EoM of a linear-damped MDoF system with n-DoFs reads

$$\mathbf{M\ddot{x}} + \mathbf{C\dot{x}} + \mathbf{Kx} = \mathbf{f}(t) \tag{1.263}$$

and can be rewritten using the modal coordinate $\mathbf{q}$ and the natural modes of the system $\boldsymbol{\phi}_r$ as

$$\sum_{r=1}^{N} \mathbf{M}\boldsymbol{\phi}_r \ddot{q}_r + \sum_{r=1}^{N} \mathbf{C}\boldsymbol{\phi}_r \dot{q}_r + \sum_{r=1}^{N} \mathbf{K}\boldsymbol{\phi}_r q_r = \mathbf{f}(t), \tag{1.264}$$

where $N \leq n$. Unlike the case of the MDoF system without damping, these modal equations may be coupled through the terms of the damping matrix $\mathbf{C}$. Therefore, after multiplying each term of the equation with the transpose vector of the i. mode shape $\boldsymbol{\phi}_i^\top$, we get

$$\hat{m}_i \ddot{q}_i + \sum_{r=1}^{N} \hat{c}_{ir} \dot{q}_r + \hat{k}_i q_i = \hat{F}_i, \tag{1.265}$$

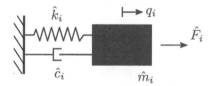

FIGURE 1.42
Generalized SDoF system with damping for the i. mode represented by the uncoupled EoM in modal coordinate q_i.

where $\hat{c}_{ir} = \phi_i^\top \mathbf{C} \phi_r$. The set of N equations, which are coupled through the damping terms, are written as

$$\hat{\mathbf{M}}\ddot{\mathbf{q}} + \hat{\mathbf{C}}\dot{\mathbf{q}} + \hat{\mathbf{K}}\mathbf{q} = \hat{\mathbf{f}}(t). \tag{1.266}$$

For certain structures with idealized damping forms, such as uncontrolled multistory buildings uniform damping distribution, these equations can be uncoupled. These systems are assumed to have *classical damping* (cf. Section 1.8) then $\hat{c}_{ir} = 0$ for $i \neq r$ and the modal equations will be uncoupled and reduced as

$$\hat{m}_i \ddot{q}_i + \hat{c}_i \dot{q}_i + \hat{k}_i q_i = \hat{F}_i, \tag{1.267}$$

where the generalized damping is given as $\hat{c}_i = \phi_i^\top \mathbf{C} \phi_i$. In matrix form, the set of N uncoupled equations are written again as

$$\hat{\mathbf{M}}\ddot{\mathbf{q}} + \hat{\mathbf{C}}\dot{\mathbf{q}} + \hat{\mathbf{K}}\mathbf{q} = \hat{\mathbf{f}}(t), \tag{1.268}$$

where, unlike Equation 1.266, this time, the damping matrix $\mathbf{C}$ is a diagonal matrix, in which the entries outside the main diagonal are zero. For further information on the construction of the damping matrix, please refer to Section 1.8.4. The generalized damped SDoF system, which is represented by the uncoupled equations, is shown in Figure 1.42. Dividing Equation 1.267 by $\hat{m}_i$ gives the *modal equation* of the generalized SDoF system with damping as

$$\ddot{q}_i + 2D_i \omega_i \dot{q}_i + \omega_i^2 q_i = \frac{\hat{F}_i}{\hat{m}_i} = f_i, \tag{1.269}$$

where ω_i is the natural angular frequency of vibration and D_i is the damping ratio of the i. mode.

The general approach for the modal analysis of MDoF systems with classical damping is summarized below.

Modal Analysis

1. Initial calculations:
   Determine mass matrix $\mathbf{M}$ and stiffness matrix $\mathbf{K}$.
   Estimate the modal damping ratios D_i.
   Determine the natural frequencies ω_i and the mode shapes $\boldsymbol{\phi_i}$ by solving the eigenvalue problem $\left[\mathbf{K} - \omega_i^2\mathbf{M}\right]\boldsymbol{\phi}_i = \mathbf{0}$.

2. Calculations for each mode $\boldsymbol{\phi}_i$:
   Solve the modal equation $\ddot{q}_i + 2D_i\omega_i\dot{q}_i + \omega_i^2 q_i = f_i$ for $q_i(t)$.
   Compute modal responses by $\mathbf{x_i}(t) = \boldsymbol{\phi}_i q_i(t)$.

3. Calculation of total system response:
   Compute the total system response by combining the modal contributions $\mathbf{x}(t) = \sum_{i=1}^n \mathbf{x}_i(t)$.

The exact value of the response of both controlled and uncontrolled systems can be only determined by including the response contributions of all modes. However, a sufficient approximation can be achieved usually with few modes. Therefore, it is not necessary to compute all modal contributions. This is especially important for MDoF systems with a large number of DoFs. Typically, in analyzing an n-DOF system, the first N modes are included, where N may be much smaller than n. Further information on this topic can be found in structural dynamics reference books, such as [2].

For a classically damped linear system, the natural modes are equal to those of the undamped case and it holds

$$\mathbf{CM}^{-1}\mathbf{K} = \mathbf{KM}^{-1}\mathbf{C}, \tag{1.270}$$

otherwise the system is *nonclassical damped* [9]. RAYLEIGH damping is a special case of this equation [10], cf. Section 1.8.4. For controlled structures, due to highly nonlinear damping behavior of the control devices, classical damping assumption is usually not accurate enough. Such systems have nondiagonal entries in $\hat{C}$, which prevents the uncoupling of the EoMs. For modal analysis of these systems, please refer to [2, 10]. Alternatively, time-domain methods can be directly applied to solve coupled EoMs, cf. Sections 1.5, 1.9 and [2, 3].

1.8 Damping Models

The internal damping behavior of structures is governed by microscopic and macroscopic mechanisms of materials. To model material damping, the classical *viscous damping model* can be used as previously introduced in Sec-

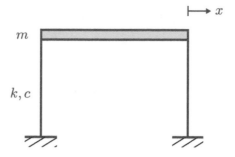

FIGURE 1.43
A simple one-story elastic structure with velocity-proportional viscous damping behavior.

tion 1.3. Furthermore, *hysteretic damping models* exist, which reflect the nonlinear damping behavior of materials based on chosen constitutive models.

Apart from material damping, structural damping is induced by friction effects within joints or supports of structural components. To model such processes commonly the COULOMB damping model is used.

1.8.1   Viscoelastic Behavior

If the responses of a system depend linearly on its deformation, the system behavior is referred to as *elastic*. However, if the responses depend linearly on the deformation rate, the system behavior is referred to as *viscous*. Accordingly, the *viscoelastic model* combines these behaviors and provides a displacement-proportional restoring force and a velocity-proportional damping force, which are defined for an SDoF system as

$$F_R = kx \qquad F_D \qquad = c\dot{x} = 2D\omega_1 m\dot{x} = \frac{\eta k}{\omega_1}\dot{x}, \qquad (1.271)$$

Here, as introduced before, k is the stiffness of the system and c is the *viscous damping coefficient*, which is equivalent to the energy dissipated in the actual system. The damping coefficient corresponds to $c = 2D\omega_1 m$ and the equation can also be formulated by using the damping ratio D and the natural angular frequency ω_1. Besides damping ratio, also the *loss factor* $\eta = 2D$ can used to represent the same damping behavior (cf. Section 4.4).

To clarify the difference of viscous and elastic responses, we consider a simple one-story structure as shown in Figure 1.43. In case of a free vibration (cf. Section 1.3.1), the harmonic response of the structure can be written as

$$x = x_0 \sin(\omega_1 t), \qquad (1.272)$$

where x_0 is the vibration amplitude and ω_1 corresponds to the natural angular

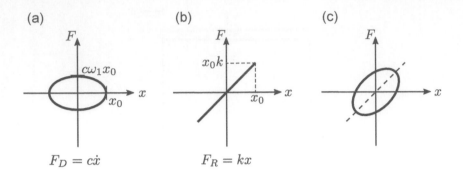

FIGURE 1.44
Viscous (a), elastic (b) and viscoelastic (c) forces of an SDoF system.

frequency of the SDoF system, if we assume that the damping of the structure is low. Using the mathematical formulation $\sin^2 a + \cos^2 a = 1$ and accordingly $\cos a = \pm(1 - \sin^2 a)^{0.5}$, we can obtain the vibration velocity as

$$\dot{x} = \pm x_0 \omega_1 \sqrt{1 - \sin^2(\omega_1 t)}. \tag{1.273}$$

Here, using the harmonic displacement equation, we replace $\sin^2(\omega_1 t)$ with x^2/x_0^2 and get

$$\dot{x} = \omega_1 \pm \sqrt{x_o^2 - x^2}. \tag{1.274}$$

The previously introduced velocity-proportional internal damping force can now be rewritten as

$$F_D = c\dot{x} = \pm c\omega_1 \sqrt{x_o^2 - x^2}, \tag{1.275}$$

which gives us after reformulating the ellipse function

$$\left(\frac{F_D}{c\omega_1 x_0}\right)^2 + \left(\frac{x}{x_0}\right)^2 = 1 \tag{1.276}$$

of the form $(y/a)^2 + (x/b)^2 = 1$. This ellipse function represents the viscous damping behavior of the system. Using this function, as shown in Figure 1.44 (a), we can compute the viscous damping force F_D over the displacement x. A further force acting on the system is the restoring force, which is depicted in Figure 1.44 (b). The sum of the viscous damping force with the restoring force represents the viscoelastic behavior of the SDoF system. The corresponding force-displacement curve is shown in Figure 1.44 (c).

The amount of damping energy of the system can be calculated from the area enclosed by the curve of the damping force plot shown in Figure 1.44 (a) and reads

$$E_D = \pi c \omega_1 x_o^2. \tag{1.277}$$

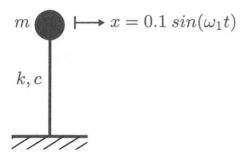

FIGURE 1.45
A viscoelastic SDoF system exhibiting harmonic displacement $x(t)$, where *omega*$_1$ is the natural angular frequency of the system.

Similarly, the elastic energy of the SDoF system corresponds to the area below the curve defining the restoring force, which is shown in Figure 1.44 (b). Accordingly, the elastic energy of the SDoF system reads

$$E_E = \frac{1}{2}x_0^2 k, \tag{1.278}$$

which is used previously in Equation 1.19 to calculate the potential energy of an SDoF system. From these two definitions, we find a relationship between the damping ratio of the SDoF system and the energy definitions by manipulating the equation of the damping ratio as

$$D = \frac{c}{2m\omega_1} \cdot \frac{\pi\omega_1 x_0^2}{\pi\omega_1 x_0^2} = \frac{E_D}{4\pi E_E}. \tag{1.279}$$

This equation shows that the damping ratio D corresponds also to the ratio of damping and elastic energies, which is scaled by $(4\pi)^{-1}$. Similarly, the relationship of the loss factor η with the energy definitions corresponds to

$$\eta = 2D = \frac{E_D}{2\pi E_E}. \tag{1.280}$$

Example 1.10 *An SDoF system with viscous damping is given in Figure 1.45. System mass and stiffness parameters as well as the damping coefficient are given as $m = 1$ t, $k = 16$ kN m^{-1} and $c = 0.4$ kNs m^{-1}, respectively. Accordingly, the natural angular frequency corresponds to $\omega_1 = 4$ rad s^{-1}. The system is assumed to exhibit a harmonic displacement response according to the function $x(t) = x_0 sin(\omega_1 t)$ with an amplitude of $x_0 = 0.1$ m.*

Due to harmonic displacement response, the velocity response of the system corresponds to

$$\dot{x} = x_0\omega_1 cos(\omega_1 t) = \pm\omega_1\sqrt{x_0^2 - x^2}, \tag{1.281}$$

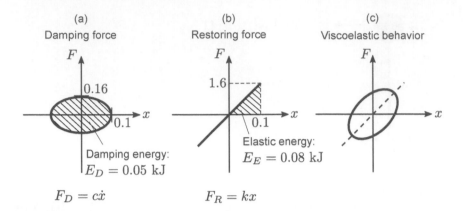

FIGURE 1.46

Damping (a) and restoring (b) forces of an SDoF system and its viscoelastic total response (c). The corresponding elastic and dissipated energies are depicted.

from which, we formulate the ellipse function of the damping force as

$$\left(\frac{F_D}{c\omega_1 x_0}\right)^2 + \left(\frac{x}{x_0}\right)^2 = 1. \tag{1.282}$$

The function is evaluated in Figure 1.46 (a). In Figure 1.46 (b), also the restoring force is evaluated. Figure 1.46 (c) shows the viscoelastic behavior of the system. We calculate from the area of the force-displacement curve of the damping the corresponding energy as

$$E_D = \pi c \omega_1 x_0^2 = 0.05 \ kNm{=}kJ \tag{1.283}$$

and from the restoring force plot the elastic energy as

$$E_E = \frac{1}{2} x_0^2 k = 0.08 \ kNm{=}kJ. \tag{1.284}$$

From both energies, the damping ratio and the loss factor can be obtained as

$$D = \frac{E_D}{4\pi E_E} = 5 \ \%, \quad \eta = 2D = 10 \ \%. \tag{1.285}$$

1.8.2   Hysteretic Damping

As introduced in the previous section, in viscoelastic material behavior, the damping force is defined by the constant damping coefficient c and the vibration velocity $\dot{x}$. Accordingly, as shown in Equation 1.277, the damping

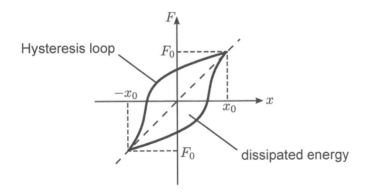

FIGURE 1.47
Hysteresis loop of a nonlinear material response plotted during a vibration cycle. The area enclosed by the curve corresponds to the dissipated energy during the cycle.

energy is proportional to the vibration frequency ω or the natural angular frequency ω_1 in case of free vibration. Consequently, in viscoelastic damping, the amount of dissipated energy increases with increasing frequency. However, the response measurements of some structures show that the energy loss can be either independent of frequency or even decrease with increasing frequency. For these systems, the frequency-dependent viscous damping mechanism is not suitable and an alternative approach is necessary.

Hysteretic damping model is frequency independent and enables, therefore, a more general representation of the damping behavior of systems. During each oscillation of the system a *hysteresis loop* is formed by the load-deformation curve, the area of which is equivalent to the total amount of dissipated energy as shown in Figure 1.47.

The EoM of an SDoF system with hysteretic damping can be formulated as

$$m\ddot{x} + F_H(x, \dot{x}) = F(t), \tag{1.286}$$

where the nonlinear force-displacement relationship of the hysteresis loop is represented by the nonlinear force F_H. The solution of this nonlinear differential equation can be carried out by numerical methods, cf. Section 1.9.

However, for low damping behavior, the equation can be linearized, in which case the computed responses of the SDoF system will correspond to the viscoelastic behavior. The linearization is performed by separating F_H into stiffness F_R and damping F_D components, which are defined as

$$F_R = kx, \quad F_D = c\dot{x}, \tag{1.287}$$

where the nonlinear stiffness of the system is approximated by an average

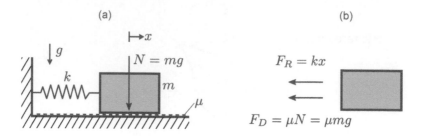

FIGURE 1.48
An SDoF system with friction (a). Restoring and friction-induced damping forces acting on the system (b).

constant stiffness k and the nonlinear damping is approximated by using the classical viscous damping formulation.

In structural control, external damping devices or materials exhibit highly nonlinear damping behavior. To model such devices and materials, specific hysteretic damping models are used. Some examples are presented in Chapter 4. For SMA-based control systems, further nonlinear modeling approaches are introduced in Chapter 8.

1.8.3   Coulomb Damping

Friction caused energy dissipation can be represented by using the COULOMB damping model. The friction resistance against sliding is defined by the damping force

$$F_D = \mu N, \tag{1.288}$$

where μ is the *friction coefficient* and N is the normal force acting on the contact surface. As SDoF system with friction is depicted in Figure 1.48. The EoM of the system reads

$$m\ddot{x} + kx + \mu N \mathrm{sgn}(\dot{x}) = 0, \tag{1.289}$$

where the friction force shows always the opposite motion direction $\dot{x}$ as considered in the equation by the signum function.

The friction-induced damping force versus system displacement is shown in Figure 1.49. From the hysteresis area, the amount of dissipated energy is determined as

$$E_D = 4\mu N x_0. \tag{1.290}$$

From the corresponding viscoelastic damping energy, the equivalent viscous damping coefficient is formulated as

$$c = \frac{4\mu N}{\pi \omega x_0}, \tag{1.291}$$

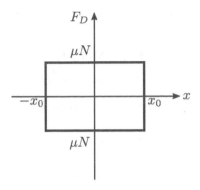

FIGURE 1.49
Damping force versus displacement of an SDoF system with friction.

which can be used to linearize SDoF systems with friction.

1.8.4 Construction of the Damping Matrix

As discussed in Section 1.7.2, unlike the case of MDoF systems without damping, the modal equations of damped systems may be coupled through the damping terms. For the i. mode of an MDoF system with n DoFs, the EoM is given as

$$\hat{m}_i\ddot{q}_i + \sum_{r=1}^{N}\hat{c}_{ir}\dot{q}_r + \hat{k}_iq_i = \hat{F}_i, \tag{1.292}$$

where N is the number of considered modes with $N \leq n$ and $\hat{c}_{ir} = \phi_i^{\top}C\phi_r$. For systems with classical damping, the modal equations can be uncoupled (cf. Section 1.7.2) and read

$$\hat{m}_i\ddot{q}_i + \hat{c}_i\dot{q}_i + \hat{k}_iq_i = \hat{F}_i. \tag{1.293}$$

The modal damping matrix of structures with classical damping is diagonal:

$$\hat{\mathbf{C}} = \begin{bmatrix} \hat{c}_1 & & & \\ & \hat{c}_2 & & \\ & & \ddots & \\ & & & \hat{c}_n \end{bmatrix}, \tag{1.294}$$

where with the generalized mass $\hat{m}_i$, the natural angular frequency of vibration ω_i and the damping ratio $\hat{D}_i$ for the i. mode, the corresponding damping coefficient is given by

$$\hat{c}_i = 2\hat{D}_i\omega_i\hat{m}_i. \tag{1.295}$$

The classical modal damping matrix can be determined using the

RAYLEIGH damping from the generalized mass and stiffness of the system as

$$\hat{\mathbf{C}} = \alpha \hat{\mathbf{M}} + \beta \hat{\mathbf{K}} \tag{1.296}$$

$$\Leftrightarrow \hat{c}_i = \alpha \hat{m}_i + \beta \omega_i^2 \hat{m}_i, \tag{1.297}$$

where the damping ratio of the i. mode is calculated by

$$\hat{D}_i = \frac{\alpha}{2}\frac{1}{\omega_i} + \frac{\beta}{2}\omega_i. \tag{1.298}$$

The α and β coefficients depend on the damping ratios D_1 and D_2, which are observed during harmonic responses with the periods T_1 and T_2, and are calculated as

$$\alpha = 4\pi \frac{T_1 D_1 - T_2 D_2}{T_1^2 - T_2^2}, \quad \beta = \frac{1}{\pi}\frac{T_1^2 T_2 D_2 - T_2^2 T_1 D_1}{T_1^2 - T_2^2}. \tag{1.299}$$

In case of only stiffness proportional damping, the coefficients are defined as

$$\alpha = 0, \quad \beta = \frac{T_1 D_1}{\pi}. \tag{1.300}$$

In case of only mass proportional damping, the coefficients read

$$\alpha = \frac{4\pi D_1}{T_1}, \quad \beta = 0. \tag{1.301}$$

Figure 1.50 compares the classical RAYLEIGH damping with the stiffness- and mass-proportional damping approaches. Here, the change of damping ratio is plotted for different natural frequencies. The stiffness proportional damping can be more realistic for most of the structures. Because the damping ratio increases for higher natural frequencies. With classical RAYLEIGH and mass proportional damping the damping ratio decreases for higher modes.

Example 1.11 *We consider again the two-story structure, which was depicted previously in Figure 1.40. This time, we assume that the structure is damped. The damping ratios are measured for the excitation frequencies $\Omega_i = \{1; 5\}$ as $D_i = \{0.01; 0.03\}$. System mass and stiffness parameters are given as $m = 1$ t, $k = 5$ kN m^{-1}.*

The mass and stiffness matrices of the system read

$$\mathbf{M} = \begin{bmatrix} m & 0 \\ 0 & m \end{bmatrix}, \quad \mathbf{K} = \begin{bmatrix} 2k & -k \\ -k & k \end{bmatrix}. \tag{1.302}$$

the corresponding periods of the excitation frequencies are

$$T_i = \frac{2\pi}{\Omega_i} = \{6.28; 1.26\} \ s. \tag{1.303}$$

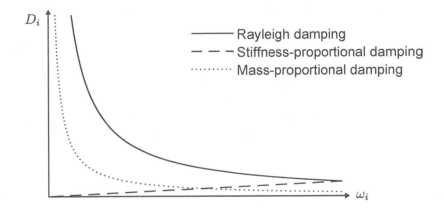

FIGURE 1.50
Damping ratio is plotted over natural frequencies for the classical RAYLEIGH damping as well as the stiffness- and mass-proportional damping.

Accordingly, for RAYLEIGH *damping, the α and β coefficients read*

$$\alpha = 4\pi \frac{T_1 D_1 - T_2 D_2}{T_1^2 - T_2^2} = 8.3 \cdot 10^{-3} \ s^{-1}, \tag{1.304}$$

$$\beta = \frac{1}{\pi} \frac{T_1^2 T_2 D_2 - T_2^2 T_1 D_1}{T_1^2 - T_2^2} = 11.7 \cdot 10^{-3} \ s. \tag{1.305}$$

Using both coefficients, we determine the damping matrix of the system as

$$\mathbf{C} = \alpha \mathbf{M} + \beta \mathbf{K} \tag{1.306}$$

$$= \alpha \begin{bmatrix} m & 0 \\ 0 & m \end{bmatrix} + \beta \begin{bmatrix} 2k & -k \\ -k & k \end{bmatrix} \tag{1.307}$$

$$= \begin{bmatrix} 0.125 & -0.058 \\ -0.058 & 0.067 \end{bmatrix} \ kNs \ m^{-1}. \tag{1.308}$$

Again, using the RAYLEIGH *damping coefficients, we calculate the damping ratios of each mode as*

$$\hat{D}_1 = \frac{\alpha}{2} \frac{1}{\omega_1} + \frac{\beta}{2}\omega_1 = 0.011, \tag{1.309}$$

$$\hat{D}_2 = \frac{\alpha}{2} \frac{1}{\omega_2} + \frac{\beta}{2}\omega_2 = 0.022, \tag{1.310}$$

with the natural angular frequencies ω_1 and ω_2, which were previously determined in Example 1.9 as 1.38 rad s^{-1} and 3.62 rad s^{-1}, respectively. Finally, *we determine the modal damping matrix of the system as*

$$\hat{C} = \begin{bmatrix} \hat{c}_1 = 2\hat{D}_1\omega_1\hat{m}_1 & 0 \\ 0 & \hat{c}_2 = 2\hat{D}_2\omega_2\hat{m}_2 \end{bmatrix} \tag{1.311}$$

$$= \begin{bmatrix} 0.031 & 0 \\ 0 & 0.161 \end{bmatrix} kNs\ m^{-1}. \tag{1.312}$$

In Section A.6, a MATLAB code is provided for the construction of the damping matrix with the calculation of the example is provided.

1.9   Nonlinear Vibrations

Systems with nonlinear material behavior respond to dynamic loads with nonlinear vibrations. The analytic solution of the EoM of these systems is usually not possible and may require numerical approaches based on time stepping methods. In Section 1.5, for linear systems, apart from the interpolation of excitation method, the NEWMARK's method and the central difference method were introduced. This section enhances these methods for nonlinear systems.

1.9.1   Newmark's Method

We consider a nonlinear SDoF system, which is represented at time step t_i by the EoM

$$m\ddot{x}_i + c\dot{x}_i + (f_K)_i = F_i, \tag{1.313}$$

where f_K is the nonlinear restoring force of the system. This force models the nonlinear material behavior of the system. At the subsequent time step t_{i+1}, the EoM reads

$$m\ddot{x}_{i+1} + c\dot{x}_{i+1} + (f_K)_{i+1} = F_{i+1}. \tag{1.314}$$

The difference of these two EoMs is given by the incremental equilibrium equation

$$m\Delta\ddot{x}_i + c\Delta\dot{x}_i + (\Delta f_K)_i = \Delta F_i. \tag{1.315}$$

As shown in Figure 1.51, the secant stiffness $k_{S,i}$ can be used to calculate the incremental restoring force $(\Delta f_K)_i$ for the displacement increment from x_i to x_{i+1}. Accordingly, the incremental restoring force reads

$$(\Delta f_K)_i = k_{S,i}\Delta x_i. \tag{1.316}$$

For small time steps, the incremental restoring force $(\Delta f_K)_i$ can be also approximated by the tangent stiffness $k_{T,i}$, cf. Figure 1.51. Accordingly, the incremental restoring force reads

$$(\Delta f_K)_i \approx k_{T,i}\Delta x_i. \tag{1.317}$$

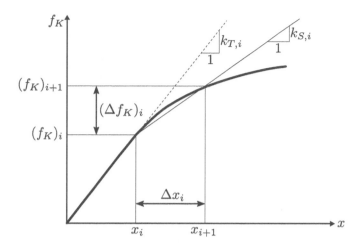

FIGURE 1.51
Restoring force f_K of a nonlinear SDoF system is plotted over its displacement x.

The equilibrium equation is then rewritten as

$$m\Delta\ddot{x}_i + c\Delta\dot{x}_i + k_{T,i}\Delta x_i = \Delta F_i. \tag{1.318}$$

From this equation, the key equation of NEWMARK's method can be obtained for nonlinear systems as

$$\hat{k}_{T,i}\Delta x_i = \Delta\hat{F}_i, \tag{1.319}$$

where the stiffness is calculated using the tangent stiffness $k_{T,i}$ by

$$\hat{k}_{T,i} = k_{T,i} + \frac{\gamma}{\beta\Delta t}c + \frac{1}{\beta\Delta t^2}m. \tag{1.320}$$

The force increment remains the same as for linear systems and reads

$$\Delta\hat{F}_i = \Delta F_i + \left(\frac{1}{\beta\Delta t}m + \frac{\gamma}{\beta}c\right)\dot{x}_i + \left(\frac{1}{2\beta}m + \Delta t\left(\frac{\gamma}{2\beta} - 1\right)c\right)\ddot{x}_i. \tag{1.321}$$

The key equation is solved in an iterative manner. The modified NEWTON-RAPHSON method can be utilized for this purpose. As shown in Figure 1.52, the tangent stiffness $\hat{k}_{T,i}$ gives the first displacement approximation $\Delta x^{(1)}$ by

$$\hat{k}_{T,i}\Delta x^{(1)} = \Delta\hat{F}_i. \tag{1.322}$$

The corresponding residual force $\Delta R^{(2)}$ is defined as the difference between the force increment $\Delta\hat{F}_i$ and the real force increment $\Delta f^{(1)}$. Accordingly, the next approximation of $\Delta x^{(2)}$ can be determined from

$$\Delta R^{(2)} = \Delta\hat{F}_i - \Delta f^{(1)} = \hat{k}_{T,i}\Delta x^{(2)}. \tag{1.323}$$

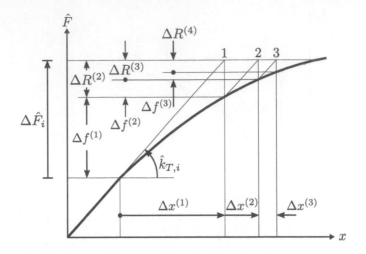

FIGURE 1.52
Iterations performed to calculate the residual force ΔR.

The iteration continues until a convergence is achieved. The calculations are terminated after n iterations when the incremental residual force $\Delta R^{(n)}$ becomes smaller than a predefined error value ϵ.

$$|\Delta R^{(n)}| < \epsilon \tag{1.324}$$

The final displacement increment is then given by

$$\Delta x_i = \sum_{j=1}^{n} \Delta x^{(j)}. \tag{1.325}$$

The general approach of the modified Newton-Raphson method is summarized below.

Modified Newton-Raphson Method

1. Initial calculations:
$$x_{i+1}^{(0)} = x_i, \quad f_K^{(0)} = (f_K)_i, \quad \Delta R^{(1)} = \Delta \hat{F}_i$$
$$\hat{k}_{T,i} = k_{T,i} + \frac{\gamma}{\beta \Delta t} c + \frac{1}{\beta \Delta t^2} m$$

2. Calculations for each iteration $j = \{1, 2, 3, \dots\}$:
$$\hat{k}_{T,i} \Delta x^{(j)} = \Delta R^{(j)} \quad \Rightarrow \Delta x^{(j)}$$
$$x_{i+1}^{(j)} = x_{i+1}^{(j-1)} + \Delta x^{(j)}$$
$$\Delta f^{(j)} = f_K^{(j)} - f_K^{(j-1)} + \left(\hat{k}_{T,i} - k_{T,i}\right) \Delta x^{(j)}$$
$$\Delta R^{(j+1)} = \Delta R^{(j)} - \Delta f^{(j)}$$

3. Repetition for the next iteration step $j + 1$ until $|\Delta R^{(n)}| < \epsilon$:
   Replace j by $j + 1$ and repeat the calculation above.

The general approach for the solution of a nonlinear SDoF system using NEWMARK's method is summarized below.

Newmark's Method for Nonlinear SDoF

1. Initial calculations:
   $\gamma = 1/2$ and $\beta = 1/4$ for average acceleration method
   $\gamma = 1/2$ and $\beta = 1/6$ for linear acceleration method
   $$\ddot{x}_0 = \frac{F_0 - c\dot{x}_0 - (f_K)_0}{m}$$
   $a = \frac{1}{\beta \Delta t} m + \frac{\gamma}{\beta} c$ and $b = \frac{1}{2\beta} m + \Delta t \left(\frac{\gamma}{2\beta} - 1\right) c$

2. Calculations for each time step t_i:
   $$\Delta \hat{F}_i = \Delta F_i + a\dot{x}_i + b\ddot{x}_i$$
   Determine the tangent stiffness $k_{T,i}$
   $$\hat{k}_{T,i} = k_{T,i} + \frac{\gamma}{\beta \Delta t} c + \frac{1}{\beta \Delta t^2} m$$
   Using the iterative modified Newton-Raphson method solve Δx_i
   from $\hat{k}_{T,i}$ and $\Delta \hat{F}_i$
   $$\Delta \dot{x}_i = \frac{\gamma}{\beta \Delta t} \Delta x_i - \frac{\gamma}{\beta} \dot{x}_i + \Delta t \left(1 - \frac{\gamma}{2\beta}\right) \ddot{x}_i$$
   $$\Delta \ddot{x}_i = \frac{1}{\beta \Delta t^2} \Delta x_i - \frac{1}{\beta \Delta t} \dot{x}_i - \frac{1}{2\beta} \ddot{x}_i$$
   $$x_{i+1} = x_i + \Delta x_i, \; \dot{x}_{i+1} = \dot{x}_i + \Delta \dot{x}_i \text{ and } \ddot{x}_{i+1} = \ddot{x}_i + \Delta \ddot{x}_i$$

3. Repetition for the next time step t_{i+1}:
   Replace i by $i + 1$ and repeat the calculation above.

Example 1.12 *A nonlinear version of the SDoF system from Example 1.6 with its nonlinear stiffness function f_k is depicted in Figure 1.53. The mass and damping parameters of the system are given as $m = 1$ t and $c = 0.316$ kNs m^{-1}, respectively. The dynamic load $F(t)$ is applied to the structure*

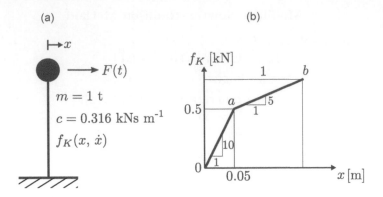

FIGURE 1.53
Nonlinear SDoF system (a) and its restoring force-displacement curve (b).

as previously shown in Figure 1.26. The system is at rest at time $t = 0$. Accordingly, the initial conditions read

$$x_0 = 0, \quad \dot{x}_0 = 0. \tag{1.326}$$

To calculate the nonlinear vibration response of the system, we use the NEWMARK's method with the modified NEWTON-RAPHSON iteration. First, we conduct the initial calculations as

$$\ddot{x}_0 = \frac{F_0 - c\dot{x}_0 - (f_K)_0}{m} = 0, \tag{1.327}$$

$$a = \frac{1}{\beta \Delta t} m + \frac{\gamma}{\beta} c = 40.63 \ t \ s^{-1}, \tag{1.328}$$

$$b = \frac{1}{2\beta} m + \Delta t \left(\frac{\gamma}{2\beta} - 1 \right) c = 2 \ t, \tag{1.329}$$

with $\gamma = 0.5$ and $\beta = 0.25$ for the average acceleration assumption. The time step size is chosen as $\Delta t = 0.1$ s corresponding to previous Example 1.6. The tangent stiffness $k_{T,i}$ is defined according to Figure 1.53 (b) with $\epsilon = 0.001$ as

$$k_{T,i} = \begin{cases} 10 \ kN \ s^{-1} & for \ 0 - a, \\ 5 \ kN \ s^{-1} & for \ a - b. \end{cases} \tag{1.330}$$

We continue with the calculations of the first time step $(i = 0)$ as

$$\hat{k}_{T,0} = k_{T,0} + \frac{\gamma}{\beta \Delta t}c + \frac{1}{\beta \Delta t^2}m = 416.32 \ kN \ m^{-1}, \tag{1.331}$$

$$\Delta \hat{F}_0 = \Delta F_0 + a\dot{x}_0 + b\ddot{x}_0 = 0.\bar{3} \ kN, \tag{1.332}$$

$$\Delta x_0 = \frac{\Delta \hat{F}_0}{\hat{k}_{T,0}} = 0.0008 \ m, \tag{1.333}$$

$$\Delta \dot{x}_0 = \frac{\gamma}{\beta \Delta t}\Delta x_0 - \frac{\gamma}{\beta}\dot{x}_0 + \Delta t\left(1 - \frac{\gamma}{2\beta}\right)\ddot{x}_0 = 0.016 \ m \ s^{-1}, \tag{1.334}$$

$$\Delta \ddot{x}_0 = \frac{1}{\beta \Delta t^2}(\Delta x_0 - \Delta t\dot{x}_0) - \frac{1}{2\beta}\ddot{x}_0 = 0.32 \ m \ s^{-2}, \tag{1.335}$$

where the initial load change is given in Figure 1.26 as $\Delta F_0 = 0.\bar{3} \ kN$. Accordingly, the system responses corresponding to the first time step read

$$x_1 = 0 + \Delta x_0 = 0.0008 \ m, \tag{1.336}$$

$$\dot{x}_1 = 0 + \Delta \dot{x}_0 = 0.016 \ m \ s^{-1}, \tag{1.337}$$

$$\ddot{x}_1 = 0 + \Delta \ddot{x}_0 = 0.32 \ m \ s^{-2}. \tag{1.338}$$

Due to linearity of the stiffness function, these steps are repeated without iteration until time step 5. The results are listed in Table 1.2. At time step 5, the displacement x reaches $0.0495 \ m$. Accordingly, for time step 6, due to a possible stiffness change an iterative solution is required. For this purpose, the modified NEWTON-RAPHSON *method is employed. First the initial calculations are conducted as*

$$x_6^{(0)} = x_5 = 0.0495 \ m, \tag{1.339}$$

$$f_K^{(0)} = (f_K)_5 = 0.5 \ kN, \tag{1.340}$$

$$\Delta R^{(1)} = \Delta \hat{F}_5 = 7.0401 \ kN, \tag{1.341}$$

$$\hat{k}_{T,5} = k_{T,5} + \frac{\gamma}{\beta \Delta t}c + \frac{1}{\beta \Delta t^2}m = 416.32 \ kN \ m^{-1}. \tag{1.342}$$

We begin with the first iteration $j = 1$ as

$$\Delta x^1 = \frac{\Delta R^{(1)}}{\hat{k}_{T,5}} = 0.0169 \ m, \tag{1.343}$$

$$x_6^{(1)} = x_6^{(0)} + \Delta x^1 = 0.0664 \ m, \tag{1.344}$$

$$\Delta f^{(1)} = f_K^{(1)} - f_K^{(0)} + \left(\hat{k}_{T,5} - k_{T,5}\right)\Delta x^{(1)} = 6.9488 \ kN, \tag{1.345}$$

$$\Delta R^{(2)} = \Delta R^{(1)} - \Delta f^{(1)} = 0.0913 \ kN. \tag{1.346}$$

with $f_K^{(1)} = 0.5 + 5 \cdot (x_6^{(1)} - 0.05) = 0.582 \ kN$, cf. Figure 1.53 (b), and $k_{T,5} =$

$10~kN~m^{-1}$. As $|\Delta R^{(2)}| > \epsilon$, the calculation continues with the second iteration with $j = 2$ as

$$\Delta x^2 = \frac{\Delta R^{(2)}}{\hat{k}_{T,5}} = 0.0002~m, \tag{1.347}$$

$$x_6^{(2)} = x_6^{(1)} + \Delta x^2 = 0.0661~m, \tag{1.348}$$

$$\Delta f^{(2)} = f_K^{(2)} - f_K^{(1)} + \left(\hat{k}_{T,5} - k_{T,5}\right)\Delta x^{(2)} = 0.0824~kN, \tag{1.349}$$

$$\Delta R^{(3)} = \Delta R^{(2)} - \Delta f^{(2)} = 0.0089~kN. \tag{1.350}$$

with $f_K^{(2)} = 0.5 + 5 \cdot (x_6^{(2)} - 0.05) = 0.583~kN$, cf. Figure 1.53 (b), and $k_{T,5} = 10~kN~m^{-1}$. As $|\Delta R^{(3)}| > \epsilon$, the calculation continues with the second iteration with $j = 3$ as

$$\Delta x^3 = \frac{\Delta R^{(3)}}{\hat{k}_{T,5}} = 0.0000~m, \tag{1.351}$$

$$x_6^{(3)} = x_6^{(2)} + \Delta x^3 = 0.0661~m, \tag{1.352}$$

$$\Delta f^{(3)} = f_K^{(3)} - f_K^{(2)} + \left(\hat{k}_{T,5} - k_{T,5}\right)\Delta x^{(3)} = 0.0081~kN, \tag{1.353}$$

$$\Delta R^{(4)} = \Delta R^{(3)} - \Delta f^{(3)} = 0.0008~kN. \tag{1.354}$$

with $f_K^{(3)} = 0.5 + 5 \cdot (x_6^{(3)} - 0.05) = 0.583~kN$, cf. Figure 1.53 (b), and $k_{T,5} = 10~kN~m^{-1}$. As $|\Delta R^{(3)}| < \epsilon$, the iteration process stops with the final displacement result $x_6 = 0.0672~m$. The corresponding velocity and acceleration values are $\dot{x}_6 = 0.150~m~s^{-1}$ and $\ddot{x}_6 = -0.632~m~s^{-2}$. Further results until $t = 1.2~s$ are listed in Table 1.2. The time histories of the calculated displacement, velocity and the acceleration responses are depicted in Figure 1.54. In Section A.7, a MATLAB code of the Newmark's method with the calculation of the example is provided.

1.9.2   Central Difference Method

Given is the EoM of a nonlinear SDoF system at time t_i as

$$m\ddot{x}_i + c\dot{x}_i + (f_K)_i = F_i, \tag{1.355}$$

Using the central difference expressions for velocity at time t_i

$$\dot{x}_i = \frac{x_{i+1} - x_{i-1}}{2\Delta t} \tag{1.356}$$

and for acceleration

$$\ddot{x}_i = \frac{x_{i+1} - 2x_i + x_{i-1}}{\Delta t^2}, \tag{1.357}$$

the EoM can be rewritten with the expression

$$m\frac{x_{i+1} - 2x_i + x_{i-1}}{\Delta t^2} + c\frac{x_{i-1} - x_i}{2\Delta t} + (f_K)_i = F_i. \tag{1.358}$$

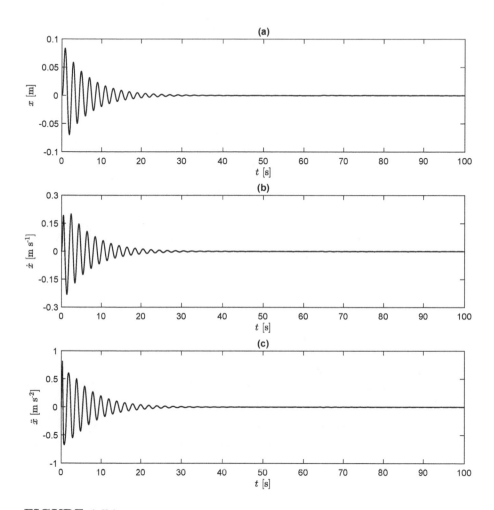

FIGURE 1.54
Displacement (a), velocity (b) and acceleration (c) responses of the nonlinear SDoF system.

TABLE 1.2
Forced vibration response of the nonlinear SDoF.

i	t_i	$\hat{k}_{T,i}$	$\Delta\hat{F}_i$	x_i	$\dot{x}_i$	$\ddot{x}_i$
-	[s]	[kN m^{-1}]	[kN]	[m]	[m s^{-1}]	[m s^{-2}]
0	0.0	416.32	0.3	0.000	0.000	0.000
1	0.1	416.32	1.625	0.001	0.016	0.320
2	0.2	416.32	4.054	0.005	0.062	0.600
3	0.3	416.32	6.686	0.014	0.133	0.814
4	0.4	416.32	7.930	0.031	0.189	0.302
5	0.5	416.32	7.041	0.050	0.193	−0.223
6	0.6	411.32	4.821	0.066	0.150	−0.632
7	0.7	411.32	2.101	0.078	0.085	−0.670
8	0.8	411.32	−0.637	0.084	0.018	−0.674
9	0.9	411.32	−3.260	0.082	−0.049	−0.645
10	1.0	411.32	−5.644	0.074	−0.110	−0.586
11	1.1	411.32	−7.681	0.060	−0.164	−0.501
12	1.2	416.32	−9.116	0.042	−0.207	−0.352

Here, compared to the linear case (cf. Section 1.5.3, only the term kx_i is replaced with the nonlinear restoring force $(f_K)_i$. Next, the system response x_{i+1} at time t_{i+1} is determined from

$$\hat{k}x_{i+1} = \hat{F}_i \qquad (1.359)$$

where the stiffness and load functions are defined as

$$\hat{k} = \frac{m}{\Delta t^2} + \frac{c}{2\Delta t}, \qquad (1.360)$$

$$\hat{F}_i = F_i - \left(\frac{m}{\Delta t^2} - \frac{c}{2\Delta t}\right)x_{i-1} - (f_K)_i + \frac{2m}{\Delta t^2}x_i. \qquad (1.361)$$

The general approach for the solution of a linear damped SDoF system using the central difference method is summarized below.

Central Difference Method

1. Initial calculations:
$$\ddot{x}_0 = \frac{F_0 - c\dot{x}_0 - (f_K)_0}{m}$$
$$x_{-1} = x_0 - \Delta t\dot{x}_0 + \frac{\Delta t^2}{2}\ddot{x}_0$$
$$\hat{k} = \frac{m}{\Delta t^2} + \frac{c}{2\Delta t}$$
$$a = \frac{m}{\Delta t^2} - \frac{c}{2\Delta t}$$

2. Calculations for each time step t_i:
$$b_i = f_{ki} - \frac{2m}{\Delta t^2} \qquad \hat{F}_i = F_i - ax_{i-1} - b_ix_i$$
$$x_{i+1} = \frac{\hat{F}_i}{\hat{k}}$$

3. Repetition for the next time step t_{i+1}:
Replace i by $i+1$ and repeat the calculation above.

2

Structural Control Systems

2.1 Introduction

This chapter is concerned with the general definition and classification of structural control systems. Section 2.2 introduces the motivation for the application of control systems and presents briefly the historical development of structural control in the context of structural engineering. Section 2.3 proposes a three-step classification scheme for control systems. In first step, control systems are classified depending on the utilized control device as tuned mass dampers (TMDs) and dissipators. In second step, the classification occurs depending on the applied operational strategy as passive, active and semi-active systems. The last step considers the incorporated materials and distinguishes between classical and smart materials.

2.2 Background

Both natural events and anthropogenic activities can induce dynamic responses on structures in the form of vibrations. Wind is a typical example, which especially for flexible tower-like buildings, imposes psychophysiological effects on occupants and deteriorates the serviceability of buildings. Further examples are onshore and offshore wind turbines. Wind and waves can induce detrimental vibrations on these structures, which lead to fatigue, limit their design lifetime and, consequently, impair energy production. Apart from economic effects, vibrations can cause collapse of structures and loss of life as unfortunately has happened several times during previous earthquakes.

Besides wind, waves and earthquake, also traffic loads, such as imposed on bridges by railways, cars and pedestrians, or further operational loads, such as caused by industrial machines, as well as extreme events, such as explosions, induce vibrations. These may also jeopardize structural integrity and human life as well as be hazardous for the operational reliability of structures.

Vibrations are particularly decisive for the design of modern structures with slender architecture. Such structures exhibit generally low inherent damping and respond highly sensitive to dynamic loads. Furthermore, vibrations play an important role for existing structures particularly during their rehabilitation and retrofitting, which can be necessary due to increased economic and environmental needs, such as rising traffic loads and extreme weather conditions.

Traditional structural design has relied so far on the improvement of strength and ductility of structures. However, this approach does not seem to be sufficient anymore to prevent vibrations and their aftereffects. Resource scarcity, architectural design demands and changing environmental conditions of the modern era require alternative solutions. In this regard, recent technological developments in computer science and particularly in cybernetics allowed the integration of structural control as a new field within the discipline of structural engineering.

Structural control involves design of solutions, which encompass methods and application of supplementary devices to improve the dynamic performance of structures. The principles of structural control originate from control engineering and have been utilized in the design of numerous engineering structures.

One of the earliest vibration control examples is the tomb of Cyrus the Great, which was built in Pasargadae, ancient Persia and dates back to 500 BC. Horizontally separated layers with relatively smooth surfaces assemble the foundation of the still existing structure. This approach allowed the massive tomb to slide during major earthquake events and dissipate its vibration energy via friction [11, 12]. Seismic protection was also considered in ancient Greece for important structures, such as in the Parthenon Temple in Athens from 400 BC. The columns of the structure are assembled by marble disks, which are connected with each other via lead dowels. Both yielding of the dowels and the friction between the disks enabled the structure to dissipate its vibration energy [13].

First modern control applications of structural engineering began by the end of 1960s, such as with the pioneering studies of Masri on two-particle impact dampers [14], Reed on hanging-chain impact dampers [15], Nordell on active systems [16] and Yao on adaptive systems [17]. These developments were followed in 1972 by the conceptual study and survey paper of Yao [18]. The first technical session in a conference related to structural control was organized with the title *Seismic Response Control of Structural Systems* during the 9th World Conference on Earthquake Engineering in Tokyo-Kyoto in 1988. In the next World Conference in Madrid in 1992, there were already several technical sessions with numerous contributions related to structural control.

Following the devastating Northridge Earthquake in California in January 1994, the International Association for Structural Control (IASC) has been established with Prof. G. W. Housner as president. In August of the same year, the first World Conference on Structural Control has been organized in Los Angeles with over 300 international participants, after which also the Journal of Structural Control and Health Monitoring has started publishing. The foundation of structural control in structural engineering has been accomplished

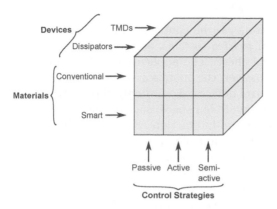

FIGURE 2.1
Classification of structural vibration control systems based on the utilized device, control strategy and material type.

by the comprehensive and foresighted paper of Housner et al. in 1997 [19], which was accompanied by further survey papers, such as by Kareem et al. [13], Soong and Spencer [20] as well as Spencer and Nagarajaiah [21]. Reference books, such as by Soong and Constantinou [22], Constantinou et al. [23], Adeli and Saleh [24], Soong and Dargush [25], Hanson and Soong [26], Petersen [27] as well as Connor [28] have accelerated this process.

2.3 Classification

Applied systems are generally classified in structural control by the type of the utilized control device, the chosen control strategy and the type of the used material as shown in Figure 2.1.

2.3.1 Control Devices

Structural control utilizes generally two approaches: reduction of the mechanical vibrations by improving the modal structural response and/or dissipation of the kinetic vibration energy by converting it to heat energy. Accordingly, two types of control devices exist. The *TMDs* are auxiliary masses, which are attached on structures to improve their modal response by exhibiting out of phase oscillations. The second approach is realized by *dissipators*, which convert the vibration energy via physical phenomena, such as friction, viscoelastic or plastic deformation and phase transformation into heat energy.

2.3.2 Control Strategies

Both TMDs and dissipators can be controlled in a passive, active and semi-active fashion. In case of *passive systems*, only the vibration of the structure itself initiates the control effect. The application of this approach is usually quite straightforward. However, the operation range is generally limited by the design parameters of control devices. Accordingly, if the structural parameters change and deviate from the design space, such as due to degradation effects, temperature and soil-structure-interaction (SSI), or if the designated usage changes, such as by an increase in traffic loads, passive systems lose their control efficiency. Performance deterioration may also arise due to uncertainties involved in the methods and data used for the estimation of the design parameters. Apart from passive systems, more advanced approaches exist and grouped as active and semi-active systems.

Active systems use external energy to generate active mechanical forces on structures. These systems were directly adopted in numerous mechanical engineering applications, where the prediction accuracy of load characteristics is usually high. However, in civil engineering, due to the random characteristics of excitation and the uncertainties in design, the response of active devices is difficult to predict and may cause a destabilization of structures. Moreover, large forces are required for the control of civil engineering structures, which can only be realized with high amounts of external energy. This can be particularly critical during major events, such as earthquakes, due to possible interruptions in power network. High capital and maintenance costs are further disadvantages of active systems.

Passive and active approaches can be also combined as *hybrid systems*. These systems offer a more robust control performance than active systems. However, they still struggle with the limitations of both passive and active control, particularly due to their lack of adaptation to the structural and environmental changes.

On the other hand, *semi-active systems* are capable of autonomous adaptation and can realize a superior control performance. So far conducted studies, such as [13, 19–21, 29], agree that semi-active systems exhibit best features of both passive and active systems. The semi-active systems generate their control forces in a passive manner and are, therefore, also called controllable passive devices. Consequently, they cannot cause any stability problems and operate with orders of magnitude less power, which can be even supplied by batteries.

2.3.3 Materials Incorporated in Control Devices

Besides the device type and the chosen control strategy, also the incorporated material is decisive for the performance of a control system. Apart from the conventional materials, *smart materials* are employed in structural control, such as piezoelectric ceramics, MR and electrorheological (ER) liquids as well as *shape memory alloys* (SMAs). Such materials exhibit unique characteristics, which can be also controlled by external stimuli, such as temperature, magnetic and electric fields as well as stress. Therefore, control devices (both TMDs and dissipators) incorporating smart materials can be applied not only in a passive but also both in active or semi-active schemes. In this context, due to their durability, the recent research focus has been particularly on the superelastic SMAs, which can recover large deformations with up to 8 % strain or even higher and efficiently dissipate vibration energy.

3

Principles of Structural Control

3.1 Introduction

In structural control, apart from classical EoMs, the state-space representation is used for the mathematical modeling of the controlled systems. The state-space representation uses first-order differential equations. Compared to traditional second-order differential equations of EoMs, the approach allows a mathematically higher efficiency, which is decisive particularly for real-time applications, such as the active and semi-active control, cf. Chapter 6. Furthermore, in structural control, numerous control algorithms have been utilized for the estimation of control parameters and the generation of required restoring and/or damping forces. To realize an optimum vibration control, these algorithms consider the conditions of structures and loads in a most efficient way.

Section 3.2 introduces the governing equations of the state-space representation and shows the application of the method on examples. Section 3.3 is concerned with control algorithms. First different operation modes are introduced, by which the control strategies are classified. Then, in particular, the control algorithms of on-off and Fuzzy logic controller (FLC) are presented.

The chapter aims to give a general overview of the methods, which are particularly suitable for semi-active systems. For further detailed information about control algorithms and in general control engineering, the interested readers are referred to reference books on the topic, such as by Lewis et al. [30]. Furthermore, control algorithms operate generally online and require an observer, which identifies the current state of structures and attached control devices, such as their accelerations, velocities and displacements, as well as their parameters, such as stiffness and damping. The topic of system identification is covered in Chapter 9.

3.2   State-Space Representation

To introduce the method, we consider the EoM of a linear dynamic system with n DoFs

$$\mathbf{M}\ddot{\mathbf{x}} + \mathbf{C}\dot{\mathbf{x}} + \mathbf{K}\mathbf{x} = \mathbf{f}, \tag{3.1}$$

where $\mathbf{M} \in \mathbb{R}^{n \times n}$, $\mathbf{C} \in \mathbb{R}^{n \times n}$ and $\mathbf{K} \in \mathbb{R}^{n \times n}$ are the mass, damping and stiffness matrices of the system. $\ddot{\mathbf{x}} \in \mathbb{R}^{n \times 1}$, $\dot{\mathbf{x}} \in \mathbb{R}^{n \times 1}$ and $\mathbf{x} \in \mathbb{R}^{n \times 1}$ are the acceleration, velocity and displacement vectors of the system. $\mathbf{f} \in \mathbb{R}^{n \times 1}$ is a force vector, which involves the external force acting on the system.

The *continuous state-space representation* of such dynamic systems consists of *transition* and *output equations*, which read

$$\dot{\mathbf{z}} = \mathcal{A}\mathbf{z} + \mathcal{B}\mathbf{u} + \mathbf{w}, \tag{3.2}$$

$$\mathbf{y} = \mathcal{C}\mathbf{z} + \mathcal{D}\mathbf{u} + \mathbf{v}. \tag{3.3}$$

Here, $\mathcal{A} \in \mathbb{R}^{2n \times 2n}$ and $\mathcal{B} \in \mathbb{R}^{2n \times n}$ are the *transition* and *input matrices* of the dynamic system:

$$\mathcal{A} = \begin{bmatrix} \mathbf{0} & \mathbf{I} \\ -\mathbf{M}^{-1}\mathbf{K} & -\mathbf{M}^{-1}\mathbf{C} \end{bmatrix}, \quad \mathcal{B} = \begin{bmatrix} \mathbf{0} \\ \mathbf{M}^{-1} \end{bmatrix}. \tag{3.4}$$

In Equation 3.3, $\mathcal{C} \in \mathbb{R}^{m \times 2n}$ and $\mathcal{D} \in \mathbb{R}^{m \times n}$ are the *output* and *feedthrough matrices*, which depend on the location and type of m sensors of the dynamic system. For instance, a system with only displacement sensors is represented by

$$\mathcal{C} = \begin{bmatrix} \mathbf{N}_y & \mathbf{0} \end{bmatrix}, \quad \mathcal{D} = \mathbf{0}, \tag{3.5}$$

where $\mathbf{N}_y \in \mathbb{R}^{m \times n}$ is the distribution matrix of the sensors. On the other hand, a system with only velocity sensors is represented by

$$\mathcal{C} = \begin{bmatrix} \mathbf{0} & \mathbf{N}_y \end{bmatrix}, \quad \mathcal{D} = \mathbf{0}. \tag{3.6}$$

In case of a system with only accelerometers, the output and feedthrough matrices read

$$\mathcal{C} = \begin{bmatrix} -\mathbf{N}_y\mathbf{M}^{-1}\mathbf{K} & -\mathbf{N}_y\mathbf{M}^{-1}\mathbf{C_D} \end{bmatrix}, \quad \mathcal{D} = \begin{bmatrix} \mathbf{N}_y\mathbf{M}^{-1} \end{bmatrix}. \tag{3.7}$$

For a dynamic system with mixed sensor types, the above given representations can be combined. In this regard, apart from Examples 3.1–3.3, Section 9.6.1 presents the state-space representation of an MDoF structure with accelerometers. Furthermore, in Section 8.7.1.2, the state-space representation is applied to an MDoF frame structure, where all displacement, velocity and acceleration responses are assumed to be observed.

In state-space representation, $\mathbf{z} \in \mathbb{R}^{2n \times 1}$ is the *state vector*, which reads

$$\mathbf{z} = \begin{bmatrix} \mathbf{x} \\ \dot{\mathbf{x}} \end{bmatrix} \tag{3.8}$$

and $\mathbf{y} \in \mathbb{R}^{m \times 1}$ is the *output vector*, which involves the measured signals. Furthermore, $\mathbf{u} \in \mathbb{R}^{n \times 1}$ is the *input vector*, which involves the external dynamic forces acting on the system. Moreover, the vectors $\mathbf{w} \in \mathbb{R}^{2n \times 1}$ and $\mathbf{v} \in \mathbb{R}^{m \times 1}$ represent the *process* and *measurement noises* of the system.

From the continuous state-space representation, it can be shown that the solution of the differential transition equation (Equation 3.2) assembles homogeneous and particular solutions as

$$\mathbf{z}(t) = e^{\mathcal{A}(t-t_0)}\mathbf{z}(t_0) + \int_{t_0}^{t} e^{\mathcal{A}(t-\tau)}\mathcal{B}\mathbf{u}(\tau)\,\mathrm{d}\tau + \mathbf{w}(t), \tag{3.9}$$

where t_0 is the initial time step. By applying the *zero-order hold* (ZOH) principle over the sampling time Δt for $t_0 = k\Delta t$ and $t = (k+1)\Delta t$ with $k \in \mathbb{Z}$, the solution reads

$$\mathbf{z}_{k+1} = e^{\mathcal{A}\Delta t}\mathbf{z}_k + \mathbf{u}_k \int_{0}^{\Delta t} e^{\mathcal{A}\tau}\mathcal{B}\,\mathrm{d}\tau + \mathbf{w}_k, \tag{3.10}$$

where according to ZOH, $\mathbf{z}(t)$, $\mathbf{u}(t)$ and $\mathbf{w}(t)$ are assumed to be piece wise constant over Δt and equal to $\mathbf{z}_k$, $\mathbf{u}_k$ and $\mathbf{w}_k$. Accordingly, we determine the discrete version of the transition equation as

$$\mathbf{z}_{k+1} = \mathcal{A}_d\mathbf{z}_k + \mathcal{B}_d\mathbf{u}_k + \mathbf{w}_k, \tag{3.11}$$

where the discrete transition matrix $\mathcal{A}_d \in \mathbb{R}^{2n \times 2n}$ and the discrete input matrix $\mathcal{B}_d \in \mathbb{R}^{2n \times n}$ are given by

$$\mathcal{A}_d = e^{\mathcal{A}\Delta t} \quad \text{and} \quad \mathcal{B}_d = \int_{0}^{\Delta t} e^{\mathcal{A}\tau}\mathcal{B}\,\mathrm{d}\tau, \tag{3.12}$$

both of which can be approximated by the TAYLOR series as

$$\mathcal{A}_d = e^{\mathcal{A}\Delta t} = \mathbf{I} + \mathcal{A}\Delta t + \frac{1}{2!}\mathcal{A}^2\Delta t^2 + \frac{1}{3!}\mathcal{A}^3\Delta t^3 + \dots, \tag{3.13}$$

$$\mathcal{B}_d = \int_{0}^{\Delta t} e^{\mathcal{A}\tau}\mathcal{B}\,\mathrm{d}\tau = \mathcal{B}\Delta t + \frac{1}{2!}\mathcal{A}\mathcal{B}\Delta t^2 + \frac{1}{3!}\mathcal{A}^2\mathcal{B}\Delta t^3 + \dots. \tag{3.14}$$

A first order approximation can also be achieved by the forward EULER method with

$$\dot{\mathbf{z}} = \frac{\mathbf{z}_{k+1} - \mathbf{z}_k}{\Delta t} = \mathcal{A}\mathbf{z}(t) + \mathcal{B}\mathbf{u}(t) + \mathbf{w}(t), \tag{3.15}$$

which yields again

$$\mathbf{z}_{k+1} = \mathcal{A}_d\mathbf{z}_k + \mathcal{B}_d\mathbf{u}_k + \mathbf{w}_k, \tag{3.16}$$

where

$$\mathcal{A}_d = \mathbf{I} + \mathcal{A}\Delta t, \tag{3.17}$$

$$\mathcal{B}_d = \mathcal{B}\Delta t. \tag{3.18}$$

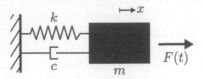

FIGURE 3.1
A damped SDoF system under the dynamic excitation $F(t)$.

As the output equation (Equation 3.3) of the state-space representation does not have a differential equation form, the discrete version can be directly written as

$$\mathbf{y}_k = \mathcal{C}\mathbf{z}_k + \mathcal{D}\mathbf{u}_k + \mathbf{v}_k, \tag{3.19}$$

where again according to ZOH, $\mathbf{y}(t)$, $\mathbf{z}(t)$, $\mathbf{u}(t)$ and $\mathbf{v}(t)$ are assumed to be piece wise constant over Δt and equal to $\mathbf{y}_k$, $\mathbf{z}_k$, $\mathbf{u}_k$ and $\mathbf{v}_k$.

Example 3.1 *We consider a damped SDoF system as shown in Figure 3.1. The system mass, stiffness and damping coefficient are given as m, k and c respectively. To the system the dynamic force $F(t)$ is applied.*
We write the transition and input matrices of the system according to Equation 3.4 as

$$\mathcal{A} = \begin{bmatrix} 0 & 1 \\ -m^{-1}k & -m^{-1}c \end{bmatrix}, \quad \mathcal{B} = \begin{bmatrix} 0 \\ m^{-1} \end{bmatrix}, \tag{3.20}$$

where $\mathcal{A} \in \mathbb{R}^{2n \times 2n}$ and $\mathcal{B} \in \mathbb{R}^{2n \times n}$ with $n = 1$ corresponding to the SDoF of the system. The state and the input vectors of the system read

$$\mathbf{z} = \begin{bmatrix} x \\ \dot{x} \end{bmatrix}, \quad \mathbf{u} = \begin{bmatrix} F(t) \end{bmatrix}, \tag{3.21}$$

which are $\mathbf{z} \in \mathbb{R}^{2n \times 1}$ and $\mathbf{u} \in \mathbb{R}^{n \times 1}$ again with $n = 1$. Accordingly, the transition equation of the system reads

$$\dot{\mathbf{z}} = \mathcal{A}\mathbf{z} + \mathcal{B}\mathbf{u} \tag{3.22}$$

corresponding to Equation 3.2 after neglecting the process noise as $\mathbf{w} = \mathbf{0}$. After substituting the transition and input matrices as well as the state and input vectors, the transition equation of the system reads

$$\begin{bmatrix} \dot{x} \\ \ddot{x} \end{bmatrix} = \begin{bmatrix} 0 & 1 \\ -m^{-1}k & -m^{-1}c \end{bmatrix} \begin{bmatrix} x \\ \dot{x} \end{bmatrix} + \begin{bmatrix} 0 \\ m^{-1} \end{bmatrix} \begin{bmatrix} F(t) \end{bmatrix}, \tag{3.23}$$

which gives

$$\dot{x} = \dot{x}, \quad \ddot{x} = -\frac{k}{m}x - \frac{c}{m}\dot{x} + \frac{1}{m}F(t), \tag{3.24}$$

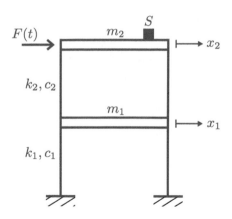

FIGURE 3.2
A damped 2-DoF system under the dynamic excitation $F(t)$. On the top floor, the horizontal vibration response of the system is observed by the velocity sensor S.

where, the first equation is a true statement and the second equation corresponds to the EoM of the SDoF system. Accordingly, the proposed transition equation of the system is valid.

Example 3.2 *We consider a damped MDoF system as shown in Figure 3.2. The system has two DoF: x_1 and x_2. To the top floor of the system the dynamic force $F(t)$ is applied. The horizontal vibration response of the system is observed by the velocity sensor S, which is attached also on the top floor. System mass, stiffness and damping parameters are given as $m_1 = m_2 = 1$ t, $k_1 = k_2 = 5$ kN m^{-1} and $c_1 = c_2 = 0.06$ kNs m^{-1}.*

The mass, stiffness and damping matrices of the system read

$$\mathbf{M} = \begin{bmatrix} m_1 & 0 \\ 0 & m_2 \end{bmatrix} = \begin{bmatrix} 1 & 0 \\ 0 & 1 \end{bmatrix} t, \tag{3.25}$$

$$\mathbf{K} = \begin{bmatrix} k_1 + k_2 & -k_2 \\ -k_2 & k_2 \end{bmatrix} = \begin{bmatrix} 10 & -5 \\ -5 & 5 \end{bmatrix} kN\ m^{-1}, \tag{3.26}$$

$$\mathbf{C} = \begin{bmatrix} c_1 + c_2 & -c_2 \\ -c_2 & c_2 \end{bmatrix} = \begin{bmatrix} 0.12 & -0.06 \\ -0.06 & 0.06 \end{bmatrix} kNs\ m^{-1}, \tag{3.27}$$

from which we get

$$\mathbf{M}^{-1} = \frac{1}{det\,\mathbf{M}}\,adj\,\mathbf{M} = \frac{1}{m_1 m_2}\begin{bmatrix} m_2 & 0 \\ 0 & m_1 \end{bmatrix} = \begin{bmatrix} 1 & 0 \\ 0 & 1 \end{bmatrix}\,t, \qquad (3.28)$$

$$-\mathbf{M}^{-1}\mathbf{K} = \begin{bmatrix} -10 & 5 \\ 5 & -5 \end{bmatrix}\,kN\,(mt)^{-1}, \qquad (3.29)$$

$$-\mathbf{M}^{-1}\mathbf{C} = \begin{bmatrix} -0.12 & 0.06 \\ 0.06 & -0.06 \end{bmatrix}\,kN\,(mt)^{-1}. \qquad (3.30)$$

Accordingly, using Equations 3.4 we determine the transition and input matrices as

$$\mathcal{A} = \begin{bmatrix} \mathbf{0} & \mathbf{I} \\ -\mathbf{M}^{-1}\mathbf{K} & -\mathbf{M}^{-1}\mathbf{C} \end{bmatrix} = \begin{bmatrix} 0 & 0 & 1 & 0 \\ 0 & 0 & 0 & 1 \\ -10 & 5 & -0.12 & 0.06 \\ 5 & -5 & 0.06 & -0.06 \end{bmatrix}, \qquad (3.31)$$

$$\mathcal{B} = \begin{bmatrix} \mathbf{0} \\ \mathbf{M}^{-1} \end{bmatrix} = \begin{bmatrix} 0 & 0 \\ 0 & 0 \\ 1 & 0 \\ 0 & 1 \end{bmatrix}, \qquad (3.32)$$

where $\mathcal{A} \in \mathbb{R}^{2n \times 2n}$ and $\mathcal{B} \in \mathbb{R}^{2n \times n}$ with $n = 2$ corresponding to the two-DoFs of the system. Furthermore, from Equation 3.6, we determine the output and feedthrough matrices of the system as

$$\mathcal{C} = \begin{bmatrix} 0 & 0 & 0 & 1 \end{bmatrix}, \quad \mathcal{D} = \begin{bmatrix} 0 & 0 \end{bmatrix}, \qquad (3.33)$$

where $\mathcal{C} \in \mathbb{R}^{m \times 2n}$ and $\mathcal{D} \in \mathbb{R}^{m \times n}$ with $n = 2$ and $m = 1$ corresponding to the two-DoFs of the system and the one sensor, respectively. The state, input and output vectors of the system read

$$\mathbf{z} = \begin{bmatrix} \mathbf{x} \\ \dot{\mathbf{x}} \end{bmatrix} = \begin{bmatrix} x_1 \\ x_2 \\ \dot{x}_1 \\ \dot{x}_2 \end{bmatrix}, \quad \mathbf{u} = \begin{bmatrix} 0 \\ F(t) \end{bmatrix}, \quad \mathbf{y} = \begin{bmatrix} \dot{x}_2 \end{bmatrix}, \qquad (3.34)$$

where $\mathbf{z} \in \mathbb{R}^{2n \times 1}$, $\mathbf{u} \in \mathbb{R}^{n \times 1}$ and $\mathbf{y} \in \mathbb{R}^{m \times 1}$ with $n = 2$ and $m = 1$ corresponding to the two-DoFs of the system and the one sensor, respectively. From Equations 3.2 and 3.3 and by neglecting the process and measurement noises as $\mathbf{w} = \mathbf{0}$ and $\mathbf{v} = \mathbf{0}$, we write the transition and output equations and determine the state-space representation of the given system as

$$\dot{\mathbf{z}} = \mathcal{A}\mathbf{z} + \mathcal{B}\mathbf{u}, \qquad (3.35)$$

$$\mathbf{y} = \mathcal{C}\mathbf{z} + \mathcal{D}\mathbf{u}. \qquad (3.36)$$

Example 3.3 *We consider again the same 2-DoF system from Example 3.2.*

We assume this time that the horizontal vibration response of the system is observed by an acceleration sensor, which is attached to the top floor of the system.

In this case, the transition and input matrices as well as the state and input vectors remain same. From Equation 3.7, we determine with $\mathbf{N}_y^\top = [0\ 1]$ the output and feedthrough matrices as

$$\mathcal{C} = \left[\begin{bmatrix} 0 \\ 1 \end{bmatrix} \begin{bmatrix} -10 & 5 \\ 5 & -5 \end{bmatrix} \right] = \begin{bmatrix} 5 & -5 & 0.06 & -0.06 \end{bmatrix}, \tag{3.37}$$

$$\mathcal{D} = \left[\begin{bmatrix} 0 \\ 1 \end{bmatrix} \begin{bmatrix} 1 & 0 \\ 0 & 1 \end{bmatrix} \right] = [0\ 1], \tag{3.38}$$

The state-space representation method will be applied to a SDoF system with a tuned mass damper (TMD) in Section 5.2.1.1. Furthermore, in Section 7.2.1.4, the representation will be applied to a system with a tuned liquid column damper.

3.3 Structural Control Algorithms

3.3.1 Operation Modes

Depending on the observed information, we distinguish between two operation modes: *Feedforward control* and *feedback control.*

In *feedforward control* mode, the system is operating as an *open-loop control* and only the excitation of the system is considered by the control algorithms. An example for the feedforward control is the control process of a seismically excited building by calculating the control forces using only the ground acceleration signal, which is measured on the foundation. Although this operation mode requires usually fewer sensors than the feedback control, it is still difficult to realize, as a very accurate knowledge about the response behavior of the system is necessary. Accordingly, the dynamics of both the structure and the control device must be known exactly. Consequently, feedforward control is only applicable for simple systems. An open-loop control system is shown in Figure 3.3.

In *feedback control* mode, the system is operating as a *closed-loop control* and only the response of the system is considered by the control algorithms. An example for the feedback control is the control process of a seismically excited building by calculating the control forces using only the structural acceleration response, which is measured on the floors of the building. This method is generally more robust and accordingly more common than the feedforward control mode. A closed-loop control system is shown in Figure 3.4.

Both operation modes can also be combined. A system, which uses both the excitation and the response of the system, is referred to as an *open/closed-loop*

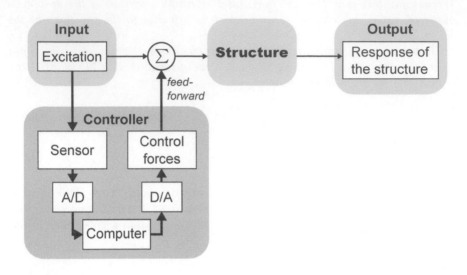

FIGURE 3.3

A control system operating in feedforward control mode (Open-loop control system).

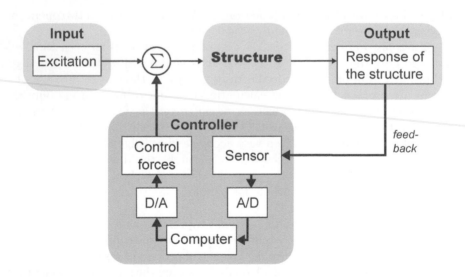

FIGURE 3.4

A control system operating in feedback control mode (Closed-loop control system).

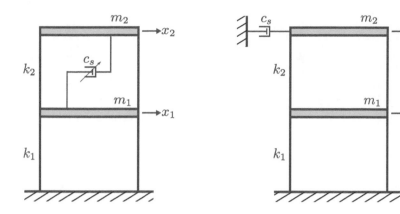

FIGURE 3.5
On-off skyhook controller applied to a 2-DoF system.

control. Compared to feedforward and feedback control modes, open/closed-loop systems are generally more robust.

3.3.2   Controller Algorithms

3.3.2.1   On-Off Controller

As a feedback operating system, the on-off controller uses the state information of the system, such as velocities and displacements. An example with the controller is the tuning of damping coefficients of semi-active devices between their high-state $c_{D,H}$ and low-state $c_{D,L}$ values.

We distinguish between the on-off *skyhook* and *groundhook* controllers. For illustration of both controllers, we consider a 2-DoF system consisting of masses $m_{1,2}$, stiffnesses $k_{1,2}$ and the damping coefficient c_s.

As shown in Figure 3.5, with a skyhook controller, the goal is to control the vibrations of the mass m_2 by tuning the damping coefficient c_s. Accordingly, this control scheme is equivalent to a system with a passive damper, which is hooked between m_2 and the sky.

The approach can be particularly applied to semi-active dissipators, which are controlling the interstory displacements of buildings. In this case, the on-off skyhook controller defines the optimum damping coefficient as

$$\dot{x}_2(\dot{x}_2 - \dot{x}_1) \geq 0 \quad c_s = c_{s,\max}, \tag{3.39}$$

$$\dot{x}_2(\dot{x}_2 - \dot{x}_1) < 0 \quad c_s - c_{s,\min}, \tag{3.40}$$

where $\dot{x}_{1,2}$ are the velocities of the system at its DoFs. $c_{s,\max}$ and $c_{s,\min}$ correspond respectively to the high and low-state damping coefficients of the semi-active device.

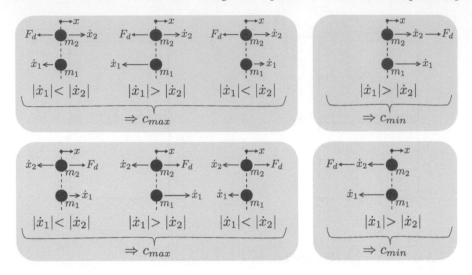

FIGURE 3.6
Case definitions for the on-off skyhook controller applied to a 2-DoF system showing the direction of the damping force F_d depending on the motion of m_1 and m_2.

When m_2 is moving in the positive direction ($\dot{x}_2 > 0$), if m_1 is moving in the opposite direction or slower in the same direction ($(\dot{x}_2 - \dot{x}_1) > 0$), the damping coefficient is maximized to $c_{s,\max}$ in order to maximize the damping force and to prevent a further deflection of m_2. Otherwise if m_1 is moving faster in the same direction ($(\dot{x}_2 - \dot{x}_1) < 0$), the damping coefficient is minimized to $c_{s,\min}$ in order to minimize the damping force, so that m_1 does not pull m_2 further away from its original position.

On the other hand, when m_2 is moving in the negative direction ($\dot{x}_2 < 0$), if m_1 is moving in the opposite direction or slower in the same direction ($(\dot{x}_2 - \dot{x}_1) < 0$), the damping coefficient is maximized to $c_{s,\max}$ in order to maximize the damping force and to prevent a further deflection of m_2. Otherwise if m_1 is moving faster in the same direction ($(\dot{x}_2 - \dot{x}_1) > 0$), the damping coefficient is minimized to $c_{s,\min}$ in order to minimize the damping force, so that m_1 does not pull m_2 further away from its original position.

Accordingly, as shown in Figure 3.6, when the damping force F_d is in opposite direction of motion of m_2, the damping coefficient c_s is maximized. Otherwise, the damping coefficient is minimized.

For the illustration of the *on-off groundhook controller*, we consider the same 2-DoF system. However, this scheme represents now a system, which is hooked to the ground as shown in Figure 3.7. Accordingly, the goal is to reduce the vibrations of m_1.

The approach can be particularly applied to semi-active TMDs, which are attached to buildings. In this case, the on-off groundhook controller defines

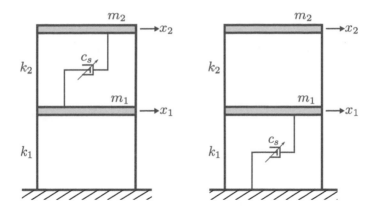

FIGURE 3.7
On-off groundhook controller applied to a 2-DoF system.

the optimum damping coefficient as

$$\dot{x}_1(\dot{x}_2 - \dot{x}_1) \leq 0 \quad c_s = c_{s,\text{max}}, \tag{3.41}$$

$$\dot{x}_1(\dot{x}_2 - \dot{x}_1) > 0 \quad c_s = c_{s,\text{min}}. \tag{3.42}$$

When m_1 is moving in the positive direction $(\dot{x}_1 > 0)$, if m_2 is moving in the opposite direction or slower in the same direction $((\dot{x}_2 - \dot{x}_1) < 0)$, the damping coefficient is maximized to $c_{s,\text{max}}$ in order to maximize the damping force and to prevent a further deflection of m_1. Otherwise if m_2 is moving faster in the same direction $((\dot{x}_2 - \dot{x}_1) > 0)$, the damping coefficient is minimized to $c_{s,\text{min}}$ in order to minimize the damping force, so that m_2 does not pull m_1 further away from its original position.

On the other hand, when m_1 is moving in the negative direction $(\dot{x}_1 < 0)$, if m_2 is moving in the opposite direction or slower in the same direction $((\dot{x}_2 - \dot{x}_1) > 0)$, the damping coefficient is maximized to $c_{s,\text{max}}$ in order to maximize the damping force and to prevent a further deflection of m_1. Otherwise if m_2 is moving faster in the same direction $((\dot{x}_2 - \dot{x}_1) < 0)$, the damping coefficient is minimized to $c_{s,\text{min}}$ in order to minimize the damping force, so that m_2 does not pull m_1 further away from its original position.

Accordingly, as shown in Figure 3.8, when the damping force F_d is in opposite motion direction of m_1, the damping coefficient c_s is maximized. Otherwise, the damping coefficient is minimized.

The case definitions of both skyhook and groundhook controllers can be also written in a *displacement-based* form. Applied to the groundhook controller, we get

$$x_1(\dot{x}_2 - \dot{x}_1) \leq 0 : \quad c_s = c_{s,\text{max}}, \tag{3.43}$$

$$x_1(\dot{x}_2 - \dot{x}_1) > 0 : \quad c_s = c_{s,\text{min}}, \tag{3.44}$$

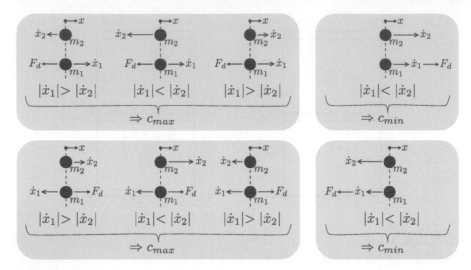

FIGURE 3.8

Case definitions for the on-off groundhook controller applied to a 2-DoF system showing the direction of the damping force F_d depending on the motion of m_1 and m_2.

which is similar to the previous velocity-based formulation with only the difference that the direction information of m_1 is defined by its displacement x_1. This displacement-based formulation covers all cases shown in Figure 3.8.

Besides tuning of the damping coefficient, the case definitions of the groundhook controller can be applied to semi-active devices with frequency adaptation capabilities, such as semi-active TMDs and TLCDs, to tune their restoring forces by adjusting their stiffness value k_d corresponding to the velocities $\dot{x}_{1,2}$ and, in case of displacement-based formulation, corresponding to the displacement x_1. For tuning of the stiffness value, the case definitions must be modified and c_s must be replaced by the stiffness value k_s, which has the maximum value $k_{s,\mathrm{max}}$ and the minimum value $k_{s,\mathrm{min}}$. Accordingly, this definition requires a semi-active device, which is capable of shifting its frequencies to generate $k_{s,\mathrm{max}}$ and $k_{s,\mathrm{min}}$ stiffnesses in real-time.

Furthermore, both the skyhook and the groundhook controllers can be also applied to devices, which can realize their parameters not only in a discrete but also in a continuous manner. In this case, the parameters, i.e., damping coefficients or stiffness values, are not limited only to two constant states. For the tuning of damping, all above presented equations of the case definitions remain unchanged. Only the maximum parameter value, which is $c_s = c_{s,\mathrm{max}}$, must be replaced by

$$c_s = \max\Big[c_{s,\mathrm{min}}, \min\big(g\dot{x}_i, c_{s,\mathrm{max}}\big)\Big], \tag{3.45}$$

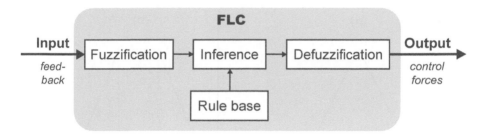

FIGURE 3.9
Processes performed by a fuzzy logic controller (FLC).

where g is a gain factor, which must be tuned corresponding to the requirements of the system. Furthermore, in Equation 3.45, we substitute $\dot{x}_i = \dot{x}_1$ for the skyhook and $\dot{x}_i = \dot{x}_2$ for the groundhook controller. With this continuous definition, the tuning gets flexible as it not limited by two discrete values anymore.

3.3.2.2 Fuzzy Logic Controller

Fuzzy logic controller (FLC) is a soft-computing-based method, which replicates the human way of thinking to control complex systems with highly nonlinear behavior. The FLC employs for control heuristic and qualitative knowledge in three subsequent processes, which are fuzzification, inference and defuzzification as shown in Figure 3.9.

The first step of the FLC *fuzzification* is the mapping of the crisp feedback, which is measured by sensors, into fuzzy values. For this purpose, the process uses fuzzy sets, which were first introduced by Zadeh [31]. Each fuzzy set A represents a linguistic value and consists of pairs of the crisp values x, as well as their degree of membership $\mu_A(x)$:

$$A = \{x, \mu_A(x)) | \forall x \in X\}, \tag{3.46}$$

where X is usually denoted as the universe of discourse [32], from which x is derived.

$\mu_A(x)$ is calculated by mathematical equations expressing the membership degree of x to A. The equations are referred to as membership functions (MFs). Several classes of MFs are defined in references, such as by Jang et al. [32] as well as Siddique and Adeli [33], for example triangular, trapezoidal, GAUSSIAN, bell-shaped and sigmoidal. A triangular and a trapezoidal MF are shown in Figure 3.10. The triangular MF reads

$$\mu(x) = \max \left[\min \left(\frac{x - x_1}{x_2 - x_1}, \frac{x_3 - x}{x_3 - x_2} \right), 0 \right], \tag{3.47}$$

where $\{x_1, x_2, x_3\}$ are parameters that are specifying the function. Similarly

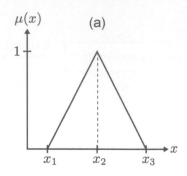 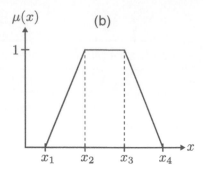

FIGURE 3.10
Triangular (a) and trapezoidal (b) membership functions of a fuzzy set.

a trapezoidal MF reads

$$\mu(x) = \max\left[\min\left(\frac{x - x_1}{x_2 - x_1}, 1, \frac{x_4 - x}{x_4 - x_3}\right), 0\right],\qquad(3.48)$$

where $\{x_1, x_2, x_3, x_4\}$ are parameters that are specifying the function.

In the second step of the FLC, the *inference* combines the fuzzified inputs by fuzzy logic operators and applies to them fuzzy rules. Fuzzy logic operators are distinguished as T-norm (triangular norm) and T-conorm operators.

T-norm operators are also referred to as AND-operators. Accordingly, they represent in Fuzzy rules the AND-condition and can be computed by minimum (MIN) or product (PROD) methods. If we consider for the two inputs x and y the membership functions $\mu_A(x)$ and $\mu_B(y)$ of fuzzy sets A and B (linguistic values), the methods read

$$\mu_R(x, y) = \min[\mu_A(x), \mu_B(y)],\qquad(3.49)$$
$$\mu_R(x, y) = \mu_A(x) \cdot \mu_B(y).\qquad(3.50)$$

On the other hand, T-conorm operators are also referred to as OR-operators. Accordingly, they represent in Fuzzy rules the OR-condition and can be computed by maximum (MAX) or algebraic sum (SUM) methods. Again for the two inputs x and y, the methods read

$$\mu_R(x, y) = \max[\mu_A(x), \mu_B(y)],\qquad(3.51)$$
$$\mu_R(x, y) = \mu_A(x) + \mu_B(y) - \mu_A(x) \cdot \mu_B(y).\qquad(3.52)$$

Fuzzy rules consist of antecedent and consequent parts and have typically the form *if x is A, then z is C*, where A and C are linguistic values. It is possible to have more than one input, which are then combined with T-norm or T-conorm operators.

MAMDANI, SUGENO and TSUKAMOTO type fuzzy models are commonly

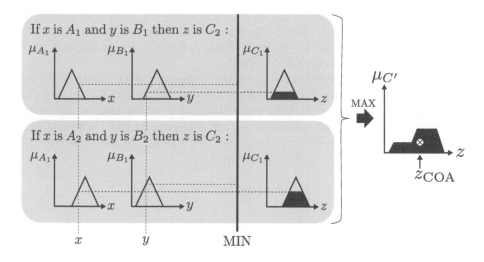

FIGURE 3.11
Two-input single-output MAMDANI type fuzzy model.

used for the inference process. The process structure of the MAMDANI fuzzy inference [34] is shown in Figure 3.11. The example fuzzy model has two inputs x and y, and one output z. Each input and the output have two triangular MFs $\mu_{i_{1,2}}$. The model involves two fuzzy if-then rules. A T-norm (MIN) is applied to the inference to calculate the firing strength of each rule. Subsequently, a T-conorm (MAX) operator is applied for aggregation of the results. The output of the MAMDANI inference process is again a fuzzy set, on which the *defuzzification* process is applied to determine a crisp value as a result. The defuzzification is carried out by calculating the centroid of the area of the fuzzy output by

$$z_{\text{COA}} = \frac{\int \mu_{C'} z \, dz}{\int \mu_{C'} \, dz}. \tag{3.53}$$

Alternatively, the defuzzification step could be carried out by calculating the bisector of the output area z_{BOA}, which satisfies

$$\int_{z_{\min}}^{z_{\text{BOA}}} \mu_{C'} \, dz = \int_{z_{\text{BOA}}}^{z_{\max}} \mu_{C'} \, dz, \tag{3.54}$$

where $z_{\min}$ and $z_{\max}$ are the lowest and highest z-values, respectively.

Furthermore, the defuzzification can be also computed as the mean of maximum of the output by

$$z_{\text{MOM}} = \frac{\int_{z'} z \, dz}{\int_{z'} \, dz}, \tag{3.55}$$

where z' is the range, in which $\mu_{C'}$ is maximum.

In the SUGENO fuzzy inference, in contrary to the MAMDANI fuzzy model, the consequent part of the fuzzy rule is represented by a crisp function [35]. On the other hand, in the TSUKAMOTO fuzzy inference, the consequent part of the fuzzy rule is a monotonic MF [36].

The design of FLCs requires expert knowledge about the system and its dynamic behavior, which can be acquired from experiments. To simplify the design process proper optimization algorithms, such as the genetic algorithm can be also applied. An alternative approach is to expand FLC- with learning capabilities using machine learning methods, such as neural networks [32, 33].

Part II

Conventional Damping Systems

4

Dissipators

4.1 Introduction

Dissipators are auxiliary devices, which are able to enhance the reliability of vibration prone structures by supplementary damping. These damping devices dissipate the vibration energy by converting it to heat energy. Various types of dissipators have been developed for this purpose, which use friction, viscoelastic or plastic deformation as well as phase transformation for the energy transformation. This chapter is concerned with the dissipators, which operate in a passive manner and encompass conventional materials. Modeling approaches are introduced and application examples are provided. Depending on the energy dissipation process and utilized material types, this chapter classifies passive dissipators as

- Metallic dampers (Section 4.2)

- Friction dampers (Section 4.3)

- Viscoelastic dampers (Section 4.4)

- Viscous fluid dampers (Section 4.5)

The chapter describes the functionality of these devices, presents governing equations and gives examples. For further information about these devices, the interested reader may refer to reference books, such as Connor [28], Constantinou et al. [23], Hanson and Soong [26] as well as Soong and Dargush [25].

As mentioned in Chapter 2, apart from passive strategy, active and semi-active strategies exist, which can be also used to control dissipators. For active and semi-active devices please refer to Chapter 6. Furthermore, dissipators can incorporate smart materials, such as SMAs. Dissipators incorporating SMAs will be described in Chapter 8.

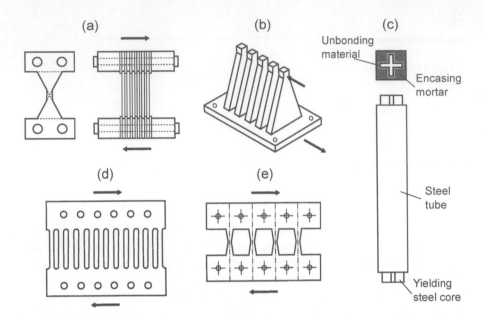

FIGURE 4.1
Examples of metallic dampers: X-shaped plate damper (a), triangular plate damper (b), buckling restrained brace (c), steel slit-type damper (d) and honeycomb damper (e). Arrows show the direction of relative displacements, which are required for the activation of the energy dissipation process.

4.2   Metallic Dampers

Metallic dampers incorporate metals, such as mild steel, lead or metal alloys. Inelastic deformation of this metallic material allows the dissipation of the vibration energy. Integrated into the primary structure, metallic dampers develop plastic hinges with large deformations associated energy dissipation capabilities. Metallic dampers can be connected in frame structures between the building floors or can be used as a part of base isolation systems between the superstructure and its foundation [37]. Figure 4.1 shows some examples of metallic dampers.

Both experimental and analytical studies have been conducted with metallic dampers since 1970s. The first mathematical modeling approaches are proposed in works of Ozdemir [38] and Bhatti et al. [39], which are accompanied by experimental investigation of Kelly et al. [40] and Skinner et al. [41].

As shown in Figure 4.2, *elastoplastic, bilinear* and *polynomial models* can be used to reproduce the behavior of metallic dampers and to predict the response of controlled structures. These models allow mathematically consistent frameworks to represent the relationship between the deformation x applied to the device and its response force F. The shown cyclic inelastic deformation paths enclose hysteresis loops, areas of which corresponds to the amount of dissipated energy. This damping process is generally referred to as *hysteretic damping*, cf. Section 1.8.

Using the elastoplastic model for $x_0 \geq x_y$, we determine the dissipated energy as

$$E_D = 4F_y(x_0 - x_y), \tag{4.1}$$

where x_0 is the maximum displacement of the device, x_y the yield displacement and F_y the corresponding yield force. Similarly, with the bilinear model, the dissipated energy per cycle corresponds to

$$E_D = 4(k_E - k_H)x_y(x_0 - x_y), \tag{4.2}$$

where the initial elastic and the strain-hardening stiffnesses are calculated as

$$k_E = F_y/x_y, \quad k_H = (F_0 - F_y)/(x - 0 - x_y), \tag{4.3}$$

where F_0 is the maximum force. With the polynomial model, the dissipated energy is calculated as

$$E_D = 4x_yF_y\frac{r-1}{r+1}\left(\frac{F_0}{F_y}\right)^{r+1}, \tag{4.4}$$

where r corresponds to the order of the polynomial. From these energies, we can now approximate the damping behavior with a linear equivalent viscous damping coefficient and the stiffness as

$$c_D = E_D/(\pi\omega x_0^2), \quad k_D = F_y/x_0, \tag{4.5}$$

where ω is the excitation frequency applied to the device.

Metallic dampers exhibit generally a stable hysteretic behavior. Their long-term reliability and insensitivity to ambient temperature are further advantages of these dampers. Furthermore, the used materials and their behavior are similar to other types of materials used in structural engineering. On the other hand, the nonlinear material behavior requires nonlinear analysis for an accurate response calculation. Furthermore, due to the residual deformation, metallic dampers require after each major event a replacement. This problem can be solved by utilizing superelastic SMAs, cf. Chapter 8.

Example 4.1 *A metallic damper is installed to a SDoF structure as shown in Figure 4.3. The hysteretic stress-strain curve of a metallic elastic-perfectly plastic damper device is obtained experimentally with an excitation frequency of $\omega = 62.83$ rad s^{-1}. The yield force and the corresponding displacement of the device are given as $F_y = 300$ kN and $x_y = 0.01$ m. The maximum possible displacement is $x_0 = 0.2$ m. The mass and stiffness parameters of the structure*

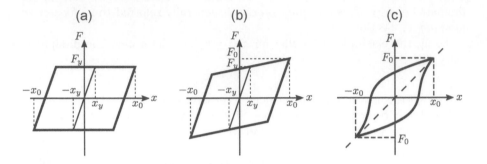

FIGURE 4.2
General material models used for the modeling of metallic dampers: Elastoplastic model (a), bilinear model (b) and polynomial model (c). F_y is the yield
force, x_y is the corresponding yield displacement, F_0 is the maximum force
and x_0 is the maximum displacement of the device.

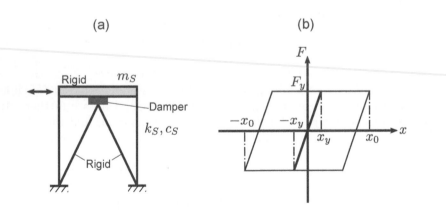

FIGURE 4.3
An SDoF system with an elastic-perfectly plastic metallic damper (a), which
has a hysteretic stress-strain curve (b).

are given as $m_S = 2$ t and $k_S = 44444.\bar{4}$ kN m^{-1}, respectively. The structure has an initial damping ratio of $D_S = 0.05$.

From the stress-strain curve, the energy dissipated by the metallic damper is determined for one load cycle as

$$E_D = 4F_y(x_0 - x_y) = 228 \ kNm. \tag{4.6}$$

The linearized damping coefficient stiffness of the damper read

$$c_D = \frac{E_D}{\pi \omega x_0^2} = 28.88 \ kNs \ m^{-1}, \tag{4.7}$$

$$k_D = \frac{F_y}{x_0} = 1500 \ kN \ m^{-1}. \tag{4.8}$$

The total stiffness, the natural frequency as well as the increment of the damping ratio of the SDoF+damper system are determined as

$$k_T = k_S + k_D = 45944.\bar{4} \ kN \ m^{-1}, \tag{4.9}$$

$$\omega_1 = \sqrt{\frac{k_T}{m}} = 151.57 \ rad \ s^{-1}, \tag{4.10}$$

$$c_S = 2Dm\omega_1 = 30.31 \ kNs \ m^{-1}, \tag{4.11}$$

$$c_T = c_S + c_D = 59.19 \ kNs \ m^{-1}, \tag{4.12}$$

$$D_T = \frac{c_T}{2m\omega_1} = 9.8 \ \%. \tag{4.13}$$

4.3   Friction Dampers

Friction dampers dissipate energy through sliding contact friction. These devices are usually attached to the bracing elements of buildings, which are connecting adjacent floors. As chevron friction dampers, they can be used to connect the bracing elements with the floor. Furthermore, as cross friction dampers, they can also connect four bracing elements together.

The *Pall device* is one of the first developed friction dampers [42], which is followed by the *Sumitomo Friction Damper* [43]. Figure 4.4 shows these dampers and a classical slotted bolted connection.

For the most of the friction dampers, the concept of COULOMB damping applies as a theoretical basis, cf. Section 1.8. The damping force reads

$$F_D = F_0 \mathrm{sgn}(\dot{x}) = \mu N \mathrm{sgn}(\dot{x}) \tag{4.14}$$

and depends on the friction coefficient μ as well as the normal force N acting on the sliding adjoining surfaces. The force direction corresponds to

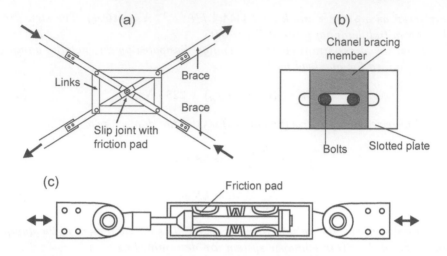

FIGURE 4.4
Examples of friction dampers: Pall device (a), slotted bolted connection (b) and Sumitomo friction damper (c). Arrows show the direction of the relative displacements, which are required for the activation of the energy dissipation process.

the deformation rate $\dot{x}$. The force–displacement relationship according to the COULOMB damping is shown in Figure 4.5. The dissipated energy per cycle is calculated from the area enclosed by the curve as

$$E_D = 4F_0 x_0 = 4\mu N x_0, \tag{4.15}$$

where x_0 is the maximum displacement of the damper.

Friction dampers can dissipate large amounts of energy during each oscillation cycle. They are insensitive to ambient temperature. However, the sliding interface conditions may deteriorate with time. As the response of the damper is strongly nonlinear, a corresponding nonlinear analysis is required for the design. Furthermore, higher modes can be excited in structures during sliding. After each excitation event, permanent displacements occur. To prevent this effect friction dampers require restoring or centering force mechanisms, such as by SMAs, cf. Chapter 8.

Example 4.2 *The cross bracings of a water tank are retrofitted using 36 friction dampers as shown in Figure 4.6. A hysteretic loop of one of the friction dampers is determined for 10 cycles.*

The added energy dissipation per cycle is determined as

$$E_D = (4x_0 F_0) \cdot 36 = 3.53 \cdot 10^3 \ kNm, \tag{4.16}$$

where the maximum displacement and load are obtained from the provided hysteretic loop as $x_0 = 0.035$ m and $F_0 = 700$ kN, respectively.

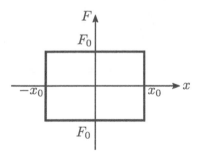

FIGURE 4.5
COULOMB damping model used for the modeling of friction dampers. F_0 is the maximum load generated by the device and x_0 is its maximum displacement.

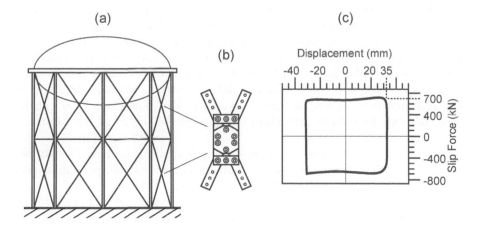

FIGURE 4.6
A water tank tower structure (a), which is retrofitted with 36 friction dampers (b). The hysteretic loop of the friction dampers (c), which is determined from a cyclic test.

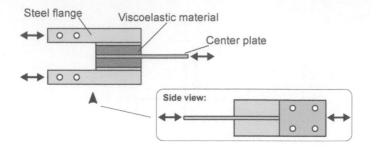

FIGURE 4.7

Example of a viscoelastic damper. Arrows show the direction of the relative displacements, which are required for the activation of the energy dissipation process.

4.4 Viscoelastic Dampers

Shear deformation initiates in viscoelastic materials, such as copolymers or elastomeric substances, energy dissipation. A typical *viscoelastic damper* consists of one or more layers of viscoelastic material bonded to metallic plates as shown in Figure 4.7. The energy dissipation begins when the structural vibration induces a relative motion between the outer layers of the damper and its center. Viscoelastic dampers should be placed in locations, which allow the largest relative displacement between attachment points of the damper. Depending on the physical constraints and the architectural requirements, these devices can be integrated to the structural components such as bracing elements or used as a part of base isolation systems. Viscoelastic dampers are activated even at low displacement amplitudes. Therefore, they can be used, not only for seismic protection but also for the control of wind-induced vibrations.

In civil engineering, the viscoelastic damper applications have already begun in 1960s. The first and the most prominent viscoelastic damper application was the towers of World Trade Center in New York [44].

Viscoelastic material models are used to describe the response of the damper, such as by Zhang et al. [45] as well as Zhang and Soong [46], cf. Section 1.8. Figure 4.8, shows the shear stress–strain relationship. For a harmonic loading with an angular excitation frequency of ω, the shear strain and its corresponding rate read

$$\gamma = \gamma_0 \sin \omega t, \quad \dot{\gamma} = \gamma_0 \omega \cos \omega t. \tag{4.17}$$

The viscoelastic material behavior is defined by two material parameters: the *shear storage modulus* G', which provides the elastic shear stiffness of

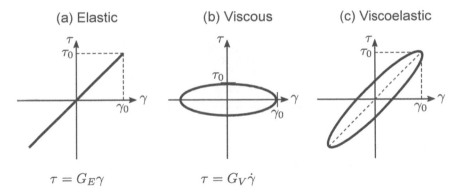

FIGURE 4.8
Viscoelastic damping model defining the relationship between shear stress τ, shear strain γ and its rate $\dot{\gamma}$. Elastic behavior (a), viscous behavior (b) and viscoelastic behavior (c). $G_{E/V}$ are shear moduli corresponding to the elastic and viscous material behaviors.

the material, and the *shear loss modulus* G'', which provides the velocity dependent viscous stiffness of the material. Accordingly, it applies

$$\tau = G'\gamma \pm \frac{G''}{\Omega}\dot{\gamma}, \tag{4.18}$$

$$= \gamma_0(G' \sin \omega t + G'' \cos \omega t), \tag{4.19}$$

where $G' = G_E$ and $G'' = G_V\omega$ can be rewritten with the moduli of elastic and viscous behavior G_E, G_V. The linearized stiffness and viscous damping coefficients of a cubic viscoelastic material can then be determined from

$$k_D = G'\frac{A}{h}, \quad c_D = \frac{G''}{\omega}\frac{A}{h}, \tag{4.20}$$

where A is the area of the shear surface and h the height of the viscoelastic element as shown in Figure 4.9.

The ratio of the shear loss G'' and storage moduli G' corresponds to the *loss factor* η, which represents the energy dissipation capacity of a viscoelastic material. Accordingly, it holds

$$\eta = \frac{G''}{G'} = \tan \delta, \tag{4.21}$$

where δ corresponds to the phase-shift between the shear stress and the corresponding strain. Depending on the type of the material behavior, for δ applies

$$
\begin{aligned}
&\delta = 0 && : \text{Elastic behavior } (\eta = \tan 0 = 0),\\
&0 < \delta < \pi/2 && : \text{Viscoelastic behavior},\\
&\delta = \pi/2 && : \text{Viscous behavior } (\eta = \tan \pi/2 = \infty).
\end{aligned}
\tag{4.22}
$$

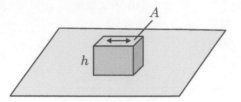

FIGURE 4.9
A cubic viscoelastic material with the shear surface A and height h.

The quantity G''/ω corresponds to the viscous damping coefficient c_D. Furthermore, G' corresponds to the stiffness k_D of the viscoelastic material. Accordingly, the damping ratio of the damper reads

$$D_D = \frac{c_D}{2\omega m} = \frac{G''}{\omega}\frac{\omega}{2G'} = \frac{\eta}{2}. \tag{4.23}$$

From the shear stress–strain relationship, the dissipated energy per unit volume and per cycle of oscillation can be also determined as

$$E_D = \int_0^T \tau\dot{\gamma}\,dt \tag{4.24}$$

$$= \int_0^{T=2\pi/\omega} \gamma_0^2\omega\cos\omega t(G'\sin\omega t + G''\cos\omega t)\,dt \tag{4.25}$$

$$= \pi\gamma_0^2 G''. \tag{4.26}$$

The material behavior can be linearized (linear damping force) as shown by Equation 4.20 and, therefore, a simplified modeling approach can be used at least for the preliminary design of the damper. On the other hand, the deformation of viscoelastic dampers is generally limited by the shear strain capacity of the material. In case of large displacements, debonding and tearing of the material can occur. Furthermore, the material properties and accordingly its control performance depend on the excitation frequency and the temperature.

Example 4.3 *A viscoelastic damper is installed to the same structure from Example 4.1 instead of the metallic damper. The material data graph of the viscoelastic damper is provided in Figure 4.10. The area of the damper is $A = 0.05$ m^2 and its height is $h = 0.1$ m. The operational material temperature of the damper is assumed to be $T = 30°$. The structure is excited in horizontal direction by a harmonic force with an excitation frequency of $\omega = 62.83$ rad s^{-1}. The load amplitude causes a static displacement of $x_{st} = 2 \cdot 10^{-3}$ m. The mass of the structure is $m_S = 3$ t. The natural frequency of the SDoF+damper systems is $\omega_1 = 2\pi10$ rad s^{-1}. The structure is assumed to be initially undamped.*

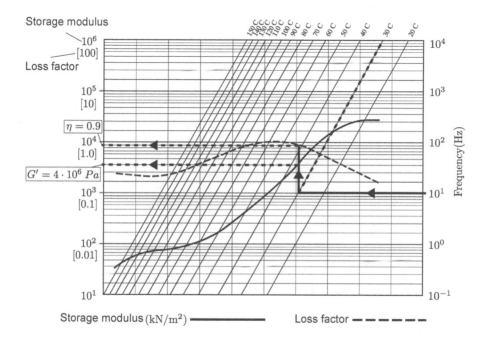

FIGURE 4.10

Material data graph of a viscoelastic damper. The material behavior depends on the excitation frequency and temperature.

First, we determine from Figure 4.10 the shear storage module G' and the loss factor η. For this purpose, we choose from the right vertical scale the corresponding excitation frequency of $\omega = 62.83$ rad s^{-1}, which corresponds to 10 Hz and follow the frequency line to the desired operational temperature isoterm at $T = 30°$. From the intersect, we go vertically until intersecting the shear modulus and loos factor curves, which gives us $G' = 4 \cdot 10^3$ kN m^{-2} and $\eta = 0.9$. Next, we calculate the shear loss module G'' and determine the linearized equivalent viscous damping coefficient c_D as

$$G'' = \eta G' = 3.6 \cdot 10^3 \ kN \ m^{-1}, \tag{4.27}$$

$$c_D = \frac{G''}{\omega}\frac{A}{h} = 28.65 \ kNs \ m^{-1}, \tag{4.28}$$

The increased damping ratio of the structure with the damper is determined as

$$D_S = \frac{c_D}{2m_S\omega_1} = 7.6 \ \%. \tag{4.29}$$

For the resonance case, the deformation response factor of the SDoF+damper system reads

$$R_{d,max} = \frac{1}{2D_S} = \frac{x_{dy}}{x_{st}}, \tag{4.30}$$

from which, we calculate the maximum dynamic displacement of the controlled structure as

$$x_{dy} = \frac{x_{st}}{2D_S} = 13.2 \cdot 10^{-3} m. \tag{4.31}$$

4.5   Viscous Fluid Dampers

In a *viscous fluid damper*, the vibration energy is dissipated by the high velocity flow of the damper fluid through an orifice. Figure 4.11 shows some examples of viscous fluid dampers. Just like other dissipators, viscous fluid dampers can be placed at any location of a structure with large relative displacements and can be integrated directly to bracing elements.

First civil engineering applications of viscous fluid dampers have begun in 1990s, such as by Constantinou et al. [47], Makris et al. [48] as well as Taylor and Constantinou [49].

Apart from orificed fluid dampers, viscous damping walls have been proposed. These devices can be installed as panels within structural frames as studied by Reinhorn et al. [50].

The damping force of an viscous fluid damper can be calculated according to Constantinou et al. [47] from

$$F_D = c_\alpha(\omega)|\dot{x}|^\alpha \mathrm{sgn}(\dot{x}), \tag{4.32}$$

where x is the relative displacement applied to the damper and $\dot{x}$ is the corresponding velocity. $c_\alpha(\omega)$ is a generalized excitation frequency–dependent

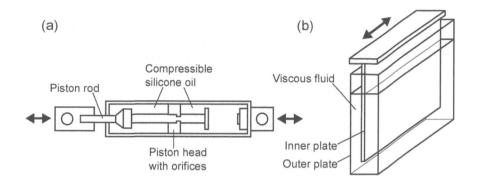

FIGURE 4.11
Examples of viscous fluid dampers: Orificed fluid damper (a) and viscous damping wall (b). Arrows show the direction of the relative displacements, which are required for the activation of the energy dissipation process.

damping coefficient. For the excitation frequencies $\omega/(2\pi) = f \leq 4\,\mathrm{Hz}$, c_α can be assumed to be constant and the effective stiffness of the device is negligible. For α applies

$$\alpha = \{0.3\ldots2.0\}, \tag{4.33}$$

where for high-velocity shocks $\alpha = 0.5$ and for wind or earthquake applications $\alpha = 1$ can be used. $\alpha = 1$ defines a linear damper behavior with the constant damping coefficient $c_\alpha = c_D$.

The energy dissipation of a viscous fluid damper is given by

$$E_D = \int F_D\,\mathrm{d}x \tag{4.34}$$

$$= 4c_\alpha(\omega)x_0^{1+\alpha}\omega^\alpha J(\alpha), \tag{4.35}$$

where $J(\alpha)$ is calculated using the gamma function

$$\Gamma(x) = \int_0^\infty u^{x-1}e^{-u}\,\mathrm{d}u, \quad \text{with} \quad J(\alpha) = 2^\alpha\frac{\Gamma^2(1+\alpha/2)}{\Gamma(2+\alpha)}. \tag{4.36}$$

An equivalent viscous damping coefficient c_D can be obtained backward from the energy equation as

$$E_D = \pi c_D\omega x_0^2 = 4c_\alpha(\omega)x_0^{1+\alpha}\omega^\alpha J(\alpha) \tag{4.37}$$

$$\Leftrightarrow c_D = \frac{4}{\pi}c_\alpha(\omega)x_0^{\alpha-1}\omega^{\alpha-1}J(\alpha), \tag{4.38}$$

which gives for $\alpha = 1$ again $c_D = c_\alpha(\omega)$.

Similar to the viscoelastic dampers also viscous fluid dampers can be activated at low displacements. However, in contrast to viscoelastic dampers, the properties of the viscous fluid dampers are largely frequency and temperature independent. They apply only a minimal restoring force. For the linear case with $\alpha = 1$, the modeling of the damper is simple. For $\alpha > 1$, the design of viscous fluid dampers require a nonlinear analysis of the damper behavior.

5

Tuned Mass Dampers

5.1 Introduction

This chapter is concerned with tuned mass dampers and provides an overview of the so far developed passive TMDs, including

- classical tuned mass dampers (TMDs) (Section 5.2)

- pendulum tuned mass dampers (P-TMDs) (Section 5.2.2)

- tuned liquid dampers (TLDs) or tuned sloshing dampers (TSDs) (Section 5.2.3)

- tuned liquid column dampers (TLCDs) (Section 5.2.4)

These auxiliary devices introduce both supplementary damping and restoring forces to structures. Methods for the modeling are presented, and tuning concepts are described. Information about the applications on real structures is provided along with calculation examples.

An overview of the so far developed actively and semi-actively controlled devices is found in Chapter 6. Furthermore, Chapter 7 introduces two semi-active TLCDs. An overview of TMDs incorporating smart materials, such as MR fluids, is found also in Chapter 6. TMDs using SMAs are mentioned in Chapter 8.

5.2 Classical Tuned Mass Dampers

Conventional TMDs are utilized in passive systems to improve the modal response of structures. These devices are usually also equipped with supplementary dissipators to dissipate simultaneously the vibration energy by converting it to heat energy.

TMDs consist of a mass, which exhibits a phase-shifted oscillation against the vibration direction of the structure. Besides solid masses, liquids can be used.

Depending on the used mass and its motion type, four general TMD formats exist:

- Classical tuned mass dampers (TMDs)

- Pendulum tuned mass dampers (P-TMDs)

- Tuned liquid/sloshing dampers (TLDs/TSDs)

- Tuned liquid column dampers (TLCDs)

The first TMD was patented by Frahm already at the beginning of 20th century [51]. Frahm's *dynamic vibration absorber* corresponds to classical TMDs without dissipators. Classical TMDs with dissipators date back to the 1930s [52]. Some of the pioneering TMD applications are the John Hancock Tower in Boston (Building height: 241 m. Built in 1976), the Canadian National Tower in Toronto (102 m, in 1976), the Citigroup Center in New York City (279 m, in 1977) and the Sydney Tower (213 m, in 1981). All these applications were designed to control wind-induced vibrations. The readers interested in these applications are referred to [13, 23, 25, 28].

For classical TMDs, spring and damper elements are used to attach the auxiliary mass to structures. The phase-shifted oscillation of this mass is enabled by tuning the stiffness and damping parameters. During the motion of the auxiliary mass, both restoring and damping forces are generated. Figure 5.1 shows a classical TMD attached to an SDoF structure. The mass of the TMD is m_D. The stiffness of the spring element is denoted as k_D, and the damping coefficient of the viscous damping element is denoted as c_D. The oscillation direction of the TMD corresponds to the DoF in horizontal x_D direction. The mass, stiffness and damping coefficients of the structure are m_S, k_S and c_S, respectively. Both the DoF x_S of the controlled structure and the DoF x_D of the TMD are in parallel direction. Accordingly, the control direction of the TMD corresponds to the vibration direction of the structure.

Simplicity and stable structural control are the most important merits of TMDs. Furthermore, they are easy to design and compute. However, classical TMDs operate uniaxial. Omnidirectional control with the TMDs requires complex bearing systems. Furthermore, the spring elements of TMDs are usually problematic for frequencies below 0.5 Hz, which limits their application field and is particularly problematic for application on high-rise structures.

TMDs can control vibrations belonging to only a single mode shape. Accordingly, they must be placed at maximum deflection points of the corresponding mode shape. For tower-like structures with a uniform mass and stiffness distribution, the ideal TMD location is usually the top of the structure. Accordingly, the required construction measures and design alterations are limited, which is advantageous compared to distributed control systems, such as dissipators. TMDs show high control performance particularly under harmonic loading at the resonance frequency, to which they are tuned. Other modes stay uncontrolled. To control them further, TMDs can be included as

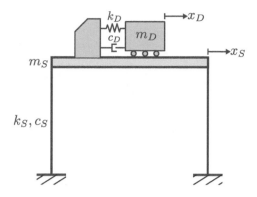

FIGURE 5.1
A classical TMD attached to an SDoF structure.

proposed by Clark [53]. However, this approach increases consequently the total damper mass. On the other hand, some of the semi-active systems can control multiple modes and realize, accordingly, a more efficient control without increasing the damper mass, cf. Chapter 6.

5.2.1 Mathematical Modeling

An SDoF structure with a TMD can be idealized by a 2-DoF system as shown in Figure 5.2 (a). To the structure, the dynamic force F is applied. As shown in Figure 5.2 (b), the TMD reacts with both restoring and damping forces, F_{D_R} and F_{D_D}, respectively. The figure shows also the generated bearing forces F_{S_R} and F_{S_D} of the structure. As an MDoF system, the corresponding EoM of the SDoF+TMD system reads

$$m_S \ddot{x}_S = F - F_{S_R} - F_{S_D} + F_{D_R} + F_{D_D}, \tag{5.1}$$
$$m_D \ddot{x}_D = -F_{D_R} - F_{D_D}, \tag{5.2}$$

where the bearing and interaction forces correspond to

$$F_{S_R} = k_s x_S, \quad F_{D_R} = k_D(x_D - x_S), \tag{5.3}$$
$$F_{S_D} = c_S \dot{x}_S, \quad F_{D_D} = c_D(\dot{x}_D - \dot{x}_S). \tag{5.4}$$

Accordingly, a matrix formulation of the EoM of the system can be introduced as

$$\Leftrightarrow \begin{bmatrix} m_S & 0 \\ 0 & m_D \end{bmatrix} \begin{bmatrix} \ddot{x}_S \\ \ddot{x}_D \end{bmatrix} + \begin{bmatrix} c_S + c_D & -c_D \\ -c_D & c_D \end{bmatrix} \begin{bmatrix} \dot{x}_S \\ \dot{x}_D \end{bmatrix}$$
$$+ \begin{bmatrix} k_S + k_D & -k_D \\ -k_D & k_D \end{bmatrix} \begin{bmatrix} x_S \\ x_D \end{bmatrix} = \begin{bmatrix} F \\ 0 \end{bmatrix}, \tag{5.5}$$

and then simplified as

$$\mathbf{M}\ddot{\mathbf{x}} + \mathbf{C}\dot{\mathbf{x}} + \mathbf{K}\mathbf{x} = \mathbf{f}, \tag{5.6}$$

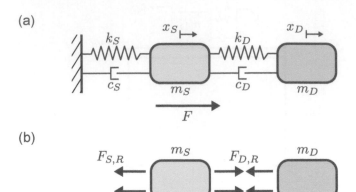

FIGURE 5.2

An SDoF structure with a classical TMD, which is idealized as a 2-DoF system (a). The interaction forces are depicted between the TMD and the structure as well as between the structure and its bearing point (b).

where the mass, stiffness and damping matrices of the SDoF+TMD system are $\mathbf{M}$, $\mathbf{K}$ and $\mathbf{C}$, respectively and the corresponding acceleration, velocity, displacement and force vectors are $\ddot{\mathbf{x}}$, $\dot{\mathbf{x}}$, $\mathbf{x}$ and $\mathbf{f}$, respectively.

5.2.1.1 State-Space Representation

The EoMs of systems with the TMDs can be formulated also using the state-space representation method. With this method the second-order differential equations are transformed to first-order equations, which are easier to solve, cf. Section 3.2.

Example 5.1 *The application of the method is presented for an SDoF+TMD system, which is shown in Figure 5.3. The system is subjected to earthquake excitation. The DoFs are denoted as x_1 and x_D and are relative to the ground motion x_g.*

The EoM of the SDoF+TMD system reads

$$m_1\ddot{x}_1 + c_1(\dot{x}_1 - \dot{x}_g) - c_D(\dot{x}_D - \dot{x}_1) + k_1(x_1 - x_g) - k_D(x_D - x_1) = 0, \quad (5.7)$$

$$m_D\ddot{x}_D + c_D(\dot{x}_D - \dot{x}_1) + k_D(x_D - x_1) = 0, \quad (5.8)$$

where the system mass, stiffness and damping parameters correspond to $m_1 = 1\ t$, $m_D = 0.05\ t$, $k_1 = 5\ kN\ m^{-1}$, $k_D = 0.23\ kN\ m^{-1}$ and $c_1 = 0.06\ kNs\ m^{-1}$ as well as $c_D = 0.03\ kNs\ m^{-1}$ respectively. In matrix form, the EoM reads

$$\mathbf{M}\ddot{\mathbf{x}} + \mathbf{C}\dot{\mathbf{x}} + \mathbf{K}\mathbf{x} = \mathbf{F}_g, \quad (5.9)$$

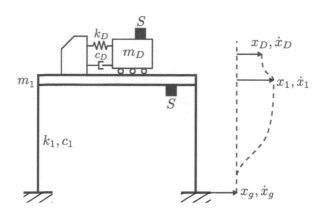

FIGURE 5.3
An SDoF system with a classical TMD. The displacement, velocity and acceleration responses are observed by two sensors S. The system is excited by earthquake.

with the following corresponding mass, stiffness and damping matrices as well as the force vector

$$\mathbf{M} = \begin{bmatrix} m_1 & 0 \\ 0 & m_D \end{bmatrix} = \begin{bmatrix} 1 & 0 \\ 0 & 0.05 \end{bmatrix} t, \tag{5.10}$$

$$\mathbf{K} = \begin{bmatrix} k_1 + k_D & -k_D \\ -k_D & k_D \end{bmatrix} = \begin{bmatrix} 5.23 & -0.23 \\ -0.23 & 0.23 \end{bmatrix} kN\ m^{-1} \tag{5.11}$$

$$\mathbf{C} = \begin{bmatrix} c_1 + c_D & -c_D \\ -c_D & c_D \end{bmatrix} = \begin{bmatrix} 0.09 & -0.03 \\ -0.03 & 0.03 \end{bmatrix} kNs\ m^{-1} \tag{5.12}$$

$$\mathbf{F}_y = \begin{bmatrix} F_g \\ 0 \end{bmatrix} = \begin{bmatrix} c_1 \dot{x}_g + k_1 x_g \\ 0 \end{bmatrix} kN. \tag{5.13}$$

The inverse mass matrix is determined and written with the stiffness as well as the damping matrices as

$$\mathbf{M}^{-1} = \frac{1}{\det \mathbf{M}}\, adj\mathbf{M} = \frac{1}{m_1 m_D} \begin{bmatrix} m_D & 0 \\ 0 & m_1 \end{bmatrix} = \begin{bmatrix} 1 & 0 \\ 0 & 20 \end{bmatrix} t, \tag{5.14}$$

$$-\mathbf{M}^{-1}\mathbf{K} = \begin{bmatrix} -5.23 & 0.23 \\ 4.6 & -4.6 \end{bmatrix} kN\ (mt)^{-1}, \tag{5.15}$$

$$-\mathbf{M}^{-1}\mathbf{C} = \begin{bmatrix} -0.09 & 0.03 \\ 0.6 & -0.6 \end{bmatrix} kNs\ (mt)^{-1}. \tag{5.16}$$

The transition and input matrices of the system are determined from

Equation 3.4 as

$$
\mathcal{A} = \begin{bmatrix} \mathbf{0} & \mathbf{I} \\ -\mathbf{M}^{-1}\mathbf{K} & -\mathbf{M}^{-1}\mathbf{C} \end{bmatrix} = \begin{bmatrix} 0 & 0 & 1 & 0 \\ 0 & 0 & 0 & 1 \\ -5.23 & 0.23 & -0.09 & 0.03 \\ 4.6 & -4.6 & 0.6 & -0.6 \end{bmatrix}, \tag{5.17}
$$

$$
\mathcal{B} = \begin{bmatrix} \mathbf{0} & \mathbf{0} \\ \mathbf{M}^{-1} & \mathbf{M}^{-1} \end{bmatrix} = \begin{bmatrix} 0 & 0 \\ 0 & 0 \\ 1 & 0 \\ 0 & 20 \end{bmatrix}, \tag{5.18}
$$

where $\mathcal{A} \in \mathbb{R}^{2n \times 2n}$ and $\mathcal{B} \in \mathbb{R}^{2n \times n}$ with $n = 2$ corresponding to the two-DoFs of the SDoF+TMD system. From Equations 3.5, 3.6 and 3.7, we determine output and feedthrough matrices of the system as

$$
\mathcal{C} = \begin{bmatrix} \mathbf{I} & \mathbf{0} \\ \mathbf{0} & \mathbf{I} \\ -\mathbf{M}^{-1}\mathbf{K} & -\mathbf{M}^{-1}\mathbf{C} \end{bmatrix} = \begin{bmatrix} 1 & 0 & 0 & 0 \\ 0 & 1 & 0 & 0 \\ 0 & 0 & 1 & 0 \\ 0 & 0 & 0 & 1 \\ -5.23 & 0.23 & -0.09 & 0.03 \\ 4.6 & -4.6 & 0.6 & -0.6 \end{bmatrix}, \tag{5.19}
$$

$$
\mathcal{D} = \begin{bmatrix} \mathbf{0} \\ \mathbf{0} \\ \mathbf{M}^{-1} \end{bmatrix} = \begin{bmatrix} 0 & 0 \\ 0 & 0 \\ 0 & 0 \\ 0 & 0 \\ 1 & 0 \\ 0 & 20 \end{bmatrix}, \tag{5.20}
$$

where $\mathcal{C} \in \mathbb{R}^{m \times 2n}$ and $\mathcal{D} \in \mathbb{R}^{m \times n}$ with $n = 2$ corresponding to the two-DoFs of the system. The two sensors of the system are measuring simultaneously displacements, velocities and accelerations, which gives $m = 6$. The state, input and output of the system read

$$
\mathbf{z} = \begin{bmatrix} \mathbf{x} \\ \dot{\mathbf{x}} \end{bmatrix} = \begin{bmatrix} x_1 \\ x_D \\ \dot{x}_1 \\ \dot{x}_D \end{bmatrix}, \quad \mathbf{u} = \begin{bmatrix} F(t) \\ 0 \end{bmatrix}, \quad \mathbf{y} = \begin{bmatrix} x_1 \\ x_D \\ \dot{x}_1 \\ \dot{x}_D \\ \ddot{x}_1 \\ \ddot{x}_D \end{bmatrix}, \tag{5.21}
$$

where $\mathbf{z} \in \mathbb{R}^{2n \times 1}$, $\mathbf{u} \in \mathbb{R}^{n \times 1}$ and $\mathbf{y} \in \mathbb{R}^{m \times 1}$ with $n = 2$ and $m = 6$ corresponding to the DoFs and the sensors of the system. From Equations 3.2 and 3.3 and by neglecting the process and measurement noises as $\mathbf{w} = \mathbf{0}$ and $\mathbf{v} = \mathbf{0}$, we write the transition and output equations and determine the state-space representation of the given system as

$$
\dot{\mathbf{z}} = \mathcal{A}\mathbf{z} + \mathcal{B}\mathbf{u}, \tag{5.22}
$$

$$
\mathbf{y} = \mathcal{C}\mathbf{z} + \mathcal{D}\mathbf{u}. \tag{5.23}
$$

5.2.1.2 Deformation Response Factor

The deformation response factor R_d of dynamic systems represents the amplification of the dynamic deflection due to a harmonic excitation compared to the static deflection with the same load amplitude, cf. Section 1.4.3. The harmonic excitation can be represented by a dynamic force, such as $F = F_0 \sin \omega t$. Here, F_0 is the load amplitude, ω is the excitation frequency and t is the time.

For TMD controlled systems, the deformation response factor is an essential parameter for the analysis of the TMD behavior and can be calculated with

$$R_d(\eta) = \frac{x_{S,dy}}{x_{S,st}} = \sqrt{\frac{f_1^2 + f_2^2}{f_3^2 + f_4^2}}, \tag{5.24}$$

where $\eta = \omega/\omega_S$ is the ratio of the angular excitation frequency to the natural angular frequency of the structure. $x_{S,dy}$ is the steady-state dynamic response of the structure, and $x_{S,st}$ is its static deformation. f_i are given as

$$f_1 = \kappa^2 - \eta^2, \tag{5.25}$$

$$f_2 = 2\eta\kappa D_D, \tag{5.26}$$

$$f_3 = \eta^4 - \eta^2(1 + \kappa^2 + \mu\kappa^2 + 4\kappa D_S D_D) + \kappa^2, \tag{5.27}$$

$$f_4 = 2\eta\big(D_S(\kappa^2 - \eta^2) + \kappa D_D(1 - \eta^2 - \mu\eta^2)\big), \tag{5.28}$$

where $\mu = m_D/m_S$ is the mass ratio of the TMD to the modal mass of the structure. $\kappa = \omega_D/\omega_S$ is the ratio of the natural angular frequency of the TMD to the natural angular frequency of the structure. D_S and D_D represent the damping ratios of the structure and the TMD, respectively. The natural angular frequencies and the damping ratios of the SDoF+TMD system can be calculated from stiffnesses and damping coefficients as

$$\omega_S = \sqrt{\frac{k_S}{m_S}}, \quad \omega_D = \sqrt{\frac{k_D}{m_D}}, \quad D_S = \frac{c_S}{2m_S\omega_S}, \quad D_D = \frac{c_D}{2m_D\omega_D}. \tag{5.29}$$

As shown in Figure 5.4, depending on the chosen TMD parameters, the deformation response of the structure changes. The SDoF structure without TMD responses at resonance, if the excitation frequency ω of the harmonic excitation coincides with the natural angular frequency of the structure ω_S. Accordingly, for $\eta = 1$, a deformation response maxima is observed.

A significant reduction of the resonance response is observed for an optimum tuning of the TMD damping ratio as $D_D = D_{D,opt}$. In such a case, as an MDoF system, the deformation response curve of the SDoF+TMD system exhibits instead two new peaks, which are adjacent to the resonance peak of the SDoF structure without the TMD.

Furthermore, for a TMD without any damping $D_D = 0$, the two adjacent peaks go infinity. On the other hand, near resonance with $\eta \approx 1$, the dynamic response of the structure approaches zero. In fact, the supplementary mass of

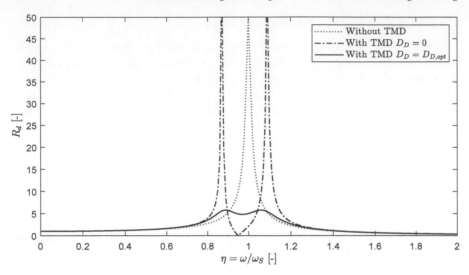

FIGURE 5.4
Deformation response factor R_d of an SDoF without TMD and with the TMD
for different tuning configurations of the damping ratio D_D. The structure has
the natural angular frequency ω_S and is under harmonic excitation with the
frequency ω.

the TMD causes a reduction of the resonance frequency and, therefore, the
dynamic response becomes zero at $\eta < 1$.

A variation of the natural angular frequency ω_D of the TMD affects the
system response as shown in Figure 5.5 (a). With the natural angular frequency ω_D, the TMD reduces the structural response optimally. However,
when the natural angular frequency of the TMD decreases to $\omega_{D,1}$, the right
peak of the response curve increases. On the other hand, when the natural
angular frequency of the TMD increases to $\omega_{D,2}$, the left peak of the response
curve increases.

The effects of the damping ratio D_D of the TMD are investigated in Figure 5.5 (b) in more detail. With the damping ratio D_D, the TMD reduces the
structural response optimally. However, when the damping ratio decreases to
$D_{D,1}$, the two peaks of the response curve increase and the response at resonance frequency simultaneously decreases. In limit state with $D_D = 0$, as
previously shown in Figure 5.4, the two peaks approach to infinity and the
response near the resonance frequency disappears. On the other hand, when
the damping ratio increases, the response around the resonance frequency
increases as well. In theoretical limit state with $D_D = 1$, the TMD does
not perform any relative motion against the structure. Accordingly, for $D_{D,2}$,
there are no restoring and damping forces generated. We observe a shift of
the resonance frequency again due to the additional TMD mass.

From the discussed variations of the natural angular frequency ω_D and the

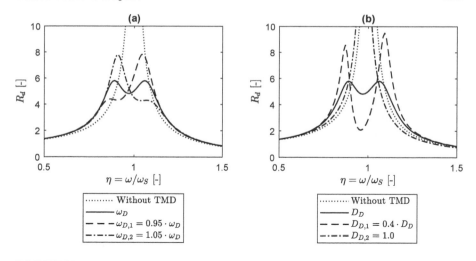

FIGURE 5.5
Deformation response factor of an SDoF without and with the TMD for varying TMD natural frequencies ω_D (a) and damping ratios D_D (b).

damping ratio D_D, it can be concluded that the TMD performs optimally, if ω_D and D_D are tuned to certain values. With these optimum natural angular frequency $\omega_{D,opt}$ and optimum damping ratio $D_{D,opt}$, the TMD is able to reduce the resonance vibrations of a harmonically excited SDoF structure very efficiently.

However, after attaching the TMD, two peaks arise in the response curve, which are adjacent to the resonance frequency of the structure. These peaks may have, even in optimal case with $\omega_{D,opt}$ and $D_{D,opt}$, a higher amplitude than the response curve of the uncontrolled structure.

In case of an MDoF structure with one TMD, the picture of the response curve stays similar. Around the natural frequency, to which the TMD is tuned, two adjacent peaks arise. Around other natural frequencies of the structure, the TMD does not affect the structural response. Accordingly, the natural frequency and damping ratio of a TMD must be tuned to the natural frequency of the structure, which is desired to be controlled.

5.2.1.3 Parameter Optimization

For the calculation of the optimum TMD parameters, several tuning methods exist. One of the most famous approaches was proposed by DEN HARTOG [52], who derived the tuning method from the response factor curve of an SDoF structure with a TMD by alternating the damper parameters. According to his criterion, the optimum natural angular frequency of the TMD is tuned to the structural natural angular frequency ω_S of the mode shape, which is aimed to be controlled, by

$$\omega_{D,opt} = \frac{\omega_S}{1+\mu},$$

(5.30)

where μ, as introduced before, is the mass ratio of the TMD to the modal mass of the structure, cf. Section 5.2.1.4. The optimum natural frequency of the TMD reads $f_{D,opt} = 2\pi\omega_{D,opt}$.

The optimum damping ratio of the TMD depends according to DEN HARTOG only on the mass ratio μ and is calculated as

$$D_{D,opt} = \sqrt{\frac{3\mu}{8(1+\mu)^3}}.$$

(5.31)

The DEN HARTOG criteria assume that the structure is undamped. However, also for structures with low damping, the criteria show only minor deviation and, therefore, apply as shown in [54].

The DEN HARTOG criteria are designed for harmonic excitation. Later, WARBURTON extended them also for random excitation with white noise spectral density [55]. For an undamped structure under random acceleration excitation, WARBURTON's criterion for the frequency tuning reads

$$\omega_{\omega,opt} = \omega_S \frac{\sqrt{1-\mu/2}}{1+\mu}.$$

(5.32)

The criterion for the optimum damping reads

$$D_{D,opt} = \sqrt{\frac{\mu(1-\mu/4)}{4(1+\mu)(1-\mu/2)}}.$$

(5.33)

As shown by WARBURTON, the influence of the structural damping on the criteria is negligible [55]. At low damping ratios $D_S \leq 0.1$ and low mass ratios $\mu \leq 0.2$, the optimum frequency and damping parameters show only a marginal change. For higher damping and mass ratios, the results start to deviate from the optimum tuning values corresponding to undamped structures.

5.2.1.4 Mass Ratio

In both DEN HARTOG and WARBURTON tuning criteria, the mass ratio is calculated with

$$\mu = m_D/m_S,$$

(5.34)

where m_D corresponds to the TMD mass and m_S to the structural modal mass of the mode shape, which should be controlled. The modal mass m_S depends on the mode shape. For tower-like structures with distributed mass, the modal mass is calculated as

$$m_S = \int_0^h \bar{m}(\xi)\varphi^2(\xi)\,d\xi,$$

(5.35)

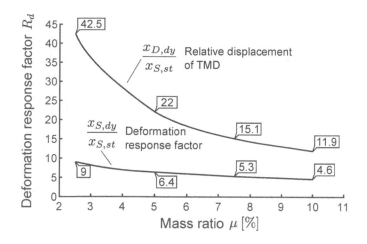

FIGURE 5.6

Change of deformation response factors of a controlled SDoF structure and its TMD for different mass ratios μ.

where m is the mass distribution along the height h of the structure. $\varphi(\xi)$ is the mode shape, which is normalized to 1 at the attachment position of the TMD. ξ is the coordinate along the structure.

Depending on the physical constraints and the architectural requirements, the mass ratio is chosen usually between 2 and 10 %. The relationships of the mass ratio with the displacements of the structure and the TMD are shown by an example in Figure 5.6. The resulting curves correspond to a negative exponential. With increasing TMD mass, both the dynamic displacement of the structure and the TMD are reducing. However, after a certain mass ratio, the slope of the curves changes only marginally and both curves converge to a certain deformation response factor.

On the other hand, for small mass ratios, the change in the TMD displacement becomes significant. Therefore, it can be concluded that for small mass ratios, generally a large TMD displacement is to be expected. This aspect can be particularly critical if the space provided for the TMD is limited. Furthermore, a large TMD displacement can be also critical for the bearing system of the damper as well as its spring and damping elements.

Moreover, large mass ratios reduce the tuning sensitivity of TMDs. This approach is applied as a general strategy in the practice to allow passive TMDs a more robust control performance. An example is shown in Figure 5.7 for two TMDs with a low mass ratio ($\mu = 2$ %) and a high mass ratio ($\mu = 10$ %). In Figure 5.7 (a), the TMDs are tuned to the natural frequency of the structure. In Figure 5.7 (b), the natural frequency of the structure decreases by 2 % and, accordingly, the TMDs get off-tuned. The performance loss for the TMD with

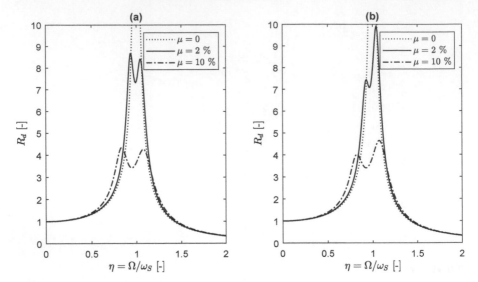

FIGURE 5.7
Deformation response factor of an SDoF without and with the TMDs of mass
ratios $\mu = 2\,\%$ and 10 %. TMDs are tuned with $\omega_D = \omega_{D,opt}$ and $D_D = D_{D,opt}$
(a). TMDs are off-tuned with $\omega_D = 0.98 \cdot \omega_{D,opt}$ and $D_D = D_{D,opt}$ (b).

the low mass ratio is more significant than for the TMD with the high mass
ratio.

Example 5.2 *An industrial chimney exhibits wind induced excessive vibrations in its first mode. Figure 5.8 shows the structure and its properties. The
natural frequency and the damping ratio of the structure are given as $f_S = 1\,Hz$
and $D_S = 0.01$ respectively.*

*From the mass distribution $\bar{m}$ and the mode shape ϕ, we determine the
modal mass of the structure as*

$$m_S = \int_0^h \bar{m}(\xi)\phi^2 \,\mathrm{d}\xi = 3.01 \ t, \tag{5.36}$$

*which corresponds to the area enclosed by the $\bar{m}\phi^2$-curve. Here, h is the total
height of the chimney and ξ is the coordinate starting from bottom of the
structure as shown in Figure 5.8.*

*For the vibration control of the structure, we design a TMD. As mass ratio,
we chose $\mu = 5\,\%$. Accordingly, a TMD mass of*

$$m_D = \mu m_S = 0.15 \ t \tag{5.37}$$

*is required for the chimney. The optimum natural frequency and the damping
ratio of the TMD can be determined from Equations 5.30 and 5.31 of the*

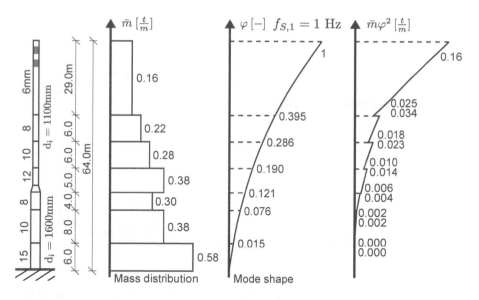

FIGURE 5.8
Industrial chimney (a). Its mass distribution $\bar{m}$ (b) and first mode shape ϕ (c) as well as $\bar{m}\phi^2$ (d).

DEN HARTOG *tuning criteria as*

$$f_{D,opt} = \frac{f_S}{1 + \mu} = 0.95 \ Hz \quad and \quad D_{D,opt} = \sqrt{\frac{3\mu}{8(1 + \mu)^3}} = 0.127. \quad (5.38)$$

The stiffness and the damping coefficient of the TMD correspond to

$$k_D = \omega_D^2 m_D = 5.362 \ kN \ m^{-1}, \quad (5.39)$$

$$c_D = 2\omega_D m_D D_D = 0.228 \ kNs \ m^{-1}. \quad (5.40)$$

The TMD mass can be realized by a steel ring, which can be attached to the structure via spring and damping elements as shown in Figure 5.9. However, it is noteworthy that for an omnidirectional operation capability, the TMD requires complex bearing systems. For a material density of $\rho = 8 \ t \ m^{-3}$, the volume of the steel ring corresponds to $V_D = 0.019 \ m^3$. For a radius of $r = 0.800 \ m$ and thickness of $d = 0.050 \ m$, the height of the steel ring is obtained as $h = 0.072 \ m$.

Figure 5.10 (a) compares the response of the structure to an initial deflection without and with the TMD. A significant improvement in the vibration dissipation capability of the structure is observed. Figure 5.10 (b) shows the corresponding frequency responses and proves the optimal tuning of the TMD.

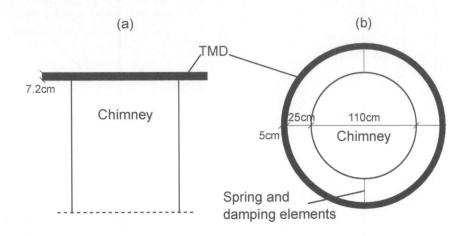

FIGURE 5.9
Side view of the TMD attached to the chimney (a). Plan view of the TMD
and the chimney (b).

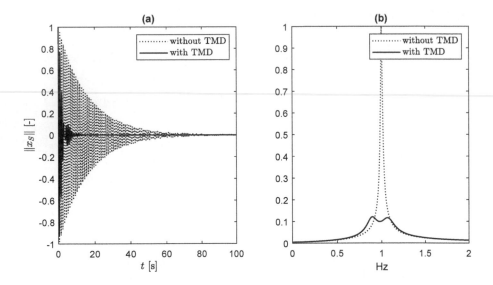

FIGURE 5.10
Normed displacement responses of the chimney to an initial deflection with
and without TMD (a). The corresponding frequency responses (b).

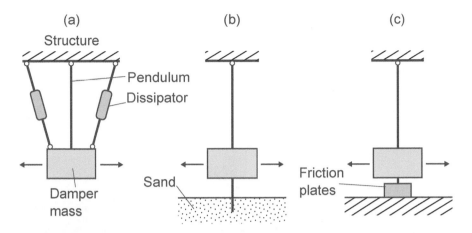

FIGURE 5.11
Examples of P-TMDs. To introduce damping, supplementary dissipators (a), sand (b) or friction plates (c) can be used.

5.2.2 Pendulum Tuned Mass Dampers

Pendulum tuned mass dampers (P-TMDs) are assembled by a solid mass, which is attached to the structure via a pendulum. The natural frequency of the P-TMDs is defined by the length of its pendulum. Therefore, the frequency tuning, contrary to the TMDs, does not require any mechanical spring elements. To improve the damping behavior, supplementary dissipators can be attached to the pendulum mass as shown in Figure 5.11 (a). Inherent damping can also be increased by friction as shown in Figure 5.11 (b,c).

P-TMDs dissipate the vibration energy of structures by restoring and damping forces, which are initiated by the motion of the supplementary damper mass. These control forces can be realized omnidirectionally. Therefore, P-TMDs are ideal for the control of multidirectional vibrations, such as for tower-like structures. Furthermore, they can easily realize natural frequencies lower than the TMDs. However, the necessary pendulum length may be long. To reduce the required space, multi-stage pendulum systems may be used.

Figures 5.12 and 5.13 show the application of a P-TMD on the skyscraper Taipei 101. The building with 101 floors is 509.2 m high and was finished in 2004. The P-TMD is installed on the 92nd floor to control wind induced vibrations. The 660 t steel damper mass is supported by supplementary viscous fluid dampers. Further information about the dynamic behavior of the structure with the P-TMD can be found in [56]. The readers interested also in other P-TMD applications are referred to [57].

As shown Figure 5.14 (a), the motion of the pendulum mass can be determined by idealizing the damper as a simple pendulum. From the tangent

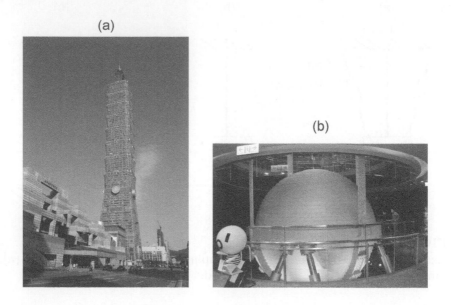

FIGURE 5.12
Taipei 101 skyscraper (a) and its P-TMD, which is attached to the 92nd floor
(b). Courtesy of O. Altay.

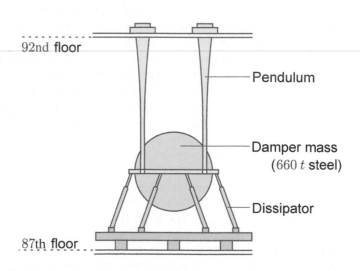

FIGURE 5.13
A drawing of the P-TMD of the skyscraper Taipei 101.

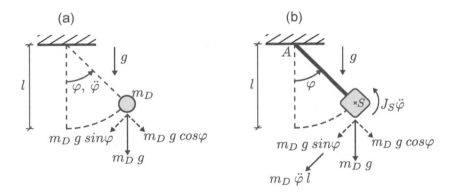

FIGURE 5.14
Idealization of the PTMD as a simple (a) and physical (b) pendulum.

component of the gravity force and the inertia force, the EoM can be determined and linearized as

$$m_D \ddot{\varphi} l + m_D g \sin \varphi = 0 \tag{5.41}$$

$$\Leftrightarrow \ddot{\varphi} + \frac{g}{l}\varphi = 0 \quad \text{for small } \varphi, \tag{5.42}$$

where φ is the rotation of the damper mass m_D around its attachment point and $\ddot{\varphi}$ is the angular acceleration. l and g are the pendulum length and acceleration of gravity respectively. The term before φ corresponds to the natural angular frequency of the P-TMD. Accordingly, it holds

$$\omega_D = \sqrt{\frac{g}{l}} \tag{5.43}$$

and the natural frequency depends consequently only on the pendulum length and the gravity.

Furthermore, P-TMDs can be mathematically described in a more accurate manner as physical pendulums as shown in Figure 5.14 (b). The EoM can then be determined from the moment equilibrium at the attachment point A as

$$\Sigma M_A: \quad m_D g \sin \varphi l + (m_D l^2 + J_{D,S})\ddot{\varphi} = 0, \tag{5.44}$$

where $J_{D,S}$ is the mass moment of inertia at the center of gravity S of the damper mass. Again for small φ, the differential equation can be linearized as

$$\ddot{\varphi} + \frac{m_D g l}{J_{D,A}}\varphi = 0, \tag{5.45}$$

where $J_{D,A}$ corresponds to the mass moment of inertia at attachment point A

of the pendulum. Accordingly, the natural angular frequency of the P-TMD is determined as

$$\omega_D = \sqrt{\frac{m_D g l}{J_{D,A}}}. \tag{5.46}$$

Example 5.3 *In this example, we design a P-TMD for the industrial chimney of Example 5.2 using the same structural properties. The structure was depicted in Figure 5.8. According to the* DEN HARTOG *tuning criteria (cf. Equations 5.30 and 5.31) and for the damper mass ratio* $\mu = 5$ *%, the following tuning parameters apply for the P-TMD*

$$f_{D,opt} = 0.95 \ Hz, \ D_{D,opt} = 0.127, \ m_D = 0.15 \ t, \ V_D = 0.019 \ m^3, \tag{5.47}$$

where $f_{D,opt}$, $D_{D,opt}$ *and* m_D *are the optimum natural frequency, optimum damping ratio and the mass of the P-TMD respectively. Here,* V_D *is the volume of the P-TMD assuming a steel ring is used with the outer radius* $r_2 = 0.825 \ m$, *inner radius* $r_1 = 0.800 \ m$ *and height* $h = 0.147 \ m$ *as shown in Figure 5.15. The pendulum length* l *is determined from the required optimum natural frequency using Equation 5.46 for physical pendulums as*

$$\omega_D = \sqrt{\frac{m_D g l}{J_{D,A}}} \tag{5.48}$$

$$\Leftrightarrow l = \frac{g}{2\omega_D^2} \pm \sqrt{\frac{g^2}{4\omega_D^4} - \frac{J_{D,S}}{m_D}} = 0.222 \wedge 0.054 \ m, \tag{5.49}$$

where $J_{D,A} = m_D l^2 + J_{D,S}$ *is the mass moment of inertia of the P-TMD at its attachment point A. Here,* $J_{D,S}$ *is the mass moment of inertia at the center of gravity and calculated for the ring form as*

$$J_{D,S} = \frac{m_D}{12} \left(3(r_2^2 - r_1^2) + h \right) = 3.169 \cdot 10^{-3} \ tm^2. \tag{5.50}$$

As $h = 0.147 > l = 0.054 \ m$ *holds,* $l = 0.222 \ m$ *is chosen for the pendulum length. The chosen dynamic parameters* (f_D, m_D) *correspond to the TMD of Example 5.2. Furthermore, we assume that the required damping ratio is realized by supplementary dissipators and corresponds to* $D_D = D_{D,opt}$. *Therefore, both the displacement time history and the frequency response of the chimney with P-TMD correspond to Figure 5.10.*

5.2.3 Tuned Liquid Dampers

The *tuned liquid dampers* (TLDs) consist of a vessel, which is partially filled with a NEWTONIAN *fluid*, such as water, as shown in Figure 5.16. The liquid motion in the vessel generates a restoring force. Furthermore, damping is induced by the formation of waves on the liquid surface and their interaction with the vessel walls. If required, inherent damping can be also increased by

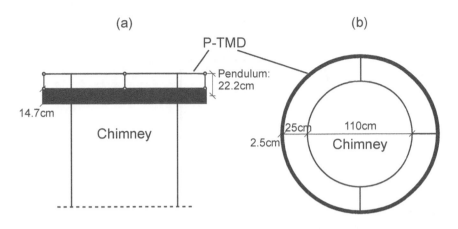

FIGURE 5.15
Side view of the P-TMD attached to the chimney (a). Plan view of the P-TMD and the chimney (b).

installing flow damping devices, such as nets, screens, baffles, poles, floating objects and further obstacles. These components introduce supplementary friction effects into the liquid flow. Because of the sloshing effect, the TLDs are also called as *tuned sloshing dampers* (TSDs).

Similar to the other TMDs, the TLDs are most efficient, when installed at locations where maximal deflections of structures occur, such as on the top floor of skyscrapers. The required damper mass can be distributed to several vessels corresponding to the available space capacity of the structure.

Some of the first TLD applications are the Nagasaki Airport Tower (42 m, in 1987), the Yokohama Marine Tower (100 m, in 1987), the Yokohama Prince Hotel (150 m, in 1992) and the Haneda Airport Tower in Tokyo (78 m, in 1993). Further information about these applications can be found in [13, 25, 58].

Pioneering studies on the damping behavior of the TLDs and their mathematical description have been conducted in 1980s, such as by Bauer [59], Fujino et al. [60], Welt and Modi [61] as well as Sun et al. [62]. These studies revealed that the natural frequency of the TLDs depends only on the geometry of their vessel. Accordingly, for the frequency tuning, the TLDs do not require any mechanical components, such as springs.

A practical approach for the design of the TLDs is to utilize a mechanical model, which approximates the damper dynamics by an equivalent modal mass m_D and a natural frequency ω_D of the first mode of sloshing response [63]. According to Ibrahim [64] and Sun et al. [65], for rectangular TLDs, these

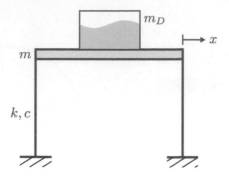

FIGURE 5.16
A TLD attached to an SDoF structure.

parameters can be determined by

$$m_D = \frac{8\rho a^2 b}{\pi^3} \tanh\left(\frac{\pi h}{a}\right), \tag{5.51}$$

$$\omega_D = \sqrt{\frac{\pi g}{a} \tanh\left(\frac{\pi h}{a}\right)}, \tag{5.52}$$

where ρ is the liquid density and a, b, as well as h represent the geometric dimensions of the TLD tank as shown in Figure 5.17 (a). According to Ibrahim [64] and Abramson [66] for circular TLDs, the modal mass and natural frequency are determined as

$$m_D = \frac{\pi \rho r^3}{2.2} \tanh\left(\frac{1.84h}{r}\right), \tag{5.53}$$

$$\omega_D = \sqrt{\frac{1.84g}{r} \tanh\left(\frac{1.84h}{r}\right)}, \tag{5.54}$$

where r and h represent the tank geometry as shown in Figure 5.17 (b). Both for rectangular and circular TLDs, the damping ratio of the first mode is given by the empirical relationship

$$D_D = C \left(\frac{\nu}{d^{3/2}\sqrt{g}}\right)^n, \tag{5.55}$$

where for the constants it holds $C \approx 1$ and $n \approx 0.5$, which changes only slightly depending on the liquid depth, cf. [64]. Here, ν is the kinematic viscosity of the liquid and g is the acceleration of gravity. Furthermore, d represents the characteristic dimension of the TLD tank with $d = a$ for rectangular and $d = r$ for circular TLDs. The supplementary damping effect of flow damping

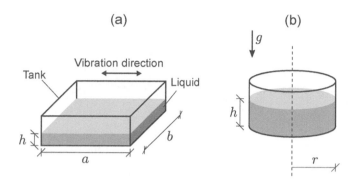

FIGURE 5.17
TLDs with rectangular (a) and circular (b) forms.

devices, such as nets and baffles, can be approximated as proposed by Konar
and Ghosh [67].

The optimum required damper parameters can be estimated by optimization methods, such as using DEN HARTOG [52] or WARBURTON [55] tuning
criteria.

Low material and maintenance costs as well as simple design are advantageous aspects of the TLDs compared to the mechanical TMDs. However,
the response of the TLD and the associated energy dissipation depend on the
sloshing effects, which are difficult to predict. A more stable type of liquid
damper will be introduced in the next section.

Example 5.4 *We consider again the industrial chimney of Example 5.2 and
design a TLD system for its vibration control. For the previously given structural parameter, the optimum damper is determined from the* DEN HARTOG
tuning criteria (cf. Equations 5.30 and 5.31) as

$$f_{D,opt} = 0.95 \ Hz, \ D_{D,opt} = 0.127, \ m_D = 0.15 \ t, \qquad (5.56)$$

*where $f_{D,opt}$, $D_{D,opt}$ and m_D are the optimum natural frequency, optimum
damping ratio and the mass of the TLD system respectively. For an omnidirectional control, a circular TLD is preferred. From Equation 5.52, the corresponding radius and height are calculated for $\omega_D = 2\pi 0.95 \ rad \ s^{-1}$ with
$h/r = 0.5$ as $r = 0.368 \ m$ and $h = 0.184 \ m$. The modal mass is calculated
from Equation 5.51 as $m_{D,i} = 0.05 \ t$. Accordingly, three TLD tanks are required to realize $m_D = 3m_{D,i} = 0.15 \ t$. The proposed TLD system is illustrated
in Figure 5.18.*

*The chosen dynamic parameters (f_D, m_D) correspond to the TMD of Ex-
ample 5.2. Furthermore, we assume that the required damping ratio is realized by flow damping devices and corresponds to $D_D = D_{D,opt}$. Therefore,*

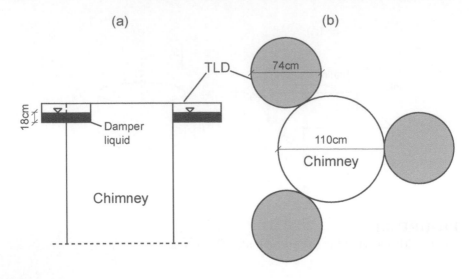

FIGURE 5.18
Side view of the TLD system attached to the chimney (a). Plan view of the
TLD system and the chimney (b).

the displacement time history and frequency response of the chimney with the
TLD system correspond to Figure 5.10 as well.

    It is noteworthy that the total mass of the TLD system is different than
its modal mass and corresponds to $m_{D,T} = 3\pi r^2 = 0.23$ t. Accordingly, the
additional passive mass of $m_{D,p} = 0.23 - 0.15 = 0.08$ t will reduce the natural
frequency of the chimney from $f_S = 1.00$ Hz to $f_S = 0.99$ Hz, due to which a
slight deterioration of the TLD performance is expected.

5.2.4 Tuned Liquid Column Dampers

Similar to the TLDs, the *tuned liquid column dampers* (TLCDs) use also a
NEWTONIAN *fluid*, such as water, which is filled into a U-shaped vessel as
shown in Figure 5.19 (a). The tuned parameters enable the liquid mass to
oscillate with a phase-shift against the motion direction of the structure. Local
friction and turbulence effects in the vessel cause the vibration energy of the
liquid to dissipate.

    The first TLCD was patented by Frahm already in 1910 [68], which was
one year before the first TMD he also patented. The design of the first TLCD
aimed to stabilize the roll motion of ships. At the end of 1980s, Sakai et al.
implemented this design for civil engineering structures [69, 70]. However,
applications of TLCDs in structural engineering are quite newer than other
types of control devices. Some of the most prominent TLCD applications are
the One Wall Centre in Vancouver (158 m, in 2001), the Random House Tower

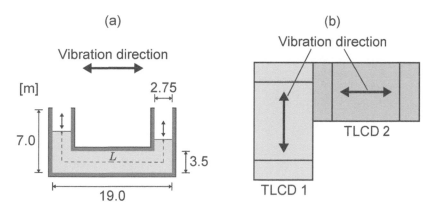

FIGURE 5.19
A TLCD example with real dimensions, which is operating uniaxially (a). Two TLCDs combined to control a structure omnidirectionally (b).

in New York City (209 m, in 2003) and the Comcast Center in Pennsylvania (297 m, in 2008). Further information about these applications can be found in [71–73].

Pioneering studies, such as by Balendra et al. [74, 75], Chang and Hsu [76], Gao et al. [77] and Hitchcock et al. [78, 79], have investigated the response of passive TLCDs. Derived mathematical descriptions show that the calculation of the natural angular frequency of the TLCD is very similar to the frequency calculation of a simple pendulum. For a vessel with a constant cross-sectional area, the natural angular frequency depends only on the length of the liquid L (cf. Figure 5.19 (a)) and reads

$$\omega_D = \sqrt{\frac{2g}{L}}. \tag{5.57}$$

The natural frequency of the TLCD is linear and allows the damper, contrary to the TLDs, a more stable damper response. TLCDs can be tuned to very low frequencies and are, therefore, especially advantageous for skyscrapers with very long natural periods. To increase the inherent damping an orifice can be installed, which can add supplementary friction effects and consequently also damping.

Classical TLCDs operate in uniaxial direction. However, they can be combined with each other to control the structural motion as shown in Figure 5.19 (b). Chapter 7 will introduce detailed information about the TLCDs and is particularly concerned with semi-active and omnidirection versions of the damper.

Part III

Advanced Damping Systems

6

Active and Semi-Active Damping Systems

6.1 Introduction

As introduced in the previous chapters, passive damping systems, such as dissipators and TMDs can be easily designed. Furthermore, their application is quite straightforward. The operation of these systems does not require any external power supply. Still, they exhibit one significant disadvantage, which is the missing adaptation capability of their control behavior to the actual requirements of the structure and the loading situation. This drawback impedes the possibility of a high performance structural control with the passive devices. Therefore, as alternative, active and semi-active systems are being researched and developed. This chapter is concerned with the description of active and semi-active damping systems.

Both systems are introduced with their relevant components in Sections 6.2.1 and 6.3.1. The chapter includes furthermore application examples of these systems. Particularly, pioneering applications are introduced in Sections 6.2.2 and 6.3.2.

6.2 Active Damping Systems

6.2.1 Description

Active damping systems generate exogenous control forces by means of actuators, which are operated by an external power. Depending on the dimensions of the controlled structure, usually large-size actuators are required. Linear motors using electro-magnetism or ball-screw mechanisms as well as hydraulic actuation are used for the force generation.

Active systems aim to realize restoring forces, which should protect both the safety of structures and the comfort requirements of the occupants. However, the dependency of the system on the external power supply is a weak point, which might become particularly dangerous during hazardous events, such as earthquakes, which are usually accompanied by power failure.

The control forces of active systems are determined in real-time from the

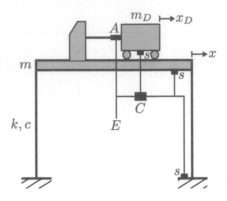

FIGURE 6.1

An AMD attached to a seismically excited SDoF structure. The actuator (A) and sensors (S) as well as the controller (C) of the system are depicted. The system requires for its operation a permanent external energy source (E)

measured dynamic response of the structure, which is excited by a natural event or an anthropogenic activity. Robust algorithms are required both for the identification of the relevant system parameters and determination of the control force characteristics. However, this can be quite challenging due to the fact that, particularly in case of wind and earthquake, structures exhibit random vibrations, the parameters of which are difficult to predict. The observability raises further challenges due to large dimensions of particularly civil engineering structures and uncertainties involved in the utilized measurement systems. Nevertheless, to tackle these limitation, besides classical control algorithms [22, 30, 80, 81], such as linear quadratic regulator (LQR), linear quadratic Gaussian controller (LQG) and model predictive control (MPC), also other methods are still being investigated and developed for active systems, such as reviewed in [82, 83]. Furthermore, soft computing-based methods are being researched as well, such the FLC, cf. Section 3.3.2.

Active damping systems can be realized in numerous manners. Figure 6.1 shows an active mass damper (AMD), which is attached to a seismically excited SDoF structure. Here, a TMD mass m_D is driven by an actuator (A). Therefore, such systems are also referred to as active-tuned mass dampers (A-TMD). The motion characteristics of the damper mass completely depend on the actuator. There are not any spring or damping elements utilized, which are common for the passive TMDs. Sensors (S) are mounted both on the structure and on the damper mass to monitor both the system response and the excitation. A controller (C) calculates the necessary actuator force in real-time from the measurement data. For the generation of the actuator force an external permanent energy source (E) is supplied.

In frame structures, active systems can also be realized by coupling the

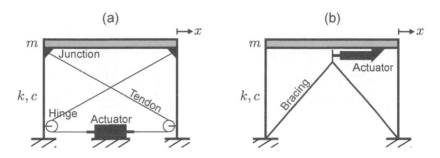

FIGURE 6.2
Active tendon system (a). Active brace system (b).

structure with actuators via tendons and bracing elements. As an example an active tendon system and an active brace system are illustrated in Figure 6.2. In both examples, the deflection of the frame structure is controlled by the actuators. Besides actuators, similar to the AMD, also these systems involve sensors as well as controllers and require a permanent external energy source, which are not depicted in the figure.

Active damping systems can also be combined with passive systems. Figure 6.3 (a) shows a hybrid mass damper (HMD). In this system, an actuator is connected to a passive TMD to control the motion of the damper mass. In case of an operational failure or an interruption of the power supply, this system continues its control function as a conventional passive TMD. Accordingly, compared to a pure active system, it can realize a more robust control.

In Figure 6.3 (b), the second system shows a small-size AMD, which is combined with a passive TMD. The required input power of this configuration is much lower than pure active systems. The main portion of the control force is generated by the natural motion of the TMD mass m_D in a passive manner. The actuator optimizes via AMD the TMD motion and increases the efficiency of the damping system. As a passive damper, the tuning frequency of the TMD corresponds to the relevant natural frequency of the structure. Furthermore, random vibrations and vibrations of other modes are covered by the AMD.

6.2.2 Application Examples

The first real-size AMD was applied in 1989 to the ten-story Kyobashi Seiwa Building in Tokyo. The wind and moderate earthquake-induced lateral and torsional vibrations of the steel frame structure are controlled by two AMDs of 4 t and 1 t masses, respectively [84].

The responses of some other famous AMD applications are investigated by Yamamoto et al. [85]. The study involves four applications from 1990s in Japan. One of the structures is the 34-story Applause Tower, for which the

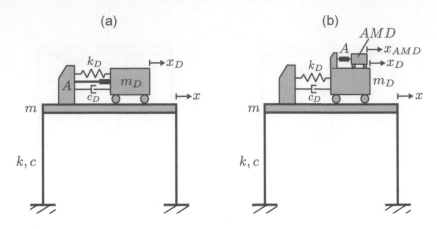

FIGURE 6.3
HMD (a) attached to an SDoF structure. The damper mass is controlled by an actuator (A). An AMD combined with a TMD (b)

480 t heliport of the building is utilized as a damper mass. The authors applied free vibration tests and analyzed structural responses to wind and earthquake. In their study, a good control performance is concluded for the control of free vibration response as well as for the control of wind-induced and low-level seismic vibrations. An overview of pioneering AMD applications in Japan is given by Nishitani and Inoue in [86].

A further AMD system was designed by Wu and Yang for the control wind-induced vibrations of the 310 m Nanjing TV transmission tower in China [87]. They investigated besides LQG also H_∞ and continuous sliding mode control algorithms using a reduced order system of the structure. The design characteristics of the damping system is reported by Cao et al. in [88].

Besides AMDs, further active systems, such as active tendons are investigated by Bossens and Preumont [89]. Rodellar et al. investigated similar systems for cable-stayed bridges [90]. Further examples can be found in [80].

Due to their robustness, HMDs dominate the so far applied active damping systems. Some examples of the most prominent HMD applications are the V-shaped HMD, which is installed in the 235 m Shinjuku Park Tower in Japan [91], and the multistep pendulum HMD, which is installed in the 296 m Yokohama Landmark Tower also in Japan [92].

For further application examples, the interested reader is referred to the review papers, such as [19] as well as [82, 93].

6.3 Semi-Active Damping Systems

6.3.1 Description

Recent significant developments in computer science and cybernetics, directed the focus of recent research in structural control to semi-active approaches. Semi-active damping systems can adapt their dynamic behavior autonomously to the requirements of structures and loading conditions. These systems involve semi-active TMDs and semi-active dissipators.

Their adaptation capability allows these devices to re-tune their parameters corresponding to the structural changes, which may be induced by external and internal effects, such as degradation and SSI. Furthermore, some semi-active systems are able to adapt themselves also in real-time during dynamic events, such as earthquake, and hence react to the loading conditions. Consequently, semi-active systems enable a more robust control compared to passive systems, which cannot adapt their parameters to the changing requirements and lose their efficiency over time due to off-tuning.

Semi-active systems generate their control forces solely from vibrations without any external energy source. Accordingly, semi-active systems cannot cause any destabilization of structures. Furthermore, for the adaptation they utilize smart materials or movable mechanisms, which can be controlled by small-size actuators and require much lower energy compared to the actuators of an active system.

For the estimation of the optimum damper parameters, semi-active systems require control algorithms. These algorithms use generally the state-space representation for the mathematical modeling of the controlled system. The governing equations of state-space representation are introduced with examples in Section 3.2. For the control of damping behavior, the on-off controller is commonly used. Furthermore, novel soft computing-based methods can be applied, such as the Fuzzy logic controller (FLC), which can handle nonlinearities and uncertainties in a more robust manner. Both approaches are introduced in Section 3.3.2.

The efficiency of semi-active systems relies on the accurate identification of the responses as well as on the parameters of the structure and the control devices. Particularly, abrupt structural changes are important and must be identified in real-time for an accurate tuning of the semi-active devices. However, the identification of such changes is challenging due to the supplementary damping effects induced by control devices. In Section 9.5, an UKF-based adaptive joint state-parameter observer is introduced for this purpose, which can accurately identify both the responses and the parameters of a controlled system under abrupt changes.

A semi-active vibration control system comprises, besides the damping device itself, also sensors and a controller. In Figure 6.4, such a system employing an adaptive TMD is illustrated. In this example, the semi-active system is

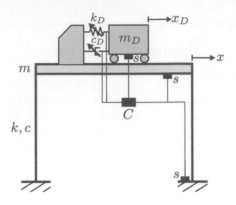

FIGURE 6.4

A semi-active vibration control system attached to an SDoF structure. An adaptive TMD is used, which can control its stiffness and damping parameters. The sensors (S) as well as the controller (C) of the system are depicted

designed to protect the SDoF structure from earthquake-induced vibrations. Sensors (S) record the responses of the structure and the TMD as well as the seismic excitation of the structure. The controller (C) evaluates the sensor signals and computes the required control forces. The semi-active TMD system can regulate both its stiffness k_D as well as the damping coefficient c_D and is, accordingly, capable of controlling both its natural frequency and damping behavior.

6.3.2 Application Examples

As introduced in Section 5.2.1.3, the stiffness and damping parameters govern the dynamic behavior of TMDs. Based on these parameters, *semi-active TMDs* with adaptive control have been developed by numerous studies. Hrovat et al. [29] and Abe [94] investigated a semi-active TMD with adjustable damping to control its position during oscillations. Variable damping was also the focus of further researchers, such as by Setareh et al. [95], Yang et al. [96] and Kang et al. [97]. On the other hand, some studies placed their focus on the variable stiffness, such as by Varadarajan and Nagarajaiah [98], Deshmukh and Chandiramani [99] and Wang et al. [100]. Furthermore, Nagarajaiah proposed a P-TMD with adaptive length for frequency adaptation [101] (cf. Section 5.2.2). These studies and further investigations, such as by Sun and Nagarajaiah [102] and Karami et al. [103], covered different application scenarios for the semi-active TMDs, such as vibration control of structures under wind and earthquake loads.

Besides classical TMDs, *semi-active TLCDs* play a significant role in semi-active structural control. As introduced in Section 5.2.4, the damping behavior of TLCDs is governed by local turbulence and friction effects. Therefore, supplementary mechanisms, such as orifices, are used to increase the inherent damping. This approach was first introduced by Sakai and Takaeda [69, 70]. On the damping effects of orifices, several studies have been conducted. Yalla and Kareem [104] tested experimentally the orifice effect by using an electro-pneumatic valve and investigated on-off and FLC-based control algorithms. La and Adam [105] investigated an on-off controller for the orifice tuning. Furthermore, the damping of TLCDs can be modulated by regulating the air flow from the columns as initially proposed by Frahm [68] and implemented in a semi-active framework by Matsuo [106].

For the control of the natural frequency of TLCDs, Nomichi and Yoshida sealed the column ends and introduced an air spring [107]. Yoshimura and Yamazaki enhanced this approach by valves to regulate the air spring effect in a semi-active manner [108]. A further development was proposed by Kagawa and Fujita [109] by dividing the air-filled portion of the columns in closable chambers. The application of this tuned liquid column gas damper (TLCGD) on structures has been investigated also by further studies, such as by Hochrainer and Ziegler [110], Reiterer and Ziegler [111], Fu [112] and by Mousavi et al. [113]. Reiterer and Kluibenschedl implemented this approach also for the semi-active control of vertical vibrations [114].

TLCDs have been also combined with mechanical spring elements to tune their frequency. Yoshimura et al. developed such a device [115] and installed it on a high-rise building [116]. Further configurations of TLCDs incorporating spring elements have been studied by Ghosh and Basu [117], Kim and Adeli [118] and Sonmez et al. [119].

Furthermore, by changing the cross section of the columns the flow speed of the liquid and the associated natural frequency of the TLCDs can be regulated. Min et al. used this effect and divided the columns in closable upright cells for the fine tuning of passive TLCDs after installation on structures [120].

Moreover, a semi-active frequency control of TLCDs is realized by installing movables panels in vertical columns and regulating them with small-size actuators [121]. By closing and opening the panels, the cross-sectional area of the columns can be regulated and the liquid flow velocity can be adapted. The proposed S-TLCD operates uniaxially. Experimental investigations [122] revealed that the real-time computational performance of the control algorithm is decisive for the frequency control. Therefore, the required calculation effort of the control algorithm must be limited. This damper is introduced in Section 7.1. An omnidirectional version of this damper, which controls its natural frequency via closable column cells is introduced in Section 7.2.

Apart from TMDs, also semi-active dissipators are being investigated. As introduced in Section 4.3, friction dampers dissipate vibration energy by sliding contact friction between two surfaces. As shown by several studies, such as [123–126], by controlling the normal force applied to both surfaces a *semi-active friction damper* can be assembled, which can adapt the generated damping force.

As introduced in Section 4.5, viscous fluid dampers consist of a hydraulic

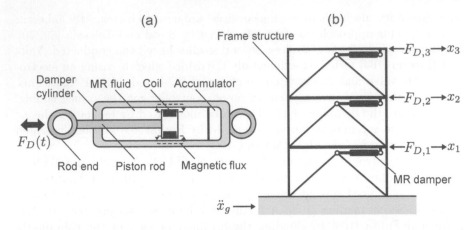

FIGURE 6.5
An MR damper (a). A seismically excited frame structure retrofitted by MR dampers (b).

cylinder with two chambers, which are filled with a viscous fluid, such as oil, and connected with each other via a small orifice. The vibration energy is dissipated by the high-speed flow of the viscous fluid through this orifice. Using a valve the opening size of the orifice can be regulated, which allows the control of generated damping in a semi-active manner. For the regulation of the valve, a small-size is used. Numerous investigations have been conducted on *semi-active viscous fluid dampers* including shaking table tests, such as [127–131].

   MR dampers operate in a similar manner to the viscous fluid dampers. However, the damper fluid (usually oil) contains 0.1–10 μm sized magnetizable particles and the viscosity can be adjusted by applying a magnetic field as shown in Figure 6.5. Several numerical and experimental studies, such as [132–138], investigated the control performance of the damper. Particular effort has been devoted on the modeling of the damper, such as based on the BOUC-WEN model [139].

   In other studies, MR dampers have been also combined with the TMDs, such as by Lin et al. [140] and Setareh et al. [141]. Some studies used also MR fluids for the semi-active control of TLCD parameters, such as by Wang et al. [142] as well as Sarkar and Chakraborty [143].

   A further semi-active strategy uses *variable stiffness devices*, which can be attached to the joints of structural elements. Variable stiffness devices have been investigated by several studies, such as [144–155]. Usually installed within bracing systems, these devices modulate the stiffness of the structure and prevent resonance phenomena. For this purpose, hydraulic cylinders have been developed, which can be locked by closing their orifice.

   The readers interested in further studies and examples are referred to review papers, such as by Housner et al. [19], Symans and Constantinou [156], Soong and Spencer [20], Fisco and Adeli [82] , and to reference books, such as by Adeli and Saleh [24], Casciati et al. [157], Cheng et al. [158] and Xu et al. [159].

7

Semi-Active Tuned Liquid Column Dampers

This chapter introduces two semi-active TLCDs with frequency and damping adaptation capabilities. The first damper operates uniaxially and uses movable panels for the frequency adaptation. This damper is referred to as the semi-active tuned liquid column damper (S-TLCD). The second damper operates omnidirectionally and encompasses closable cells in its columns for the frequency adaptation. This damper is referred to as the semi-active omnidirectional tuned liquid column damper (O-TLCD). Both dampers use orifices for the damping control.

The S-TLCD is introduced in Section 7.1. The O-TLCD is introduced in Section 7.2. For both the S-TLCD and the semi-active O-TLCD, the analytical modeling approaches are presented and validated by experiments. Furthermore, the chapter includes the numerical application of the semi-active O-TLCD to a structure. The chapter is concluded is Section 7.3.

7.1 Semi-Active Uniaxial Tuned Liquid Column Damper

Semi-active damping systems can adapt themselves to the changing structural conditions and loading situations based on the data captured by their sensors, cf. Section 6.3. This autonomous tuning capability allows semi-active structural control devices to outperform passive vibration mitigation methods. Furthermore, semi-active systems can alternate their parameters also in real-time ensuring a load correspondent, superior, stable and low-energy control performance compared to active systems. In the last decades, numerous semi-active systems have been proposed. However, most of these devices can adapt only their damping parameters. Besides the active variable stiffness (AVS) system developed by Kobori et al. [144], only a few semi-active systems exist with stiffness adaptation capabilities. Some rare examples include the adaptive length P-TMD developed by Nagarajaiah et al., which are also applied to real structures in the USA, Japan and China [101, 102, 160].

Semi-active systems with tunable natural frequency capabilities are particularly necessary for slender high-rise structures, such as wind turbines. During their service time, these structures change their natural frequencies due to degradation as well as soil and operational effects. Thereby, passive systems lose their efficiency by frequency off-tuning. In this context, Sun developed for wind turbines subjected to multi-hazards a sophisticated simulation environment, where degradation and soil effects can be simulated for the investigation of semi-active control systems [161, 162]. He proposed and investigated for monopile offshore wind turbines an S-TMD, which can adapt its damping and frequency parameters. The results show a significant improvement of the control system due to its semi-active adaptation capability.

Besides these mechanical TMDs, the so far proposed liquid-based TMDs, which possess both natural frequency and damping adaptation capabilities, are still in the very early development stage. Due to their geometric versatility, low prime costs and stable dynamic behavior, the TLCDs are regarded as one of the most promising solutions.

First invented by Frahm [68] already in 1910 for ships and then proposed by Sakai et al. [69, 70] in 1990s for civil engineering structures, the TLCD is one of the first structural control devices. As introduced in Section 5.2.4, the TLCDs consist of a U-shaped tube with two vertical columns, which are open at one end and connected at the other end by a horizontal segment. The tube is attached on a structure, the vibrations of which are to be controlled. For high-rise structures the TLCDs are usually attached at the top of the structure, where the maximum lateral motion is expected. The tube is partially filled with a Newtonian fluid, which is prescribed to flow with phase shift with respect to the structure. The phase shift must be tuned by the natural frequency and the damping parameters of the liquid motion. The oscillating liquid mass evokes restoring forces. Furthermore, during the oscillation, local friction and turbulence effects in the tube dissipate vibration energy, allowing a control of the structural vibrations.

To tune the damping of the TLCDs, Sakai et al. installed an orifice in the horizontal segment. This orifice can be closed manually to introduce additional damping by increasing the local friction [70]. Alternatively, Frahm and Matsuo closed the open ends of the vertical columns with orifices, which influence the air flow and reduce the speed of liquid level motion [68, 106]. As this method influences also the natural frequency, Sakai's approach counts as an established method for the damping tuning of the TLCDs.

Several studies have been conducted considering the calculation and control of the damping effects introduced by the orifice, such as by Yalla and La et al. In their study, Yalla et al. equipped a TLCD with an electro-pneumatic valve to tune its inherent damping in a semi-active manner [104]. They developed a control strategy and conducted experimental investigations. The results show that the semi-active tuning of the inherent damping provides the TLCD an up-to 25 % performance improvement. Furthermore, La et al. derived in their study a general on-off controller for the damping tuning of the TLCDs using the orifice [105]. They documented the robust performance of the controller numerically using a 5-DoF shear frame, which is subjected to earthquake. Both of these investigations use TLCDs without frequency adaptation capabilities.

For the semi-active tuning of the natural frequency, Csupor divided the horizontal segment into two channels, each with a different cross-sectional area [163]. As the fluid flow velocity in the tube depends on the cross-sectional area, this approach allows the TLCD to switch between two natural frequencies. However, for the mitigation of lateral vibrations the volume of the horizontal segment is decisive. Therefore, this TLCD configuration alternates unfavorably also its restoring force level depending on the volume of the chosen channel.

As reported in Section 6.2.2, further configurations were developed, which use air spring for the adaptation of the natural frequency, such as Nomichi and Yoshida [107], Yoshimura and Yamazaki [108], Kagawa and Fujita [109], Hochrainer and Ziegler [110], Reiterer and Ziegler [114], Fu [112], Mousavi et al. [113] as well as Reiterer and Kluibenschedl [114].

A further method for the frequency tuning is proposed by Yoshimura et al. [115]. They integrated in TLCDs mechanical spring elements influencing the flow velocity of the liquid. This device is referred to as the liquid column damper with period adjustment (LCD-PA) and has been operating since 1994 in a hotel building in Tokyo [13]. Also Ghosh, Kim and Sonmez et al. propose to control the natural frequency of TLCDs by interposing spring elements between the TLCD and the structure [117–119]. Both of these hybrid solutions control the natural frequency mechanically using spring elements while the TLCD continues its operation in a passive manner.

Min et al. divided the vertical columns in upright cells allowing a frequency tuning of the TLCD [120]. They applied this TLCD on four skyscrapers in South Korea to tune the natural frequency after installation of the damper in a passive adaptive manner. By closing the chambers, the cross-sectional area of the columns can be adjusted, which influences the liquid flow speed and enables an incremental frequency tuning.

This section is concerned with a semi-active tuned liquid column damper (S-TLCD), which can change both its natural frequency and damping continuously. Inspired by the passive TLCD, the S-TLCD is developed as a tank, which is filled with a Newtonian liquid, such as a mixture of water and antifreeze. The tank has a U-shaped geometry consisting of two rectangular liquid columns, which are arranged at a distance from each other and communicating by a horizontal passage as shown in Figure 7.1. As described also in [122], in the tank, mechanisms for tuning both the natural frequency and the damping are provided, which can adjust the parameters of the damper in real time. For this purpose, perpendicular to the flow direction, movable vertical panels are attached to the column walls, which allow to change the cross-sectional area of the columns without changing the horizontal distance between the liquid columns. By changing the cross-sectional area, the velocity of the liquid flow is controlled enabling S-TLCD to tune its natural frequency. Each movable panel consists of a transition and an upright segment. The liquid

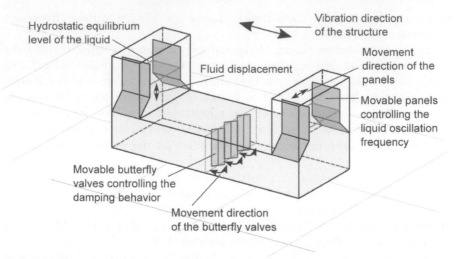

FIGURE 7.1
The S-TLCD with natural frequency and damping adaptation capabilities.

level oscillates in the upright segment range allowing the S-TLCD to keep its natural frequency constant during its operation. The gaps between the movable panels and the column walls are sealed by waterproof elastic membranes. For the tuning of the damping, several butterfly valves are installed in the connection passage of the liquid columns. By closing these valves, local friction effects can be introduced allowing an increase in damping. The motion of the panels and the butterfly valves is controlled by actuators. The liquid motion is measured by a capacitance transducer.

7.1.1 Mathematical Modeling

In fluid dynamics, the absolute acceleration **a** of a single liquid particle with infinitesimal mass dm and volume dV is determined by the surrounding force field **f**, such as gravity, and the pressure gradient $\operatorname{grad} p$, as defined by the EULER impulse equation for homogeneous frictionless incompressible fluids in Equation 7.1, [164]. In Equation 7.2, the terms are divided by the liquid density ρ. Apart from gravity, also magnetism and electricity can cause a force field. This property is used to control MR or ER liquids, cf. Section 6.3.2.

$$dm\,\mathbf{a} = dm\,\mathbf{f} - dV\operatorname{grad} p \tag{7.1}$$

$$\Leftrightarrow \mathbf{a} = \mathbf{f} - \frac{1}{\rho}\operatorname{grad} p \tag{7.2}$$

Figure 7.2 shows a liquid particle, which is moving through a tube between

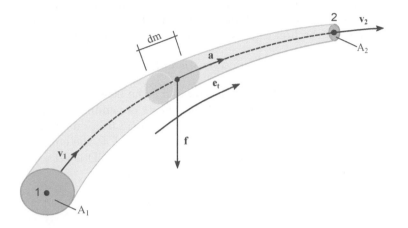

FIGURE 7.2
Homogeneous frictionless incompressible fluid particle moving through a tube
from (1) to (2).

the ends (1) and (2) with the corresponding cross-sectional areas A_1 and A_2.
The starting and ending velocities of the particle are $\mathbf{v_1}$ and $\mathbf{v_2}$. The direction
vector is tangent to the tube streamline and denoted as $\mathbf{e_t}$. The force field $\mathbf{f}$
acting on the particle and the absolute acceleration $\mathbf{a}$ of the liquid particle are
also shown in Figure 7.2.

Based on Equation 7.2, Sakai et al. [69, 70] derived the EoM of passive
TLCDs attached to horizontally vibrating high-rise structures. Xue et al. [165]
enhanced this equation for pitching structures. Hochrainer et al. [110] derived
the EoM of TLCGDs attached to horizontally vibrating high-rise structures
according to the approach developed by Sakai. Reiterer et al. [111] enhanced
this equation for TLCGDs attached to vertically and rotationally oscillating
structures, such as footbridges. In this section, the governing equations of the
S-TLCD, which is attached to a horizontally vibrating high-rise structure,
are derived in a similar manner based on the BERNOULLI equation of the
non-stationary incompressible flow of the liquid.

7.1.1.1 Equation of Motion of the S-TLCD

The total acceleration of a liquid column in a tank can be calculated in a
general form by integrating the Equation 7.2 along the streamline for the
force field of gravity $\mathbf{g}$ leading to

$$\int \mathbf{a}\,\mathbf{e_t}\,ds = \int \mathbf{g}\,\mathbf{e_t}\,ds - \frac{1}{\rho}\int \operatorname{grad} p\,\mathbf{e_t}\,ds. \qquad (7.3)$$

The integral of the pressure gradient along the streamline of the S-TLCD equals to the pressure loss Δp, which is caused by turbulence and local friction effects in the tank. Analytical formulations for the calculation of pressure loss can be found in the literature, such as according to [165], which reads

$$\Delta p = \delta \rho |\dot{u}| \dot{u}, \tag{7.4}$$

where $\dot{u}$ is the velocity of the liquid flow and δ the head-loss coefficient, which is a function representing the turbulence effects depending on the REYNOLDS number of the liquid stream together with the local friction effects. The absolute liquid acceleration $\mathbf{a}$ in Equation 7.3 is defined as the sum of the acceleration $\mathbf{a}_\mathrm{M}$ of the main structure, on which the S-TLCD is attached, and the relative acceleration $\mathbf{a}'$ of the liquid column yielding

$$\mathbf{a}\,\mathbf{e_t} = \mathbf{a_M}\,\mathbf{e_t} + \mathbf{a}'\,\mathbf{e_t}. \tag{7.5}$$

The relative acceleration $\mathbf{a}'$ of the fluid is separated into local and convective accelerations as shown below in Equation 7.6. The local acceleration describes the force field induced time-dependent change in the velocity $\dot{u}$ of the liquid particle. The convective acceleration describes the position-dependent change in the velocity of the liquid stream.

$$\mathbf{a}'\mathbf{e_t} = \frac{\partial \dot{u}}{\partial t} + \dot{u}\frac{\partial \dot{u}}{\partial s} \tag{7.6}$$

From Equation 7.3, by integrating the scalar product $\mathbf{g}\,\mathbf{e_t}$ and using the pressure loss Δp formulation from Equation 7.4, a modified version of the BERNOULLI equation can be determined for the S-TLCD as shown in Equation 7.7, where u is the displacement of the liquid in the tank.

$$\int \mathbf{a_M}\,\mathbf{e_t}\,\mathrm{d}s + \int \frac{\partial \dot{u}}{\partial t}\,\mathrm{d}s + \int \frac{\partial}{\partial s}\left(\frac{\dot{u}^2}{2}\right)\mathrm{d}s = -2gu - \frac{1}{\rho}\Delta p \tag{7.7}$$

As shown in Figure 7.3, the main parameters describing the geometry of the S-TLCD tank are in vertical direction the height V and in horizontal direction the length H of the streamline. A further parameter is d_i, where $i = [1, 4]$ giving the cross-sectional areas of the vertical columns and the horizontal passage. As seen in the side view, the cross-sectional area of the vertical columns can change depending on the inclination angle β of the movable panels. The vertical columns are hereby divided into three sections: beginning section, transition section and end section, V_1, V_2 and V_3, respectively, where the height V_2 is a function of β. Furthermore, for S-TLCDs with constant liquid mass, also the height V_3 changes depending on β.

The cross-sectional areas of the tank are calculated using Equation 7.8. Hereby, A_{V1} is the cross-sectional area of the column-top, A_{V2} is the cross-sectional area of the column-beginning and A_H is the cross-sectional area of the connecting passage of the vertical columns. As d_3 is a function of the panel position β, A_{V1} can be modified by changing β.

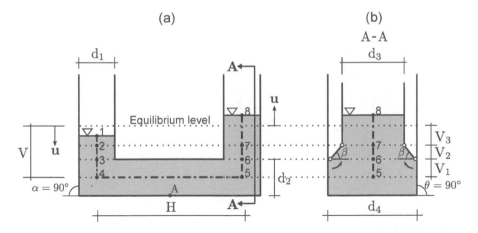

FIGURE 7.3
Mechanical model showing the governing parameters of the S-TLCD. Front
(a) and side (b) views of the S-TLCD are depicted.

$$A_{V1}(\beta) = d_1 \cdot d_3(\beta), \quad A_{V2} = d_1 \cdot d_4, \quad A_H = d_2 \cdot d_4, \quad (7.8)$$

The acceleration effects, which are induced by the oscillation of the main
structure, are calculated in Equation 7.9 from integral along the streamline
through the connection points 1–8 of the sections of the damper tank, cf.
Figure 7.3. Here, the acceleration of the main structure is assumed to occur
only in horizontal x-direction, which is usually the relevant vibration direction
for high-rise structures subjected to wind and earthquake loads. Therefore,
for the acceleration, it is assumed that $\mathbf{a_M} = \ddot{x}\,\mathbf{e_x}$. The scalar product of
the horizontal direction vector $\mathbf{e_x}$ with $\mathbf{e_t}$ vector, which is parallel to the
streamline, is conducted in Equations 7.10–7.12. The result corresponds to
the length of the horizontal passage of the damper H multiplied with the
acceleration $\ddot{x}$ of the main structure.

$$\int \mathbf{a_M}\,\mathbf{e_t}\,ds = \ddot{x}\left(\int_1^2 \mathbf{e_x}\,\mathbf{e_t}\,ds + \int_2^3 \mathbf{e_x}\,\mathbf{e_t}\,ds + \cdots + \int_7^8 \mathbf{e_x}\,\mathbf{e_t}\,ds\right) \quad (7.9)$$
$$= H\ddot{x}$$

$$1 \longrightarrow 4 : \quad \mathbf{e_x}\,\mathbf{e_t} = 0 \quad (7.10)$$
$$4 \longrightarrow 5 : \quad \mathbf{e_x}\,\mathbf{e_t} = 1 \quad (7.11)$$
$$5 \longrightarrow 8 : \quad \mathbf{e_x}\,\mathbf{e_t} = 0 \quad (7.12)$$

Figure 7.4 illustrates the velocity profile of the liquid stream in the S-
TLCD. The velocity changes in dependence with the cross-sectional area of

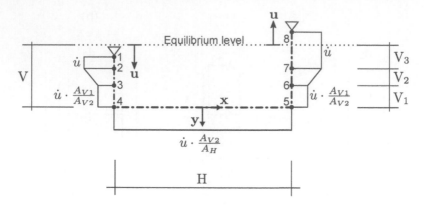

FIGURE 7.4
Velocity profile of the liquid stream in an S-TLCD tank.

each segment. This is applied in Equation 7.13–7.21 to determine the local acceleration of the liquid flow. The result corresponds to the acceleration $\ddot{u}$ of the liquid scaled by L_1, which is named as the first effective length of the S-TLCD. Equation 7.14 determines L_1 by solving the integral in Equation 7.13. For a TLCD with equal horizontal and vertical cross-sections $A_H = A_V$, the effective length corresponds to the total length of the liquid streamline.

$$\int \frac{\partial \dot{u}}{\partial t}\, ds = \sum_{i=1}^{7} \int_{i}^{i+1} \frac{\partial \dot{u}}{\partial t}\, ds = L_1 \ddot{u} \tag{7.13}$$

$$\Leftrightarrow L_1 = 2V_3 + V_2 + (2V_1 + V_2)\frac{A_{V1}}{A_{V2}} + H\frac{A_{V1}}{A_H} \tag{7.14}$$

$$1 \longrightarrow 2: \quad \frac{\partial \dot{u}}{\partial t} = \ddot{u} \tag{7.15}$$

$$2 \longrightarrow 3: \quad \frac{\partial \dot{u}}{\partial t} = \frac{\ddot{u}}{2}\left(1 + \frac{A_{V1}}{A_{V2}}\right) \tag{7.16}$$

$$3 \longrightarrow 4: \quad \frac{\partial \dot{u}}{\partial t} = \ddot{u}\frac{A_{V1}}{A_{V2}} \tag{7.17}$$

$$4 \longrightarrow 5: \quad \frac{\partial \dot{u}}{\partial t} = \ddot{u}\frac{A_{V1}}{A_H} \tag{7.18}$$

$$5 \longrightarrow 6: \quad \frac{\partial \dot{u}}{\partial t} = \ddot{u}\frac{A_{V1}}{A_{V2}} \tag{7.19}$$

$$6 \longrightarrow 7: \quad \frac{\partial \dot{u}}{\partial t} = \frac{\ddot{u}}{2}\left(1 + \frac{A_{V1}}{A_{V2}}\right) \tag{7.20}$$

$$7 \longrightarrow 8: \quad \frac{\partial \dot{u}}{\partial t} = \ddot{u} \tag{7.21}$$

As shown in Equation 7.22, due to the symmetry of the damper the integral of the convective acceleration becomes zero:

$$\int_1^8 \frac{\partial}{\partial s} \left(\frac{\dot{u}^2}{2} \right) ds = \frac{1}{2}(\dot{u}_8^2 - \dot{u}_1^2) = 0. \tag{7.22}$$

The sum of Equations 7.9, 7.13 and 7.22 gives according to Equation 7.7 the expanded nonlinear EoM of the S-TLCD for a liquid motion along the streamline as shown in Equation 7.23 and simplified in Equation 7.24, where ω_D is the adaptable natural circular frequency of the semi-active damper, which is calculated according to Equation 7.25. In Equation 7.24, we name the parameter γ_1 as the first geometric factor of the S-TLCD, which is given by Equation 7.26. Both ω_D and γ_1 depend on the first effective length of the streamline L_1. By changing the cross-sectional area of the vertical liquid columns, the effective length and following that the circular frequency ω_D and the geometric factor γ_1 change as well. As a result, the EoM of the S-TLCD reads

$$L_1\ddot{u} + 2gu + \frac{1}{\rho}\Delta p = -H\ddot{x}, \tag{7.23}$$

$$\Leftrightarrow \ddot{u} + \omega_D^2 u + \delta|\dot{u}|\dot{u} = -\gamma_1\ddot{x}, \tag{7.24}$$

where the natural frequency of the S-TLCD corresponds to

$$\omega_D = \sqrt{\frac{2\,g}{L_1}} \tag{7.25}$$

and the first geometric factor of the S-TLCD is expressed as

$$\gamma_1 = \frac{H}{L_1}. \tag{7.26}$$

As shown in Equation 7.25, the natural frequency ω_D of the S-TLCD depends only on the first effective length L_1. This equation and the calculation of the first effective length L_1 will be validated experimentally in Section 7.1.2.2.

In the EoM of the S-TLCD, the nonlinear term $\delta|\dot{u}|\dot{u}$ can be linearized according to Gao et al. [77] into $2D_D\omega_D\dot{u}$, where D_D is the equivalent viscous damping ratio of the S-TLCD. The D_D can be calculated by demanding for both the linear and nonlinear cases that the energy dissipation is for each cycle equal. This yields

$$\int_0^T |(\delta|\dot{u}|\dot{u} + \omega_D^2 u)\,\dot{u}|\,dt = \int_0^T |(2D_D\omega_D\dot{u} + \omega_D^2 u)\,\dot{u}|\,dt, \tag{7.27}$$

where T is the vibration period. Here, by substituting the time harmonic function

$$u = U_0 \cos\omega_D t \tag{7.28}$$

the equivalent viscous damping ratio D_D can be determined proportional to the vibration amplitude U_0 as

$$D_D = \frac{4U_0\delta}{3\pi}. \tag{7.29}$$

This approach linearizes the damping term for the harmonic excitation, cf. also Section 7.2. Furthermore, for random excitations, Xu et al. [166] propose

$$D_D = \delta\sqrt{\frac{2}{\pi}\frac{\sigma_{\dot{u}}}{\omega_D}}, \tag{7.30}$$

where $\sigma_{\dot{u}}$ is the standard deviation of the liquid velocity for a zero-mean GAUSSIAN process. Using these approaches, the EoM of the S-TLCD can be linearized as

$$\ddot{u} + \omega_D^2 u + 2D_D\omega_D\dot{u} = -\gamma_1\ddot{x}. \tag{7.31}$$

Generally, the nonlinear term $\delta|\dot{u}|\dot{u}$ introduced in Equation 7.24 is expected to give more accurate results. However, an accurate tuning of this parameter can be challenging. In such cases, soft-computing methods, such as FLC (cf. Section 3.3.2), neural networks and evolutionary algorithms, can circumvent the tuning problem. The interested readers can find further information about these methods in references, such as [32, 33].

7.1.1.2 Restoring Force

The restoring force of the S-TLCD on a horizontally oscillating structure is derived from the momentum equation. The momentum **I** induced by the liquid motion is computed in Equation 7.32 using the integral of the resulting liquid velocity over the liquid mass m_f. The resulting velocity is assembled by the velocity of the structure $\mathbf{v_A}$ at the attachment point A of the S-TLCD to the structure and the liquid velocity $\mathbf{v}$.

$$\mathbf{I} = \int (\mathbf{v_A} + \mathbf{v})\,\mathrm{d}m_f = \int_1^8 \left(\dot{x}\mathbf{e_t} + \dot{u}\frac{A_{V1}}{A(s)}\mathbf{e_t}\right)\rho A(s)\,\mathrm{d}s \tag{7.32}$$

The momentum projected in the lateral x-direction is computed in Equation 7.33. The restoring force is determined in Equation 7.34 from the time derivative of the momentum equation. In Equation 7.35, the liquid mass m_f is calculated. The second geometric factor γ_2 is calculated in Equation 7.36 and the corresponding second efficient length L_2 is given in Equation 7.37.

$$I_x = \rho\bigg(A_{V1}2V_3\dot{x} + (A_{V1} + A_{V2})V_2\dot{x}$$

$$+ A_{V2}2V_1\dot{x} + A_H H\left(\dot{x} + \dot{u}\frac{A_{V1}}{A_H}\right)\bigg) \tag{7.33}$$

$$F_x = \frac{dI_x}{dt} = m_f(\ddot{x} + \gamma_2 \ddot{u}) \tag{7.34}$$

$$m_f = \rho A_{V1} L_2 \tag{7.35}$$

$$\gamma_2 = \frac{H}{L_2} \tag{7.36}$$

$$L_2 = 2V_3 + V_2 + (2V_1 + V_2)\frac{A_{V2}}{A_{V1}} + H\frac{A_H}{A_{V1}} \tag{7.37}$$

The EoM of an SDoF system with an S-TLCD subjected to a time-dependent load $F(t) = f(t)m_S$ is shown in Equation 7.38, where $\mu = m_f/m_S$ is the ratio of the liquid mass to the modal mass of the structure. In the same equation, the damper interaction forces are represented by $\mu\ddot{x} + \mu\gamma_2\ddot{u}$, which is obtained from Equation 7.34. In Equation 7.38, w_S is the natural frequency of the structure and D_S is its damping ratio.

$$\ddot{x} + 2D_S\omega_S\dot{x} + \omega_S^2 x = -\mu\ddot{x} - \mu\gamma_2\ddot{u} + f \tag{7.38}$$

If the structure is subjected to a ground acceleration $\ddot{x}_g$, the EoM of the SDoF system with the S-TLCD is expanded as

$$\ddot{x} + 2\,D_S\omega_S\dot{x} + \omega_S^2 x = -\ddot{x}_g - \mu\,(\ddot{x} + \ddot{x}_g) - \mu\gamma_2\ddot{u} + f. \tag{7.39}$$

Based on this equation, the EoM of an MDoF system with several S-TLCDs will be derived in Section 7.1.1.4 reformulated by using the state-space representation.

7.1.1.3   Geometric Factors

The geometric factors γ_1 and γ_2 determine the ratio of the horizontal length H of the S-TLCD to its efficient lengths L_1 and L_2. For an S-TLCD with a constant liquid mass, apart from the cross-sectional area A_{V1}, also the height V_2 and the height V_3 depend on β. As a result, L_1 and L_2, as well as γ_1 and γ_2, become a function of β. For the sake of convenience, the equations of the efficient lengths and geometric factors representing this general case are repeated below:

$$L_1(\beta) = 2V_3(\beta) + V_2(\beta) + (2V_1 + V_2(\beta))\frac{A_{V1}(\beta)}{A_{V2}} + H\frac{A_{V1}(\beta)}{A_H}, \tag{7.40}$$

$$L_2(\beta) = 2V_3(\beta) + V_2(\beta) + (2V_1 + V_2(\beta))\frac{A_{V2}}{A_{V1}(\beta)} + H\frac{A_H}{A_{V1}(\beta)}, \tag{7.41}$$

$$\gamma_1(\beta) = \frac{H}{L_1(\beta)}, \tag{7.42}$$

$$\gamma_2(\beta) = \frac{H}{L_2(\beta)}. \tag{7.43}$$

For a TLCD with uniform cross-sectional area, the total length of the liquid column L equals to both L_1 and L_2. For a TLCD with the constant cross-sectional area of the vertical columns A_V and the cross-sectional area of the horizontal passage A_H (cf. Liquid column vibration absorber (LCVA) of Hitchcock et al. [78, 79]), both efficient lengths L_1 and L_2 are given by the sum of the total liquid length in vertical direction and the horizontal length, which is scaled by the ratio of the cross-sectional areas A_V and A_H. For this type of TLCDs, if the liquid mass is constant, a change of A_V would influence the first efficient length L_1 linearly and the second efficient length L_2 in a nonlinear manner. In case of the S-TLCD with a constant liquid mass, a change of the panel position β influences both L_1 and L_2 nonlinearly. As the natural frequency ω_D depends also on L_1, a change of the panel position β will also cause ω_D and consequently the geometric factors γ_1 and γ_2 to change in a nonlinear manner. This effect is presented in Sections 7.1.2.2 and 7.1.2.3 from the experimental results.

The interaction forces between the S-TLCD and the structure are scaled by γ_1 and γ_2. Therefore, the geometric factors are expected to effect the efficiency of the damper depending on β as well. The first geometric factor γ_1 scales the acceleration of the structure $\ddot{x}$, as given in the EoM of the S-TLCD in Equation 7.31. Accordingly, together with the acceleration $\ddot{x}$ of the structure, an increase in the geometric factor γ_1 amplifies the liquid motion amplitude u.

In the EoM of the structure, as written in Equation 7.39, the second geometric factor γ_2 scales the acceleration of the liquid motion with the term $-\mu \gamma_2 \ddot{u}$, where the minus sign represents the restoring character of the interaction force. Consequently, an increase in the geometric factor γ_2 mitigates the lateral motion x of the structure.

From the arguments above, it can be expected that γ_2 will dominate the efficiency of the S-TLCD. However, γ_1 correlates directly with L_1 and, consequently, also with the natural frequency ω_D, which is vital for the efficiency of the damper. This effect can also be observed from the experimental results presented in Section 7.1.2.2.

Furthermore, to increase the active mass fraction of the S-TLCD, both γ_1 and γ_2 must be maximized. Analogous to the classical TLCD, the active mass fraction for the S-TLCD defines the amount of liquid mass, which participates to the structural control and generates the restoring force. The rest of the liquid mass is necessary for the tuning of the natural frequency of the damper. The amount of active mass fraction can be determined according to Hochrainer et al. [110] by introducing the scaled liquid motion $\bar{u} = u/\gamma_1$ in the EoM of the structure and the S-TLCD, and by multiplying these equations with $\mathrm{diag}\,(1/(1 + \mu\,(1 - \gamma_1\,\gamma_2)), 1/\gamma_1)$, which yields

$$\begin{bmatrix} 1 + \bar{\mu} & \bar{\mu} \\ 1 & 1 \end{bmatrix} \begin{bmatrix} \ddot{x} \\ \ddot{\bar{u}} \end{bmatrix} + \begin{bmatrix} 2\,\bar{D}_S\bar{\omega}_S & 0 \\ 0 & 2\,D_D\omega_D \end{bmatrix} \begin{bmatrix} \dot{x} \\ \dot{\bar{u}} \end{bmatrix} + \begin{bmatrix} \bar{\omega}_S^2 & 0 \\ 0 & \omega_D^2 \end{bmatrix} \begin{bmatrix} x \\ \bar{u} \end{bmatrix}$$
$$= -\begin{bmatrix} 1 + \bar{\mu} \\ 1 \end{bmatrix} \ddot{x}_g + \begin{bmatrix} 1/\bar{m}_S \\ 0 \end{bmatrix} f, \quad (7.44)$$

where the ratio between the active mass fraction $\bar{m}_D$ of the S-TLCD and the sum of the modal mass of the structure m_S and the rest liquid mass of the damper is denoted by $\bar{\mu}$, which corresponds to

$$\bar{\mu} = \frac{\bar{m}_D}{\bar{m}_S} = \frac{m_D \gamma_1 \gamma_2}{m_S \left(1 + \mu \left(1 - \gamma_1 \gamma_2\right)\right)} = \frac{\mu \gamma_1 \gamma_2}{1 + \mu \left(1 - \gamma_1 \gamma_2\right)}. \tag{7.45}$$

From the numerator, it can be clearly seen, that the active mass fraction $\bar{m}_D$ equals to the total liquid mass of the damper, which is scaled by γ_1 and γ_2:

$$\bar{m}_D = m_D \gamma_1 \gamma_2. \tag{7.46}$$

In Equation 7.44, $\bar{\omega}_S$ and $\bar{D}_S$ are obtained by dividing the natural frequency ω_S and the damping ratio D_S of the structure by $\sqrt{1 + \mu \left(1 - \gamma_1 \gamma_2\right)}$.

7.1.1.4 State-Space Representation

In this section, the state-space representation of an n-DoF structure subjected to a horizontal ground acceleration $\ddot{x}_g$ and horizontal excitation forces $\mathbf{F(t)}$ with $k \times$ S-TLCDs is derived. The general aspects of the state-space representation method is introduced in Section 3.2. The EoM of the system is given in Equation 7.47. In this equation, $\mathbf{M_M}$ is a hypermatrix consisting of the mass of the structure and the S-TLCDs. The lateral displacement vector of the system and the liquid motion vector of the S-TLCDs are denoted by $\mathbf{x}$ and $\mathbf{u}$, respectively. $\mathbf{C_S}$ and $\mathbf{C_D}$ are the damping matrices of the structure and the S-TLCDs. $\mathbf{K_S}$ and $\mathbf{K_D}$ are the stiffness matrices. $\mathbf{M_S}$ is the modal mass matrix of the structure and $\mathbf{M_D}$ is the mass matrix of the S-TLCDs. The vector $\mathbf{r}$ distributes the ground excitation $\ddot{x}_g$ to the DoFs of the structure. As shown in Equation 7.48, $\mathbf{r} \in \mathbb{R}^{n \times 1}$ is an identity vector for a horizontally excited structure with DoFs only in horizontal direction.

$$\mathbf{M_M} \begin{bmatrix} \ddot{\mathbf{x}} \\ \ddot{\mathbf{u}} \end{bmatrix} + \begin{bmatrix} \mathbf{C_S} & \mathbf{0} \\ \mathbf{0} & \mathbf{C_D} \end{bmatrix} \begin{bmatrix} \dot{\mathbf{x}} \\ \dot{\mathbf{u}} \end{bmatrix} + \begin{bmatrix} \mathbf{K_S} & \mathbf{0} \\ \mathbf{0} & \mathbf{K_D} \end{bmatrix} \begin{bmatrix} \mathbf{x} \\ \mathbf{u} \end{bmatrix}$$
$$= - \begin{bmatrix} \mathbf{M_S}\,\mathbf{r} + \mathbf{L}\,\mathbf{M_D}\,\mathbf{i} \\ \mathbf{\Gamma_1}\,\mathbf{i} \end{bmatrix} \ddot{x}_g + \begin{bmatrix} \mathbf{f} \\ \mathbf{0} \end{bmatrix} \tag{7.47}$$

$$\mathbf{r}^\mathsf{T} = \begin{bmatrix} 1 & 1 & \dots & 1 \end{bmatrix} \tag{7.48}$$

The S-TLCDs are distributed to the DoFs of the structure by using an incidence matrix $\mathbf{L} \in \mathbb{R}^{n \times k}$. Equation 7.49 below shows an example of an incidence matrix for a structure with 3-DoF and two S-TLCDs, which are attached to the second and third DoFs of the structure.

$$\mathbf{L} = \begin{bmatrix} 0 & 0 \\ 1 & 0 \\ 0 & 1 \end{bmatrix} \tag{7.49}$$

For a horizontally excited system with all S-TLCDs also in the horizontal direction, the vector $\mathbf{i} \in \mathbb{R}^{k \times 1}$ is an identity vector:

$$\mathbf{i}^\top = [1\ 1\ \dots\ 1] \tag{7.50}$$

Furthermore, in Equation 7.47, $\mathbf{\Gamma}_1$ is a matrix containing the geometric factors of the S-TLCDs. The excitation forces acting on the structure are represented by $\mathbf{f}$.

The hypermatrix $\mathbf{M_M} \in \mathbb{R}^{(n+k) \times (n+k)}$ is given in Equation 7.51. It involves besides the mass matrices $\mathbf{M_S}$, $\mathbf{M_D}$ and the incidence matrix $\mathbf{L}$ also the matrices $\mathbf{\Gamma}_1$ and $\mathbf{\Gamma}_2$ with the geometric factors as given in Equation 7.52. The hypermatrix includes furthermore an identity matrix $\mathbf{I} \in \mathbb{R}^{k \times k}$, as shown in Equation 7.53.

$$\mathbf{M_M} = \begin{bmatrix} \mathbf{M_S} + \mathbf{L}\,\mathbf{M_D}\,\mathbf{L}^\top & \mathbf{L}\,\mathbf{M_D}\,\mathbf{\Gamma_2} \\ \mathbf{\Gamma_1}\,\mathbf{L}^\top & \mathbf{I} \end{bmatrix} \tag{7.51}$$

$$\mathbf{\Gamma_i} = \mathrm{diag}\left(\gamma_{i,1}, \gamma_{i,2}, \dots \gamma_{i,k}\right) \tag{7.52}$$

$$\mathbf{I}^\top = (1, 1, \dots, 1 \tag{7.53}$$

In the state-space representation, the displacements $\mathbf{x}$, $\mathbf{u}$ and the velocities $\dot{\mathbf{x}}$, $\dot{\mathbf{u}}$ of the system are composed in the state vector $\mathbf{z} \in \mathbb{R}^{2(n+k) \times 1}$ and its first time derivative $\dot{\mathbf{z}}$ as

$$\mathbf{z}^\top = (\mathbf{x}, \mathbf{u}, \dot{\mathbf{x}}, \dot{\mathbf{u}}), \tag{7.54}$$

$$\dot{\mathbf{z}}^\top = (\dot{\mathbf{x}}, \dot{\mathbf{u}}, \ddot{\mathbf{x}}, \ddot{\mathbf{u}}). \tag{7.55}$$

Using the state vector $\mathbf{z}$ and its first derivative $\dot{\mathbf{z}}$, a general representation of a horizontally excited MDoF system with several S-TLCDs is derived. The associated distribution vector $\mathbf{e_g} \in \mathbb{R}^{2(n+k) \times n}$ and the distribution matrix $\mathbf{E} \in \mathbb{R}^{2(n+k) \times n}$ are provided in Equation 7.57:

$$\dot{\mathbf{z}} = \mathcal{A}_r\,\mathbf{z} - \mathbf{e_g}\,\ddot{x}_g + \mathbf{E}\,\mathbf{f}, \tag{7.56}$$

$$\mathbf{e_g} = \begin{bmatrix} \mathbf{0} \\ \mathbf{0} \\ \mathbf{M_M^{-1}} \begin{bmatrix} \mathbf{M_S}\,\mathbf{r} + \mathbf{L}\,\mathbf{M_D}\,\mathbf{i} \\ \mathbf{\Gamma_1}\,\mathbf{i} \end{bmatrix} \end{bmatrix}, \quad \mathbf{E} = \begin{bmatrix} \mathbf{0} \\ \mathbf{0} \\ \mathbf{M_M^{-1}} \begin{bmatrix} \mathbf{I} \\ \mathbf{0} \end{bmatrix} \end{bmatrix} \tag{7.57}$$

The hypermatrix $\mathcal{A}_r$ defines dynamic properties of the system and consists of two further components $\mathcal{A} \in \mathbb{R}^{2(n+k) \times 2(n+k)}$ and $\mathcal{B} \in \mathbb{R}^{2(n+k) \times 2(n+k)}$ as well as $\mathbf{R} \in \mathbb{R}^{2(n+k) \times 2(n+k)}$:

$$\mathcal{A}_r = \mathcal{A} + \mathcal{B}\,\mathbf{R}, \tag{7.58}$$

$$\mathcal{A} = \begin{bmatrix} 0 & I \\ -M_M^{-1} \begin{bmatrix} K_S & 0 \\ 0 & 0 \end{bmatrix} & -M_M^{-1} \begin{bmatrix} C_S & 0 \\ 0 & 0 \end{bmatrix} \end{bmatrix}, \tag{7.59}$$

$$\mathcal{B} = \begin{bmatrix} 0 & I \\ -M_M^{-1} \begin{bmatrix} I & 0 \\ 0 & I \end{bmatrix} & -M_M^{-1} \begin{bmatrix} I & 0 \\ 0 & I \end{bmatrix} \end{bmatrix}, \tag{7.60}$$

$$\mathbf{R} = \begin{bmatrix} 0 & 0 & 0 & 0 \\ 0 & K_D & 0 & 0 \\ 0 & 0 & 0 & 0 \\ 0 & 0 & 0 & C_D \end{bmatrix}. \tag{7.61}$$

7.1.2 Experimental Investigations

To verify the natural frequency equation derived in Section 7.1.1.1 and to show the proof of concept of the S-TLCD as a damping device, experimental investigations are conducted at the control engineering laboratory of the RWTH Aachen University. The study is concerned with the natural frequency adaptation capability, the effects on the inherent damping and the vibration control performance of the damper particularly considering its frequency tuning as well as with the influences of the geometric factors.

7.1.2.1 Experimental Setup

The experimental setup consists of a prototype of the S-TLCD with full adaptation capabilities and a model structure, on which the S-TLCD is attached. The structure is subjected to ground motion, which is generated by a uniaxial shaking table.

The S-TLCD prototype is shown in Figure 7.5. The proposed adaptation features are included and the damper can change its natural frequency and damping behavior. For the U-shaped tank, acrylic glass is chosen to allow visual observation of the liquid motion and position change of the movable components. The tank is built by rectangular glass elements, which are glued with each other, compressed by screws and sealed with silicone. The tank is 430 mm length, 570 mm height and 130 mm width. The material thickness of the acrylic glass is 15 mm. The cross-sectional area of the vertical columns of the tank is 50 mm × 100 mm. The horizontal passage has a cross-sectional area of 100 mm × 100 mm.

The S-TLCD prototype has in each vertical column two movable panels, which are labeled as (1) in Figures 7.5 and 7.6. The panels can regulate the cross-sectional area of the vertical columns and tune the natural frequency. Each panel consists of two segments, which are connected by joints. The first transition segment is 60 mm long and connects the second segment with the vertical column of the tank. The second segment is vertical and has a length of 385 mm. In each vertical column, the panels are connected together with a threaded rod, which allows the panels to change their position simultaneously perpendicular to the liquid flow and accordingly to the operation direction of the S-TLCD. Each threaded rod is connected to an actuator, which allows an electronic adjustment of the panel positions, (2) in Figures 7.5 and 7.6. The gaps between the panels and the tank walls are sealed with water-resistant and highly-expandable plastic membranes. The construction allows the panels to change their position within a range of $\beta = 45°$–$72°$.

For the tuning of the inherent damping, two butterfly valves are installed in the middle of the horizontal passage, Figure 7.5 (3) and Figure 7.6 (4). Each valve consists of a rectangular plate with the dimensions 98 mm × 45 mm. Each plate is rotatable around a vertical axis perpendicular to the liquid flow direction. By opening the valves (rotation angle of the plate $\varphi = 90°$) and closing them ($\varphi = 0°$), the inherent damping can be changed due to the modulation of the local friction effects. Both valves are connected by gear wheels to an actuator, which allows a simultaneous electronic adjustment of the valves, Figure 7.6 (5).

For the measurement of the liquid motion, a capacitance transducer is developed and installed in one of the vertical columns of the S-TLCD prototype, which can dynamically measure the liquid level change, Figure 7.6 (3). The liquid motion is recorded with a sample rate of 100 Hz allowing the identification of relevant dynamic parameters, such as the maximum liquid amplitude and the natural frequency of the S-TLCD. The measured liquid motion is also utilized to investigate the inherent damping effects of the S-TLCD.

During the first part of the experimental investigations, the S-TLCD is directly mounted on the shaking table to study its natural frequency and inherent damping. During further tests regarding the performance of the damper, the S-TLCD is attached to a model structure, which is excited by the shaking table, Figure 7.7. The uniaxial horizontal shaking table is operated by a linear actuator and can generate harmonic vibrations in a frequency range of 0.1–50 Hz with 100 mm (±50 mm) stroke. The motion of the shaking table is measured by a microelectromechanical system (MEMS) accelerometer.

The model structure consists of a pendulum attached to a rigid frame with four steel wires, Figure 7.7 (2). The rigid frame is built by aluminum elements and has its natural frequency at 9.20 Hz. The pendulum consists of a platform, where the S-TLCD can be attached and has its natural frequency at $f_S = 1.15$ Hz, which can be changed by modifying the length of the wires. The damping ratio of the pendulum is about $D = 1.0\%$. The mass of the pendulum together with the mass of the S-TLCD is $m_{Pe} + m_{De} = 10.86$ kg. The damper liquid mass is $m_D = 4.95$ kg. The motion of the pendulum is recorded also by a MEMS accelerometer. The model structure with the pendulum simulates a horizontally oscillating SDoF system with a low natural frequency. This setup allowed the design of a very compact S-TLCD utilizing reasonable actuators, which are adequate for the investigations in the laboratory environment.

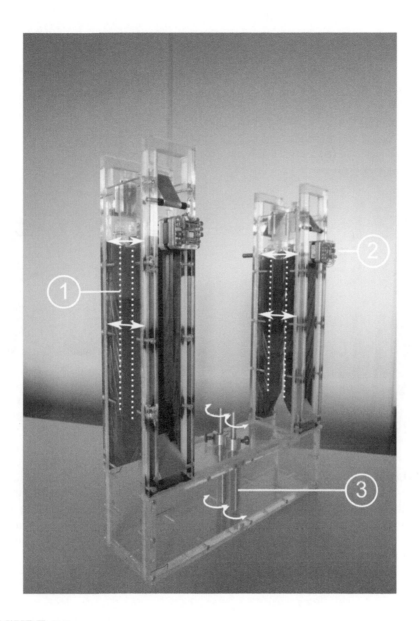

FIGURE 7.5
S-TLCD prototype. Movable panels for natural frequency tuning (1). Actuators adjusting the panels (2). Butterfly valves for damping tuning (3). Courtesy of O. Altay.

(a)   (b)

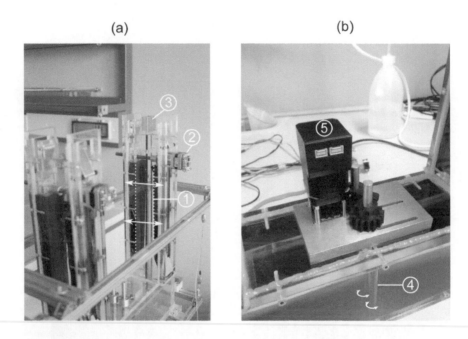

FIGURE 7.6

The S-TLCD prototype attached on the pendulum of the model structure (a)
and filled with liquid (water and colorant) (b). Movable panels for frequency
tuning (1). Actuators adjusting the movable panels (2). Sensor measuring the
liquid motion (3). Butterfly valves for damping tuning (4). Actuator adjusting
the valves (5). Courtesy of O. Altay.

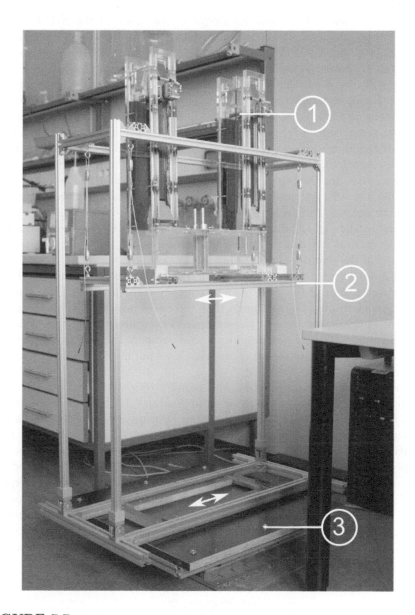

FIGURE 7.7
The S-TLCD prototype (1) is attached on a pendulum (2), which is connected
to a rigid frame. The frame structure is subjected to ground motion by a
uniaxial shaking table (3). Courtesy of O. Altay.

7.1.2.2   Investigations on the Natural Frequency

For the verification of the natural frequency equation of the S-TLCD, four damper configurations with the liquid volumes 4.69, 4.77, 4.83 and 4.95 L with variable panel positions are investigated. During these investigations, the S-TLCD prototype is directly attached on the shaking table and excited harmonically by excitation frequencies 0.50–2.00 Hz with an increment of 0.10 Hz. For each damper configuration, the experiments are repeated for the panel positions between 45° and 72° with an increment of 0.5°. The damper response is determined by measuring the liquid motion. From the maximum values of the liquid motion, the frequency response curves for the conducted frequency sweeps are determined. Figure 7.8 shows the measured frequency response curves for the tested four damper configurations and selected panel positions $\beta = 45°$, 58° and 72°. Here, Ω is the excitation frequency and u the liquid motion. The peak values of the frequency curves correspond to the resonant response of the damper, in which the excitation frequency matches with the natural frequency. From the corresponding frequency the natural frequency of the damper is determined.

In Figure 7.8, the change of the natural frequency is clearly observed from the horizontal shift of the peak of the frequency response curve. The shape of the frequency response curve shows that the panel position has also a certain effect on the damping ratio of the damper. The wide frequency curve at the panel position 45° indicates more damping compared to the panel positions 58° and 72°. Further discussion regarding this damping effect can be found in Section 7.1.2.3.

For each damper configuration and panel position, the mathematical parameters are determined, as introduced in Section 7.1.1.1. From these parameters, the natural frequencies are calculated by using equations of the natural frequency ω_D and the first efficient length L_1, which are repeated below in Equations 7.62 and 7.63. Table 7.1 shows the damper parameters for two damper configurations with 4.69 and 4.95 L, and compares the calculated and measured natural frequencies. In Figure 7.9, the calculated natural frequencies $f_{D,c}$ are compared with the measured natural frequencies $f_{D,m}$ for the tested four damper configurations within the operation range of the panels between 45° and 72°. The results correspond to each other very well proving the applicability of the mathematical equation of the natural frequency of S-TLCD.

$$f_{D,c} = \frac{1}{2\pi} \sqrt{\frac{2g}{L_1}} \qquad (7.62)$$

$$L_1 = 2V_3 + V_2 + (2V_1 + V_2)\frac{A_{V1}}{A_{V2}} + H\frac{A_{V1}}{A_H} \qquad (7.63)$$

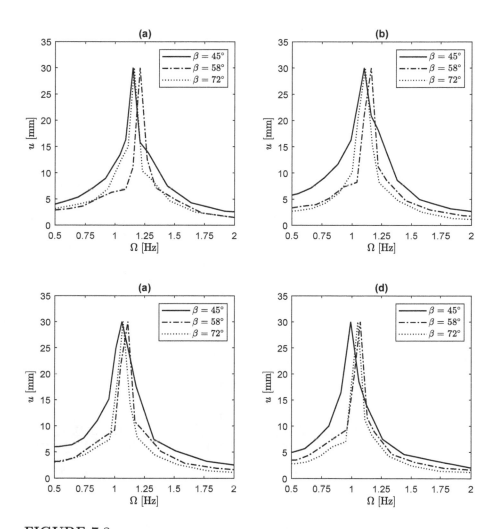

FIGURE 7.8
Frequency response spectra of the four damper configurations with the liquid volumes 4.69 L (a), 4.77 L (b), 4.83 L (c) and 4.95 L (d) at the panel positions $\beta = 45°$, 58° and 72°. The butterfly valve position is $\varphi = 0°$ (closed).

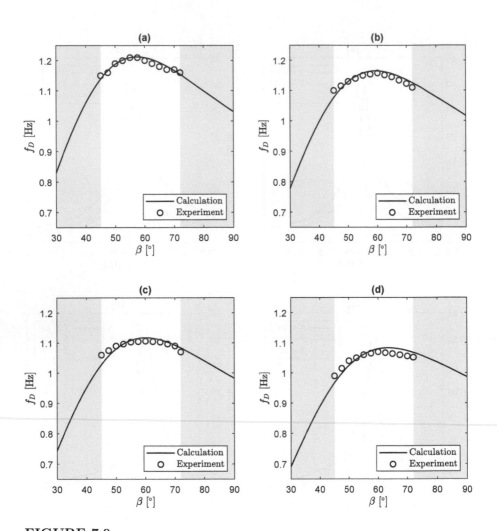

FIGURE 7.9

Comparison of the analytical and experimental results showing the relationship between the natural frequency f_D and the panel position β for four damper configurations with liquid volumes: 4.69 L (a), 4.77 L (b), 4.83 L (c) and 4.95 L (d). The non-gray areas are showing the operation range of the panels. The butterfly valve position is $\varphi = 0°$ (closed).

TABLE 7.1

Parameters of the S-TLCD configuration with 4.69 and 4.95 L liquid volume and comparison of the calculated natural frequency $f_{D,c}$ with the measured natural frequency $f_{D,m}$.

β [°]	V_1 [cm]	V_2 [cm]	V_3 [cm]	H [cm]	A_{V1} [cm²]	A_{V2} [cm²]	A_H [cm²]	L_1 [cm]	$f_{D,c}$ [Hz]	$f_{D,m}$ [Hz]
					4.69 L					
45	5	4.2	11.5	35	18	50	100	38.3	1.14	1.15
58	5	5.1	5.2	35	28	50	100	32.8	1.21	1.21
72	5	5.7	2.0	35	42	50	100	37.3	1.15	1.16
					4.95 L					
45	5	4.2	18.9	35	18	50	100	53.1	0.97	0.99
58	5	5.1	9.8	35	28	50	100	43.0	1.07	1.07
72	5	5.7	5.2	35	42	50	100	43.6	1.07	1.06

7.1.2.3 Investigations on the Inherent Damping

Study 1: Effects of the Butterfly Valves on the Inherent Damping. Four different valve positions are tested: $\varphi = \{0°, 30°, 60°, 90°\}$, where $\varphi = 0°$ corresponds to a closed position and $90°$ to open. Each test is repeated for four times to increase the statistical accuracy of the results. The tests are performed on the S-TLCD configuration with 4.69 L at a panel position of $\beta = 45°$. The S-TLCD is attached directly on the shaking table and a free vibration of the liquid is induced by a sudden start and stop of the table after a certain initial velocity. For this purpose, the motion of the shaking table is tuned in such a way that the first peak of the liquid motion is approximately always $u_{max} \approx 15$ mm. As shown in Figure 7.10, the damping ratios are determined from the recorded decay curves of the liquid motion beginning with this peak.

In Figure 7.10 (a), the diagram shows the comparison of the time histories of liquid motion u for the open and closed valve positions. The measured initial liquid motion peaks of the shown data sets are $u = 14.97$ mm for the valve in the closed position and $u = 15.24$ mm for the valve in the open position. The decay of the liquid deflection increases by closing the valve, which shows the increased damping behavior. In Figure 7.10 (b), the diagram shows the determined damping ratios D, which vary between 4.6 % and 7.4 % for the tested valve positions φ. The test results prove that the damping ratio of the S-TLCD can be modified efficiently by changing the valve position.

For the investigation of the effects arising from the nonlinearity of the S-TLCD damping and its coherence with the deflection amplitude, further tests are conducted with higher liquid deflection amplitudes. For this purpose, the motion of the shaking table is increased to induce an initial liquid motion peak of $u = 75.12$ mm, as shown in Figure 7.11. During these tests, the valve position is set as $\varphi = 90°$ (open). The damping ratios are determined from several decay curves using different amplitude combinations. For the

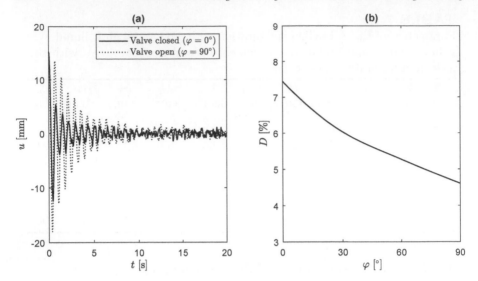

FIGURE 7.10
Time histories of the liquid motion for the valve positions $\varphi = 0°$ and $\varphi = 90°$ (a). Relationship between the damping ratio D and the valve position φ determined from time series with $u_{max} \approx 15$ mm and five oscillations (b). The panel position is $\beta = 45°$.

amplitude pairs 75.12–9.64 mm and 15.12–3.60 mm the determined damping ratios are 6.5 % and 4.6 % as evaluated in Table 7.2, where u_i and u_{i+n} are the amplitude pairs, n is the number of oscillations between the amplitude pairs, Λ is the logarithmic decrement and D is the damping ratio of the S-TLCD. It is evident from these results, that for larger amplitudes the damping ratio increases slightly in a nonlinear manner.

Due to the nonlinearity of the inherent damping, for real applications, an optimal tuning of the damping can be realized by a mapping of the relationship between the liquid stream velocity amplitude and the damping term $\delta|\dot{u}|\dot{u}$. Furthermore, sophisticated control algorithms, such as using soft-computing methods, can circumvent this challenge as well, cf. [32, 33]. These approaches

TABLE 7.2
Parameters of the experimental
investigations on the nonlinear inherent
damping effects of the S-TLCD.

	u_i [mm]	u_{i+n} [mm]	n [-]	Λ [-]	D [%]
Data set 1	75.12	9.64	5	0.41	6.5
Data set 2	15.12	3.60	5	0.29	4.6

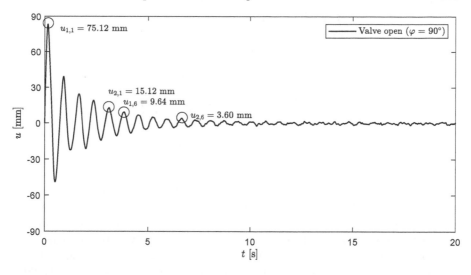

FIGURE 7.11
Investigation of the nonlinearity depending on the damping ratio and the liquid motion amplitude. The panel position is $\beta = 45°$.

can consider the nonlinearity of the damper and allow an accurate tuning of the damping depending on the liquid velocity. On the other hand, as explained in Section 7.1.1.1, the nonlinear damping term of the EoM can also be linearized by introducing the equivalent damping ratio D_D. This approximation approach [77, 166] has sufficient accuracy at least for low liquid amplitudes. At higher amplitudes, the linearization underestimates the inherent damping and can cause detuning effects, which will deteriorate the performance of the damper.

Study 2: Effects of the Movable Panels on the Inherent Damping. Further tests are conducted for the three different panel positions at 45°, 58° and 72° using the damper configuration with 4.69 L and a constant valve position at 90° (open). Further parameters are shown in Table 7.3. Similar to the previous investigations on the valve effects, the damping ratios are determined from the deflection decay curves of the liquid motion. Figure 7.12 (a) shows the time histories of the liquid deflection for the tested three panel positions and the amplitude pairs, which are used for the determination of the damping ratios. The highest damping ratio with 6.5 % was determined for 45°. The lowest damping ratio is 4.8 % at panel position 72°. By closing the panels, the damping ratio increases. This result matches with the previous results from frequency sweep tests, which are presented in Section 7.1.2.2. During the frequency sweeps, the determined frequency response curve of the panel position 45° is wider than of the panel positions 58° and 72° showing that the damping behavior is affected not only by the valves but also by the panel position, cf. Figure 7.8.

TABLE 7.3

Parameters of the experimental
investigations on the damping
effects caused by the panels.

β	u_i	u_{i+n}	n	Λ	D
[°]	[mm]	[mm]	[-]	[-]	[%]
45	75	10	5	0.41	6.5
58	75	15	5	0.32	5.1
72	75	16	5	0.30	4.8

As mentioned previously, the inherent damping of the S-TLCD is caused by local friction effects in the tank. Therefore, every tank component, which is introducing additional friction, leads to a complementary increase in the inherent damping. This effect is also observed in Study 1 by changing the position of the butterfly valves. At the panel position $\beta = 45°$, the inherent damping decreases by just opening the butterfly valves from $\varphi = 0°$ to $90°$. This effect is known also from the conventional TLCDs. In Study 2, now also the panel position is varied. The inherent damping decreases further by opening the movable panels from $\beta = 45°$ to $72°$. This effect is unique for the S-TLCD.

To investigate the interaction between the valve and panel positions φ and β in more detail, further tests are conducted. During these tests, the damping ratio is determined for different φ and β from decay curves with $u_{max} \approx 15$ mm using 5 oscillations. The results are shown in Figure 12 7.12 (a). As determined also previously, the S-TLCD reaches the maximum damping ratio at a valve position of $\varphi = 0°$ (closed) with a panel position of $\beta = 45°$. By opening the valves to $\varphi = 90°$ and the panels to $\beta = 72°$ the S-TLCD reaches its minimum damping ratio. The determined maximum and minimum damping ratios are 7.4% and 3.4%, respectively. Other variations, as also shown in Figure 7.12, lead to damping ratios, which are between these two limit values. It is worth noting that the inherent damping of the S-TLCD, as previously discussed in Study 1 and shown in Figure 7.11, also depends on the amplitude of the liquid motion.

7.1.2.4   Investigations on the Vibration Control Performance

As mentioned at the beginning of Section 7.1, the damping adaptation using an orifice, as proposed by [69, 70], counts as an established method, which is also utilized in the S-TLCD in a similar manner by the butterfly valves. Studies on the performance effects of this approach can be found in the literature such as in [104, 167]. The major novelty of the proposed S-TLCD is its frequency adaptation capability. Therefore, and for the sake of brevity, this section focuses on the effects of the frequency adaptation on the performance of the S-TLCD. In this regard, Study 1 discusses the effects of the natural frequency tuning on the damper performance. Study 2 reports the side effects

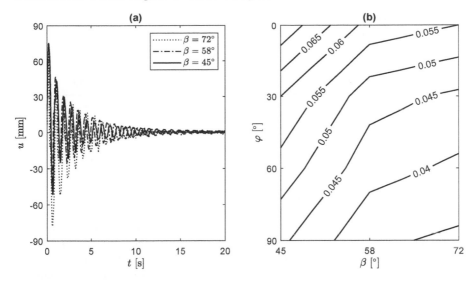

FIGURE 7.12

Investigation of the effect of the panel position on the damping behavior. Valve position is $\varphi = 90°$ (open) (a). Damping ratio [%] of the S-TLCD for different valve/panel (φ/β) positions determined from time series with $u_{max} \approx 15$ mm using 5 oscillations (b)).

of the frequency adaptation, which are initiated by the change of geometric factors.

Study 1: Effects of the Natural Frequency Tuning on the Damper Performance. The natural frequency tuning is decisive for the efficiency of structural control systems utilizing auxiliary masses. As a semi-active system, the S-TLCD is able to change its natural frequency and is therefore expected to reach a higher performance compared to passive dampers, which can be easily detuned. This effect is investigated during an experimental campaign with the S-TLCD.

During these tests, the S-TLCD with 4.95 L liquid amount is attached on the pendulum. As experimentally determined in Section 7.1.2.2, this S-TLCD configuration can change its natural frequency within the range of 0.97 Hz to 1.07 Hz. For a 1.15 Hz pendulum natural frequency, the optimum natural frequency is determined using the system parameters documented in Table 7.4 according to the DEN HARTOG criteria [52] as given in Equations 7.64–7.67 for the S-TLCD. In Table 7.4, f_D is the natural frequency of the damper, which is controlled by the panel position β, and f_S is the natural frequency of the pendulum. Further reported parameters are the liquid mass of the damper m_D, the mass of the pendulum m_{Pe} and the mass of the damper prototype m_{De}. As m_{Pe} and m_{De} are "dead masses", the vibration of which is aimed to

TABLE 7.4

Parameters of the performance study of the S-TLCD.

β [°]	$f_D(\beta)$ [Hz]	f_M [Hz]	m_D [kg]	m_{Pe} [kg]	m_{De} [kg]	m_S [kg]
$45-72$	$0.97-1.07$	1.15	4.950	1.500	9.360	10.860

be controlled, their sum is denoted as m_S and referred to as the main mass, whereas the damper mass m_D is the secondary mass.

As derived in Section 7.1.1.3 and shown in Equation 7.64, the amount of liquid mass, which is actively participating to the vibration control, is calculated by scaling the total liquid mass m_D with the geometric factors γ_1 and γ_2. The rest liquid mass is also a "dead mass", which is necessary only for the tuning of the natural frequency of the damper. Therefore, this mass is added in Equation 7.65 to the main mass m_S. In Equation 7.65, the mass ratio $\bar{\mu}$, which is necessary for the application of the DEN HARTOG tuning criteria, is calculated from the actively participating liquid mass $\bar{m}_D$ and the total amount of the dead mass $\bar{m}_S$. This mass ratio together with the natural frequency of the system f_S is substituted in Equation 7.67 to determine the optimum natural frequency of the damper. The pendulum, on which the S-TLCD prototype is attached, can be assumed to be a mathematical pendulum. Accordingly, its natural frequency is assumed to be independent from the mass. Therefore, in Equation 7.67, the natural frequency f_S does not changed by the additional mass.

$$\bar{m}_D = m_D\,\gamma_1\,\gamma_2 \tag{7.64}$$

$$\bar{\mu} = \frac{\bar{m}_D}{\bar{m}_S} = \frac{m_D\,\gamma_1\,\gamma_2}{m_S + m_D\,(1 - \gamma_1\,\gamma_2)} = \frac{\mu\,\gamma_1\,\gamma_2}{1 + \mu\,(1 - \gamma_1\,\gamma_2)} \tag{7.65}$$

$$\mu = \frac{m_D}{m_S} \tag{7.66}$$

$$f_{D,opt} = \frac{f_S}{1 + \bar{\mu}} \tag{7.67}$$

The optimum inherent damping of the S-TLCD can be approximated, similar to the frequency tuning, using the DEN HARTOG criteria. The result is the linearized damping ratio as given in Equation 7.68, where $\bar{\mu}$ is again the mass ratio of actively participating liquid mass $\bar{m}_D$ and the total amount of the dead mass $\bar{m}_M$.

$$D_{D,opt} = \sqrt{\frac{3\bar{\mu}}{8(1 + \bar{\mu})^3}} \tag{7.68}$$

As stated before, $\bar{\mu}$ depends on the geometric factors γ_1 and γ_2. As shown in Figure 7.13, a position change of the movable panels initiates a change of the geometric factors, which will be discussed in Study 2 more in detail. This change in the geometric factors has according to Equation 7.68 also an influence on the optimum damping ratio. However, in this study, the influence of geometric factors on the optimum damping ratio is neglected and the damping ratio kept constant without further tuning by adjusting the butterfly valves to the closed position with $\varphi = 0°$.

For the investigation of the frequency tuning effect, the pendulum and S-TLCD system are excited harmonically by the shaking table with the natural frequency of the pendulum inducing resonant vibrations. During the excitation, the panel position β is changed gradually in its operation range of 45°-72°. Figure 7.13 (a) shows the time history of β for a test sequence of 550 seconds. Figure 7.13 (b) shows for the same test sequence the time history of the measured natural frequency f_D and the calculated optimum natural frequency $f_{D,opt}$ of the damper according to DEN HARTOG. As illustrated in this diagram, f_D matches $f_{D,opt}$ exactly with 1.083 Hz at $\beta = 62°$. At $\beta = 72°$, f_D comes also very close to $f_{D,opt}$ with 1.068 Hz compared to 1.065 Hz. Figure 7.13 (c) shows for the same test sequence the time history of the calculated geometric factors γ_1, γ_2 and their multiplication $\gamma_1 \cdot \gamma_2$. As discussed in Section 7.1.1.3, with changing panel position the natural frequency and the geometric factors change in a nonlinear manner.

Figure 7.14 (a) shows the time history of the measured acceleration of the pendulum. The maximum acceleration is observed at the beginning of the test sequence due to the detuned natural frequency of the damper. At $\beta = 45°$, the natural frequency of the damper is 0.97 Hz and the optimum frequency is 1.12 Hz. As the natural frequency comes closer to the optimum frequency, the acceleration reduces. However, the minimum acceleration 0.032 m s^{-2} is reached at the panel position $\beta = 72°$, rather than at $\beta = 62°$, where the natural frequency is exactly equal the optimum frequency. At $\beta = 62°$, the acceleration is measured as 0.037 m s^{-2}. This effect is caused by the geometric parameters γ_1 and γ_2 due to increased amount of active mass fraction $\bar{m}_D$ and is also discussed in Study 2.

A sample of the liquid motion time history, which is measured during the investigations, is shown in Figure 7.14 (b). The liquid oscillation decreases as the pendulum vibration decreases. Also this effect is further discussed in Study 2.

The effects of the frequency tuning is determined by comparing the measured maximum and minimum pendulum accelerations shown in Figure 7.14 (a). As a result, during the experimental investigations, compared to the passive case, the frequency adaptation capability of the S-TLCD enabled a performance improvement of over 77 %.

Study 2: Effects of the Geometric Factors on the Damper Performance. As derived in Section 7.1.1.1, the geometric factors, γ_1 and γ_2 determine the amount of active liquid mass $\bar{m}_D$ participating to the vibration control. Therefore, besides frequency tuning effects, they are both expected to have an influence on the performance of the damper. Furthermore, γ_1 is directly connected with the effective length L_1, which determines the natural

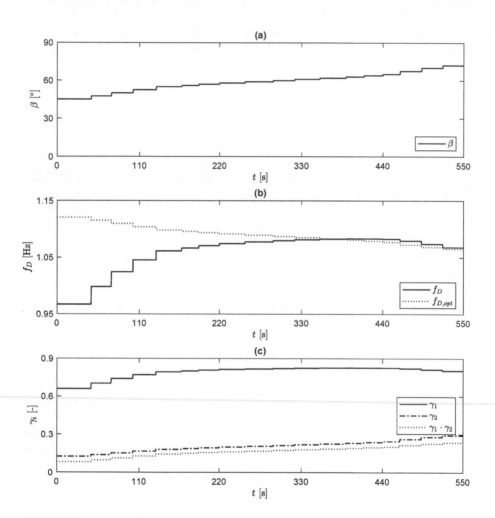

FIGURE 7.13

Time history of the panel position β (a) influencing the natural frequency f_D and the geometric factors γ_i. Change of f_D compared with the optimum natural frequency $f_{D,opt}$ (b). Change of γ_1 and γ_2 (c).

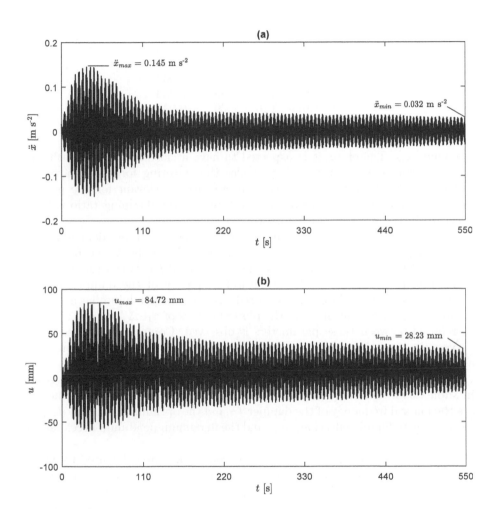

FIGURE 7.14
Time history of the pendulum acceleration induced by a resonant harmonic excitation (a). Time history of the liquid motion of the damper (b).

TABLE 7.5

Change of damper performance
depending on the panel position β.

β [°]	f_D [Hz]	$f_{D,opt}$ [Hz]	u_{max} [mm]	$\ddot{x}_{max}$ [m s^{-2}]
45	1.12	0.97	143	0.145
50	1.11	1.02	75	0.110
55	1.10	1.06	57	0.059
60	1.09	1.08	36	0.040
65	1.08	1.08	34	0.037
70	1.07	1.07	33	0.033
72	1.07	1.07	30	0.032

frequency ω_D. Therefore, it is expected to have a close correlation with the damper performance. Moreover, γ_2 scales the restoring force, as shown in Equation 7.39. Therefore, it is also expected to be relevant for the performance of the damper. Analogous to the Study 1, the damping ratio is kept constant by the butterfly valves with $\varphi = 0°$.

Figure 7.15 (a) shows the change of the geometric factors depending on the panel position β. The diagram in Figure 7.15 (b) shows the frequency detuning determined from the frequency difference Δf between the optimum damper frequency $f_{D,opt}$ and the natural frequency of the damper f_D. In Figure 7.15 (c), the S-TLCD-induced reduction of the pendulum acceleration is illustrated. From the shape of the plotted curves of γ_1, Δf and R, a direct correlation between these parameters is observed. Compared to these three curves, the two other curves, γ_2 and $\gamma_1 \cdot \gamma_2$, are showing a different progress as the panel position changes.

Parameters, which are relevant for the investigation of this effect are given in Table 7.5 for each corresponding discrete panel position β. These parameters are the natural frequency of the damper f_D and the optimum frequency $f_{D,opt}$, the maximum liquid deflection u_{max} and the maximum pendulum acceleration $\ddot{x}_{max}$.

From these parameters the correlation coefficients are calculated to investigate the relationship between the parameters. The correlation coefficient of the frequency detuning $\Delta f = f_{D,opt} - f_D$ with the acceleration reduction R is -0.992, which shows an almost full correlation. The correlation between the first geometric factor γ_1 and R is about 0.971. The correlation between γ_2 and R is 0.893. Comparing these results, it can be seen, that the frequency detuning has the highest effect on the damper performance. The standard deviations of the compared data sets of Δf, γ_1 and γ_2 are 0.052, 0.056 and 0.059 respectively, which are almost equal showing a comparable data distribution for the investigated parameters. These results match with the mathematical background derived in Section 7.1.1.1 and with the interpretations at the beginning of this section.

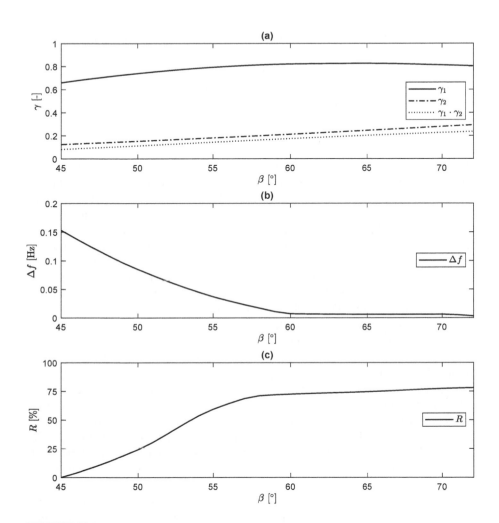

FIGURE 7.15
Change of the geometric parameters γ_1 and γ_2 depending on the panel position β (a). Frequency detuning $\Delta f = f_{D,opt} - f_D$ in relevance to the panel position β (b). Comparison of the panel position β induced natural frequency tuning effect and reduction of pendulum acceleration (c).

7.2   Semi-Active Omnidirectional Tuned Liquid Column Damper

This section introduces the semi-active O-TLCD with frequency and damping adaptation capabilities. Although several passive TLCD layouts have been proposed for the control of multidirectional vibrations, semi-active TLCDs with multidirectional control capabilities are missing (cf. Section 6.3.2).

In passive multidirectional control, one of the pioneering designs was the plus-shaped TLCD developed by Sakai and Takaeda [69, 70]. This configuration combines two TLCDs with a common joint. Zhang et al. proposed another layout by connecting multiple columns by horizontal passages, which are arranged in a crossed layout [168]. Ding et al. proposed recently a toroidal TLCD, which is assembled by dividing a ring-shaped vessel with L-shaped baffles and operates as several equivalent TLCDs, which are oriented in arbitrary directions [169]. By connecting four columns with a horizontal conduit of a rectangular layout Hitchcock et al. assembled a bi-directional TLCD [79]. Rozas et al. [170] investigated this damper analytically for translational vibrations and conducted shaking table tests. Tong et al. [171] implemented the bi-directional TLCD for the control of rotational vibrations of wind turbine platforms. Coudurier et al. investigated also the control of platform rotations using star-shaped TLCDs [172]. The mathematical model of this damper involves $n - 1$ equations corresponding to its $n \geq 3$ columns.

For the multidirectional control of translational vibrations of SDoF structures, Mehrkian and Altay [173] proposed recently also a star-shaped TLCD layout. This damper operates omnidirectionally and is termed as the O-TLCD. The O-TLCD consists of multiple columns, which are distributed around an origin and possess point symmetry as shown in Figure 7.16. Each column is connected with a horizontal passage assembling L-arms, which are communicating with each other at the origin of the layout. However, contrary to the rotational damper of Coudurier et al., the proposed mathematical approach reduces the O-TLCD, irrespective of the number of columns, to an SDoF system.

Based on this star-shaped layout, the semi-active O-TLCD is developed. The frequency adaptation is realized by regulating the column cross-sections using closable cells. These are integrated in columns of the damper and dividing the columns in vertical channels. Small-size actuators control valves to close (respectively open) the upper ends of cells and deactivate (respectively activate) them as shown in Figure 7.17. A similar approach was previously proposed by Min et al. [120] without actuators for the re-tuning of passive TLCDs after their installation. Furthermore, the semi-active O-TLCD utilizes closable orifices to regulate its inherent damping behavior, which are also controlled by small-size actuators, cf. Figure 7.16.

For the mathematical representation of the frequency control effect, a

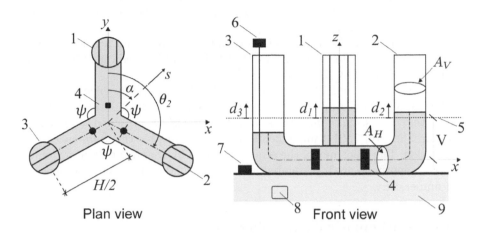

FIGURE 7.16
The semi-active O-TLCD with three columns and non-identical cross-sectional areas. Columns subdivided into cells (1)–(3), orifice (4), hydrostatic equilibrium level (5), liquid level sensor (6), vibrations sensor (7), computer/controller (8), structure (9).

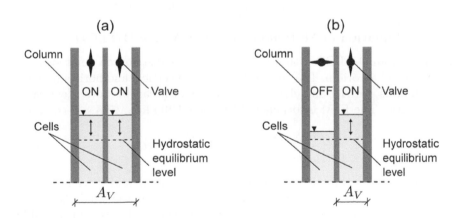

FIGURE 7.17
Closable cells integrated in columns of the semi-active O-TLCD. Small-size actuators are controlling valves. Both cells are activated (a). One cell is active (b). The liquid motion is stopped in closed cells. Liquid mass participating in vibration control changes depending on the number of active cells.

model is proposed, which considers both identical and non-identical cross-sectional areas of columns and horizontal passages. The proposed approach holds for $n \geq 2$. Accordingly, it covers not only the semi-active O-TLCD but also both classical passive TLCDs with two columns and further developments, such as the plus-shaped TLCD of Sakai and Takaeda [69, 70] and all other star-shaped TLCDs with $n \geq 3$.

Sections 7.2.1.1 and 7.2.1.2 present the EoMs of a system encompassing a semi-active O-TLCD and an SDoF structure. Section 7.2.1.3 derives the EoMs of a system consisting of multiple semi-active O-TLCDs attached to an MDoF structure. Section 7.2.1.4 implements these equations for the state-space representation. The efficiency of the semi-active O-TLCD and the accuracy of the modeling approach are investigated in Sections 7.2.2 and 7.2.3 experimentally and numerically.

7.2.1 Mathematical Modeling

As shown in Section 7.1, for the mathematical modeling of the TLCDs, the BERNOULLI equation of the non-stationary flow of the liquid can be employed. However, for the semi-active O-TLCD, the EULER-LAGRANGE equation is preferred considering the 3-dimensional complex geometry of the damper. Similar to the BERNOULLI equation, also this formulation neglects the compressibility of fluid and still holds for TLCD applications, which generally use NEWTONIAN fluids, such as water. The EULER-LAGRANGE equation has been proposed and validated before for classical TLCDs, such as by Xue et al. [165], Taflanidis et al. [174], Coudurier et al. [175] and by Ding et al. [169].

7.2.1.1 Equation of Motion of the Semi-Active O-TLCD

We consider an O-TLCD with n L-arms, which are distributed point symmetrically around an origin. The cross-sections of the vertical and horizontal parts of the L-arms are A_V and A_H, respectively, which can be both identical or non-identical. The EoM is formulated from the EULER-LAGRANGE equation

$$\frac{\mathrm{d}}{\mathrm{dt}} \left(\frac{\partial E_{kin}}{\partial \dot{d}} \right) - \frac{\partial E_{kin}}{\partial d} + \frac{\partial E_{pot}}{\partial d} = Q_d, \tag{7.69}$$

where E_{kin} is the kinetic energy of the moving fluid and E_{pot} is the potential energy caused by the gravity forces acting on the fluid. Q_d is the non-conservative force, which corresponds to the energy dissipation caused by friction and turbulence effects.

As shown in Figure 7.18, the liquid deflection in each column of the O-TLCD is represented by a separate DoF d_i, where $i = \{1, \ldots, n\}$. The sign of d_i represents the motion direction with positive sign corresponding to an upward liquid motion. Due to the fact that the L-arms are communicating with each other and the fluid mass is conservative, a generalized DoF must exist for the representation of the liquid motion. This DoF is referred to as d in the

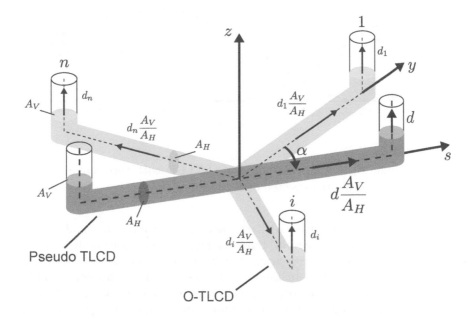

FIGURE 7.18
Reduction of the DoFs (d_i) of the semi-active O-TLCD by introducing an equivalent TLCD, which is operating parallel to the excitation direction w and has the DoF d.

aforementioned EULER-LAGRANGE equation. With an appropriate transfer function, d_i can be then represented by the generalized DoF d. Consequently, the O-TLCD can be reduced to an SDoF system. This is advantageous for the formulation and solution of the EoM of the damper. If the transfer function is chosen correctly, the resulting EoM will correspond to a TLCD, which is equivalent to the O-TLCD and operating parallel to the dominating excitation direction s as shown in Figure 7.18. This TLCD is referred to as *pseudo TLCD*.

To determine the transfer function, we consider an O-TLCD with three columns as previously shown in Figure 7.16. In Figure 7.19, the expected liquid deflection of each column d_i is normalized by the maximum expected deflection d_{max} and plotted over the excitation direction α. The angle between L-arms is $\psi = 2\pi/3$. Accordingly, the orientations of the three L-arms of the O-TLCD corresponding to the y-direction are referred to as $\theta_1 = 0$, $\theta_2 = 2\pi/3$ and $\theta_3 = 4\pi/3$. Accordingly, for $\alpha = \{0, 2\pi/3, 4\pi/3\}$, the excitation is parallel to the L-arms $1, 2$ and 3, respectively. In these cases, the highest liquid deflection d_{max} is expected in the column, which is parallel to the excitation direction. Therefore, the estimated normalized deflection value for these cases correspond to 1 and -1 for the opposite excitation direction as shown in the plot. For the other two columns, we expect, due to symmetry, the maximum

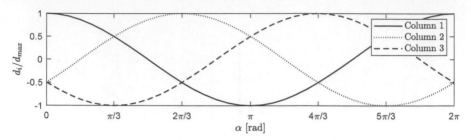

FIGURE 7.19
Liquid deflection d_i in each column of an O-TLCD with three columns normalized by the maximum expected deflection $d_{\max}$.

liquid deflection to be halved with the normalized value -0.5 and 0.5 for the opposite excitation direction. Furthermore, if the excitation direction is perpendicular to one of the L-arms, the liquid deflection in that L-arm is expected to be zero. In the remaining two L-arms, the liquid deflection will be equal to each other with opposite signs and lower than maximum deflection $\pm d_{\max}$. Consequently, we obtain for the normalized liquid deflection of each column a periodic plot, which can be represented by a harmonic function.

As a transfer function, for the general case of an O-TLCD with n columns, the harmonic function

$$d_i = d\cos(\theta_i - \alpha) \quad \text{with} \quad \theta_i = (i-1)\frac{2\pi}{n} \tag{7.70}$$

is proposed. As previously introduced, α and θ_i are the orientation angle of the excitation direction and the ith L-arm, respectively. The sum over n L-arms yields

$$\sum_{i=1}^{n} d_i = d\sum_{i=1}^{n} \cos(\theta_i - \alpha) = 0 \tag{7.71}$$

corresponding to the conservation of liquid mass. When applied to the O-TLCD with three columns and after replacing in Equation 7.70 d with $d_{\max}$, we obtain for the normalized liquid deflection, the plot shown in Figure 7.19, which matches with the expected liquid deflection. Moreover, mathematically the proposed function projects the L-arms in the horizontal plane to the dominating excitation direction s. Accordingly, the function has a physical meaning and computes for each L-arm the corresponding length of the active liquid portion participating in vibration control. Based on a similar approach, Hochrainer and Ziegler [110] as well as Reiterer and Ziegler [111] computed for TLCDs with inclined columns the active liquid portion resulting from the vertical inclination of the columns.

For an O-TLCD with n columns, the kinetic energy of the moving fluid reads

$$E_{kin} = \frac{1}{2}\rho \sum_{i=1}^{n} \left(A_H \frac{H}{2}|\dot{\mathbf{d}}_{H,i}|^2 + A_V V |\dot{\mathbf{d}}_{V,i}|^2 \right), \tag{7.72}$$

where ρ is the density of the damper fluid. The velocity vectors of the liquid flow are computed for the vertical and horizontal parts of the O-TLCD as

$$\dot{\mathbf{d}}_{V,i} = \begin{bmatrix} \dot{s}\sin\alpha \\ \dot{s}\cos\alpha \\ \dot{d}_i \end{bmatrix}, \quad \dot{\mathbf{d}}_{H,i} = \begin{bmatrix} \dfrac{A_V}{A_H}\dot{d}_i \sin\theta_i + \dot{s}\sin\alpha \\ \dfrac{A_V}{A_H}\dot{d}_i \cos\theta_i + \dot{s}\cos\alpha \\ 0 \end{bmatrix} \tag{7.73}$$

where $\dot{s}$ is the excitation velocity applied to the damper and $\dot{d}_i$ is the fluid velocity in each O-TLCD column, which can be computed using the generalized DoF d as

$$\dot{d}_i = \dot{d}\cos(\theta_i - \alpha) - \dot{\alpha}d\sin(\theta_i - \alpha). \tag{7.74}$$

For the sake of brevity, we assume that during the excitation event the excitation direction α does not change and accordingly α is time-invariant with $\dot{\alpha} = 0$. The kinetic energy formulation yields

$$E_{kin} = \frac{1}{2}\rho A_V \left(\left(\frac{H}{2}\left(\frac{A_V}{A_H}\dot{d}^2 + 2\dot{d}\dot{s} \right) + V\dot{d}^2 \right)\kappa + n\dot{s}^2 \left(\frac{A_H}{A_V}\frac{H}{2} + V \right) \right), \tag{7.75}$$

where $\kappa = \sum_{i=1}^{n}\cos^2(\theta_i - \alpha)$. In EULER-LAGRANGE equation, the corresponding partial derivative terms of the kinetic energy read

$$\frac{\mathrm{d}}{\mathrm{dt}}\left(\frac{\partial E_{kin}}{\partial \dot{d}} \right) = \frac{1}{2}\rho A_V \kappa \left(H\left(\frac{A_V}{A_H}\ddot{d} + \ddot{s} \right) + 2V\ddot{d} \right), \quad \frac{\partial E_{kin}}{\partial d} = 0. \tag{7.76}$$

The potential energy of the damper fluid reads

$$E_{pot} = \frac{1}{2}\rho g A_V \sum_{i=1}^{n}(V + d_i)^2, \tag{7.77}$$

where g is the acceleration of gravity. After introducing the generalized DoF d, the potential energy reads

$$E_{pot} = \frac{1}{2}\rho g A_V \left(nV^2 + \kappa d^2 \right), \tag{7.78}$$

with the corresponding partial derivative term

$$\frac{\partial E_{pot}}{\partial d} = \rho g A_V \kappa d, \tag{7.79}$$

where the level of zero potential energy is assumed to coincide with the horizontal sections of the L-arms.

The non-conservative term corresponding to the energy dissipation can be formulated as

$$Q_d = -\frac{1}{4}\rho\lambda A_V \sum_{i=1}^{n}|\dot{d}_i|\dot{d}_i = -\frac{1}{4}\rho\lambda A_V \kappa|\dot{d}|\dot{d}, \qquad (7.80)$$

where λ is the loss factor. As previously mentioned, energy dissipation is introduced by friction and turbulence effects, which can be caused by orifices or changes in area of flow cross-section. Corresponding real λ values can be determined according to the literature, such as from Idelchik [176] as well as Cengel and Cimbala [177]. For $n = 2$ and $\alpha = 0$, the non-conservative force formulation yields to

$$Q_d = -\frac{1}{2}\rho\lambda A_V|\dot{d}|\dot{d} \quad \text{for } n = 2 \text{ (TLCD) and } \alpha = 0, \qquad (7.81)$$

corresponding to the commonly known representation of the non-conservative force of a classical TLCDs, which is operating in excitation direction.

After introducing the partial derivative terms and the non-conservative force in the EULER-LAGRANGE equation, the EoM of the semi-active O-TLCD reads

$$\ddot{d} + \delta|\dot{d}|\dot{d} + \omega_d^2 d = -\gamma_1\ddot{s}, \qquad (7.82)$$

where δ, ω_d and γ_1 are the head-loss coefficient, the natural angular frequency and the geometric factors, which are defined as

$$\delta = \frac{\lambda}{2L_1}, \quad \omega_d = \sqrt{\frac{2g}{L_1}}, \quad \gamma_1 = \frac{H}{L_1}. \qquad (7.83)$$

The effective length L_1 reads

$$L_1 = 2V + H\frac{A_V}{A_H} \qquad (7.84)$$

and corresponds for identical cross-sectional areas $A_V = A_H$ to the total liquid length of the pseudo TLCD.

Furthermore, the EoM can be expanded for the case of a seismic excitation with the ground acceleration $\ddot{s}_g$ as

$$\ddot{d} + \delta|\dot{d}|\dot{d} + \omega_d^2 d = -\gamma_1\left(\ddot{s} + \ddot{s}_g\right). \qquad (7.85)$$

After linearizing the nonlinear term $\delta|\dot{d}|\dot{d}$, the EoM reads

$$\ddot{d} + 2D_D\omega_D\dot{d} + \omega_d^2 d = -\gamma_1\left(\ddot{s} + \ddot{s}_g\right), \qquad (7.86)$$

where D_D is the equivalent damping ratio, which is proposed for harmonically excited classical TLCDs by Gao et al. [77] as

$$D_D = \delta\frac{4}{3\pi}\frac{\Omega\, d_{max}}{\omega_D}. \qquad (7.87)$$

Here, Ω and $d_{\max}$ are the angular excitation frequency and the maximum expected displacement of the damper liquid respectively. For a stationary random excitation, the equivalent damping ratio D_D reads

$$D_D = \delta \sqrt{\frac{2}{\pi}} \frac{\sigma_{\dot{d}}}{\omega_D}, \tag{7.88}$$

as proposed for classical TLCDs by Xu et al. [166], where $\sigma_{\dot{d}}$ is the standard deviation of the velocity of the damper liquid considering a zero-mean GAUSSIAN process.

7.2.1.2 Equation of Motion of an SDoF Structure with a Semi-Active O-TLCD

We consider a structure under a dynamic load $\mathbf{F}$ and the ground acceleration $\ddot{\mathbf{s}}_{\mathbf{g}}$ in arbitrary directions as shown in Figure 7.20 (a). The structure is torsional rigid and exhibits translational vibrations in the x, y-plane. The resulting vibration direction is along the s-axis, which is inclined from the y-axis by α. The corresponding components of the dynamic load and ground acceleration in s-direction are F and $\ddot{s}_g$, respectively. m_S, c_S and k_S are the mass, damping coefficient and stiffness of the structure corresponding to the vibration direction s. As shown in Figure 7.20 (a), a semi-active O-TLCD, which has n symmetrically distributed L-arms, is attached to the structure in order to control its vibrations. The cross-sections of the vertical and horizontal parts of the L-arms are A_V and A_H, respectively. Figure 7.20 (b) shows the SDoF structure with the corresponding pseudo TLCD, which is assumed to be operating in vibration direction and is represented by the DoF d.

The linear momentum applied to each L-arm of the O-TLCD in vibration direction s reads

$$I_{s,i} = \rho \left(A_V V (\dot{s} + \dot{s}_g) + A_H \frac{H}{2} \left(\dot{s} + \dot{s}_g + \dot{d}_i \frac{A_V}{A_H} \right) \right) \cos(\theta_i - \alpha). \tag{7.89}$$

from which, the corresponding force is determined as

$$F_{s,i} = \frac{\mathrm{d}I}{\mathrm{d}t} = m_{D,i} \left(\ddot{s} + \ddot{s}_g + \gamma_2 \ddot{d}_i \right) \cos(\theta_i - \alpha), \tag{7.90}$$

where $m_{D,i}$ is the liquid mass in i^{th} L-arm. $\gamma_2 = H/L_2$ is introduced as a further geometric factor with the corresponding efficient length $L_2 = 2V + H(A_H/A_V)$. For the sake of brevity, analogous to the derivation of the EoM of the O-TLCD, α is assumed to be time-invariant during the excitation event. After replacing d_i with the generalized DoF d as $\ddot{d}_i = \ddot{d}\cos(\theta_i - \alpha)$, the resulting force for all L-arms in excitation direction yields

$$F_s = m_D \left(\ddot{s} + \ddot{s}_g + \frac{\gamma_2 \kappa}{n} \ddot{d} \right), \tag{7.91}$$

where m_D is the total liquid mass and, as previously introduced, $\kappa = \sum_{i=1}^{n} \cos^2(\theta_i - \alpha)$.

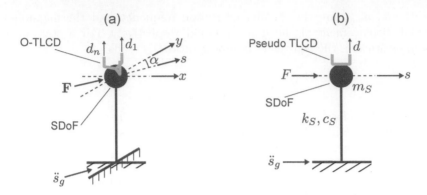

FIGURE 7.20
A torsional rigid SDoF structure with a semi-active O-TLCD subjected to the dynamic load **F** and the ground acceleration $\ddot{s}_{\mathbf{g}}$ (a). The structure exhibits translational vibrations in the x, y-plane in s direction. The corresponding components of the dynamic load and seismic excitation are F and $\ddot{s}_g$, respectively. The SDoF structure with the corresponding pseudo TLCD (b).

Finally, from the equilibrium of the forces, the EoM of the structure is defined as

$$m_S \ddot{s} + c_S \dot{s} + k_S s = -m_S \ddot{s}_g - m_D \left(\ddot{s} + \ddot{s}_g + \frac{\gamma_2 \kappa}{n} \ddot{d} \right) + F, \qquad (7.92)$$

where m_S, c_S and k_S are the mass, damping coefficient and stiffness of the structure corresponding to the vibration direction s. By introducing ω_S and D_S as the natural angular frequency and the damping ratio of the structure as well as $\mu = m_D/m_S$ as the mass ratio, the EoM can be rewritten as

$$\ddot{s} + 2D_S \omega_S \dot{s} + \omega_S^2 w = -\ddot{s}_g - \mu \left(\ddot{s} + \ddot{s}_g + \frac{\gamma_2 \kappa}{n} \ddot{d} \right) + f, \qquad (7.93)$$

where $f = F/m_S$. It is important to mention that it holds $\kappa = 2\cos^2 \alpha$ for a classical TLCD with $n = 2$ and $\kappa = n/2$ for O-TLCD with $n \geq 3$. Accordingly, in case of a classical TLCD, the restoring force in the EoM changes depending on the excitation direction α. As expected, it is zero if the excitation direction is perpendicular to the operation direction of the TLCD with $\cos \pi/2 = 0$, whereas, in case of an O-TLCD, the restoring force remains constant and independent of the excitation direction α. The EoMs reads for both cases

$$\ddot{s} + 2D_S \omega_S \dot{s} + \omega_S^2 s$$
$$= \begin{cases} -\ddot{s}_g - \mu \left(\ddot{s} + \ddot{s}_g + \gamma_2 \ddot{d} \cos^2 \alpha \right) + f & \text{for } n = 2 \text{ (TLCD)}, \\ -\ddot{s}_g - \mu \left(\ddot{s} + \ddot{s}_g + \frac{\gamma_2}{2} \ddot{d} \right) + f & \text{for } n \geq 3 \text{ (O-TLCD)}. \end{cases} \qquad (7.94)$$

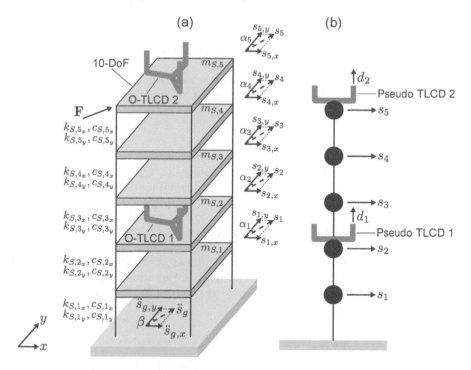

FIGURE 7.21

A torsional rigid 10 DoF structure with two semi-active O-TLCDs subjected to the dynamic load $\mathbf{F}$ and the ground acceleration $\ddot{\mathbf{s}}_{\mathbf{g}}$ (a). Each jth floor of the structure exhibits translational vibrations in s_j direction under the dynamic load $\mathbf{F}$ and the ground acceleration $\ddot{\mathbf{s}}_{\mathbf{g}}$. The MDoF structure with the corresponding pseudo TLCDs, which are assumed to perform along the resulting vibration direction of the floors (b).

7.2.1.3 Equation of Motion of an MDoF Structure with Multiple Semi-Active O-TLCDs

We consider a torsional rigid structure with m floors under a dynamic load $\mathbf{F}$ and the ground acceleration $\ddot{\mathbf{s}}_{\mathbf{g}}$, which are acting in arbitrary directions as shown in Figure 7.21 (a). The structure exhibits translational vibrations and, consequently, discretized by $2m$ DoFs in x and y directions. The resulting vibration direction of each jth floor is along the s_j-axis, which is inclined from the y-axis by α_j.

To reduce the vibrations, the structure is equipped with k semi-active O-TLCDs. Each ith damper has cross-sectional areas with $A_{V,i}$, $A_{H,i}$ and n_i L-arms distributed symmetrically around an origin. As shown in Figure 7.21 (b), each O-TLCD has a corresponding pseudo TLCD, which is assumed to operate along the resulting vibration direction s_i. The EoMs of the pseudo TLCDs are formulated by means of matrices as

$$\ddot{\mathbf{d}} + \mathbf{C}_D\dot{\mathbf{d}} + \mathbf{K}_D\mathbf{d} = -\boldsymbol{\Gamma}_1\mathbf{N}^\top\left(\boldsymbol{\nu}\ddot{s}_g + \ddot{s}\right), \tag{7.95}$$

where the vectors $\{\ddot{\mathbf{d}}, \dot{\mathbf{d}}, \mathbf{d}\} \in \mathbb{R}^{k \times 1}$ represent the liquid motions of the corresponding pseudo TLCDs, which are described as

$$\ddot{\mathbf{d}} = [\ddot{d}_1 \ldots \ddot{d}_k]^\top, \quad \dot{\mathbf{d}} = [\dot{d}_1 \ldots \dot{d}_k]^\top, \quad \mathbf{d} = [d_1 \ldots d_k]^\top. \tag{7.96}$$

Here, $\ddot{d}_i$, $\dot{d}_i$ and d_i are respectively the acceleration, velocity and displacement of the liquid motion of the ith pseudo TLCD corresponding to the ith O-TLCD. Furthermore, in Equation 7.95, the diagonal matrices $\mathbf{C}_D \in \mathbb{R}^{k \times k}$ and $\mathbf{K}_D \in \mathbb{R}^{k \times k}$ gather the damping and stiffness properties as

$$\mathbf{C}_D = \text{diag}[\delta_1|\dot{d}_1|, \ldots, \delta_k|\dot{d}_k|], \quad \mathbf{K}_D = \text{diag}[\omega_{D,1}^2, \ldots, \omega_{D,k}^2], \tag{7.97}$$

where δ_i and $\omega_{D,i}$ are respectively the head-loss coefficient and the natural angular frequency of the ith O-TLCD as given in Equation 7.83. As introduced in Section 7.2.1.1, the damping terms of the O-TLCDs can be linearized. In this case, the damping matrix reads

$$\mathbf{C}_D = \text{diag}[2D_{D,1}\omega_{D,1}, \ldots, 2D_{D,k}\omega_{D,k}], \tag{7.98}$$

where $D_{D,i}$ is the equivalent damping ratio of the ith O-TLCD calculated from Equation 7.87 for the harmonic excitation and from Equation 7.88 for the stationary random excitation of the system.

On the right side of Equation 7.95, the diagonal matrix $\boldsymbol{\Gamma}_1 \in \mathbb{R}^{k \times k}$ contains the geometric factors as

$$\boldsymbol{\Gamma}_1 = \text{diag}[\gamma_{1,1}, \ldots, \gamma_{1,k}], \tag{7.99}$$

where $\gamma_{1,i}$ is the geometric factor corresponding to the ith O-TLCD as given in Equation 7.83.

Furthermore, in Equation 7.95, $\mathbf{N} \in \mathbb{R}^{2m \times k}$ represents the distribution of the O-TLCDs corresponding to the DoFs of the structure and reads

$$\mathbf{N} = \begin{bmatrix} \tilde{\mathbf{N}} \circ \mathbf{T}_x \\ \tilde{\mathbf{N}} \circ \mathbf{T}_y \end{bmatrix}. \tag{7.100}$$

The operator $\circ$ constitutes the HADAMARD product of the incidence matrix $\tilde{\mathbf{N}} \in \mathbb{R}^{m \times k}$ with the transformation matrices $\mathbf{T}_x \in \mathbb{R}^{m \times k}$ and $\mathbf{T}_y \in \mathbb{R}^{m \times k}$. The elements of the incidence matrix $\tilde{\mathbf{N}}$ are equal to 1 for floors with O-TLCD, and 0 for floors without O-TLCD. The transformation matrices read

$$\mathbf{T}_x = \begin{bmatrix} \sin\alpha_1 & \ldots & \sin\alpha_1 \\ \vdots & \ldots & \vdots \\ \sin\alpha_m & \ldots & \sin\alpha_m \end{bmatrix}, \quad \mathbf{T}_y = \begin{bmatrix} \cos\alpha_1 & \ldots & \cos\alpha_1 \\ \vdots & \ldots & \vdots \\ \cos\alpha_m & \ldots & \cos\alpha_m \end{bmatrix} \tag{7.101}$$

with $\alpha_j = \arctan(s_{j,x}/s_{j,y})$ corresponding to the DoFs of the jth floor in x and y directions.

Accordingly, for the structure shown in Figure 7.21 with 5 floors and 2 O-TLCDs, which are attached to the second and fifth floors, the incidence matrix and the transformation matrices read

$$\tilde{\mathbf{N}} = \begin{bmatrix} 0 & 0 \\ 1 & 0 \\ 0 & 0 \\ 0 & 0 \\ 0 & 1 \end{bmatrix}, \quad \mathbf{T}_x = \begin{bmatrix} \sin\alpha_1 & \sin\alpha_1 \\ \sin\alpha_2 & \sin\alpha_2 \\ \sin\alpha_3 & \sin\alpha_3 \\ \sin\alpha_4 & \sin\alpha_4 \\ \sin\alpha_5 & \sin\alpha_5 \end{bmatrix}, \quad \mathbf{T}_y = \begin{bmatrix} \cos\alpha_1 & \cos\alpha_1 \\ \cos\alpha_2 & \cos\alpha_2 \\ \cos\alpha_3 & \cos\alpha_3 \\ \cos\alpha_4 & \cos\alpha_4 \\ \cos\alpha_5 & \cos\alpha_5 \end{bmatrix}. \quad (7.102)$$

Moreover, in Equation 7.95, $\boldsymbol{\nu} \in \mathbb{R}^{2m \times 1}$ is an incidence vector, which is distributing the ground acceleration to the DoFs of the structure as

$$\boldsymbol{\nu} = [\mathbf{t}_x | \mathbf{t}_y]^\top, \quad (7.103)$$

with the transformation vectors $\mathbf{t}_x \in \mathbb{R}^{m \times 1}$ and $\mathbf{t}_y \in \mathbb{R}^{m \times 1}$ as

$$\mathbf{t}_x = [\sin\beta \ldots \sin\beta]^\top, \quad \mathbf{t}_y = [\cos\beta \ldots \cos\beta]^\top, \quad (7.104)$$

corresponding to the x and y directions.

Finally, in Equation 7.95, $\ddot{s}_g$ and $\ddot{\mathbf{s}} \in \mathbb{R}^{2m \times 1}$ are the ground acceleration and the acceleration responses of the structure at its DoFs.

Accordingly, the EoM of the $2m$-DoF structure with k-O-TLCDs reads

$$\mathbf{M}_S\ddot{\mathbf{s}} + \mathbf{C}_S\dot{\mathbf{s}} + \mathbf{K}_S\mathbf{s} = -\mathbf{M}_S\boldsymbol{\nu}\ddot{s}_g - \tilde{\mathbf{N}}\mathbf{M}_D\tilde{\mathbf{N}}^\top (\ddot{\mathbf{s}} + \boldsymbol{\nu}\ddot{s}_g) + \mathbf{N}\mathbf{M}_D\boldsymbol{\Gamma}_2\ddot{\mathbf{d}} + \mathbf{f} \quad (7.105)$$

where $\mathbf{M}_S \in \mathbb{R}^{2m \times 2m}$, $\mathbf{C}_S \in \mathbb{R}^{2m \times 2m}$ and $\mathbf{K}_S \in \mathbb{R}^{2m \times 2m}$ are the mass, damping and stiffness matrices of the structure. $\mathbf{M}_D \in \mathbb{R}^{k \times k}$ is the mass matrix of the dampers. $\ddot{\mathbf{s}} \in \mathbb{R}^{2m \times 1}$, $\dot{\mathbf{s}} \in \mathbb{R}^{2m \times 1}$ and $\mathbf{s} \in \mathbb{R}^{2m \times 1}$ are the acceleration, velocity and displacement vectors of the structure, which are defined according to the DoFs of the structure as

$$\ddot{\mathbf{s}} = [\ddot{\mathbf{s}}_x | \ddot{\mathbf{s}}_y]^\top, \quad \dot{\mathbf{s}} = [\dot{\mathbf{s}}_x | \dot{\mathbf{s}}_y]^\top, \quad \mathbf{s} = [\mathbf{s}_x | \mathbf{s}_y]^\top, \quad (7.106)$$

where $\ddot{\mathbf{s}}_{x/y}$, $\dot{\mathbf{s}}_{x/y}$ and $\mathbf{s}_{x/y}$ contain the acceleration, velocity and displacements of the DoFs corresponding to the x and y directions. As the structure is assumed to be torsional rigid, the mass, damping and stiffness matrices read

$$\mathbf{M}_S = \text{diag}[m_{S,1}, \ldots, m_{S,m}, m_{S,1}, \ldots, m_{S,m}], \quad (7.107)$$

$$\mathbf{C}_S = [\mathbf{C}_{S,x}\, 0 \,|\, 0\, \mathbf{C}_{S,y}]^\top \quad (7.108)$$

$$\mathbf{K}_S = [\mathbf{K}_{S,x}\, 0 \,|\, 0\, \mathbf{K}_{S,y}]^\top, \quad (7.109)$$

where $\mathbf{K}_{S,x} \in \mathbb{R}^{m \times m}$ and $\mathbf{K}_{S,y} \in \mathbb{R}^{m \times m}$ constitute the stiffness components of the floors corresponding to the x and y directions. The damping matrix components $\mathbf{C}_{S,x} \in \mathbb{R}^{m \times m}$ and $\mathbf{C}_{S,y} \in \mathbb{R}^{m \times m}$ can be determined according to the RAYLEIGH damping as

$$\mathbf{C}_{S,x/y} = \alpha_{x/y}\mathbf{M}_{S,x/y} + \beta_{x/y}\mathbf{K}_{S,x/y} \quad (7.110)$$

where the RAYLEIGH damping coefficients $\alpha_{x/y}$ and $\beta_{x/y}$ are calculated from

$$\begin{bmatrix} \alpha_{x/y} \\ \beta_{x/y} \end{bmatrix} = \frac{2\omega_{S,1_{x/y}}\omega_{S,2_{x/y}}}{\omega_{S,2_{x/y}}^2 - \omega_{S,1_{x/y}}^2} \begin{bmatrix} \omega_{S,2_{x/y}} & -\omega_{S,1_{x/y}} \\ -1/\omega_{S,2_{x/y}} & 1/\omega_{S,1_{x/y}} \end{bmatrix} \begin{bmatrix} D_{S,1_{x/y}} \\ D_{S,1_{x/y}} \end{bmatrix}, \quad (7.111)$$

with the natural angular frequencies $\omega_{S,1/2_{x/y}}$ and damping ratios $D_{S,1/2_{x/y}}$ of the structure corresponding to the directions x and y.

On the right side of Equation 7.105, the diagonal matrix $\boldsymbol{\Gamma}_2 \in \mathbb{R}^{k \times k}$ contains the geometric factors of the O-TLCDs as

$$\boldsymbol{\Gamma}_2 = \mathrm{diag}[\gamma_{2,1}\kappa_1/n_1, \ldots, \gamma_{2,k}\kappa_k/n_k], \quad (7.112)$$

where $\gamma_{2,i}$ is the geometric factor of the ith O-TLCD.

Furthermore, it holds $\kappa_i = n_i/2$ for the semi-active O-TLCDs with $n_i \geq 3$. For the case of classical TLCDs with $n_i = 2$ attached to the jth floor, it holds $\kappa_i = 2\cos^2\alpha_j$. Finally, in Equation 7.105, the load vector $\mathbf{f} \in \mathbb{R}^{2m \times 1}$ contains the components of the dynamic load corresponding to the vibration directions of the structure.

7.2.1.4   State-Space Representation of an MDoF Structure with Multiple Semi-Active O-TLCDs

For the real-time application of semi-active control strategies, highly efficient mathematical models are required. For this purpose, the state-space representation method is used in common practice. As introduced in Section 3.2, this approach formulates the dynamic systems by first order differential equations.

For the derivation of the state-space representation, we consider the MDoF structure with $2m$ DoFs and k semi-active O-TLCDs with n L-arms, which are symmetrically distributed around an origin and have the horizontal and vertical cross-sections A_H and A_V, respectively. We rewrite the EoMs from previous Section 7.2.1.3 as

$$\mathbf{M}_M \begin{bmatrix} \ddot{\mathbf{s}} \\ \ddot{\mathbf{d}} \end{bmatrix} + \begin{bmatrix} \mathbf{C}_S & \mathbf{0} \\ \mathbf{0} & \mathbf{C}_D \end{bmatrix} \begin{bmatrix} \dot{\mathbf{s}} \\ \dot{\mathbf{d}} \end{bmatrix} + \begin{bmatrix} \mathbf{K}_S & \mathbf{0} \\ \mathbf{0} & \mathbf{K}_D \end{bmatrix} \begin{bmatrix} \mathbf{s} \\ \mathbf{d} \end{bmatrix}$$
$$= -\begin{bmatrix} \mathbf{M}_S\boldsymbol{\nu} + \tilde{\mathbf{N}}\mathbf{M}_D\tilde{\mathbf{N}}^\top\boldsymbol{\nu} \\ \boldsymbol{\Gamma}_1\mathbf{N}^\top\boldsymbol{\nu} \end{bmatrix} \ddot{s}_g + \begin{bmatrix} \mathbf{f} \\ \mathbf{0} \end{bmatrix} \quad (7.113)$$

where the matrix $\mathbf{M}_M \in \mathbb{R}^{(2m+k) \times (2m+k)}$ reads

$$\mathbf{M}_M = \begin{bmatrix} \mathbf{M}_S + \tilde{\mathbf{N}}\mathbf{M}_D\tilde{\mathbf{N}}^\top & \mathbf{N}\mathbf{M}_D\boldsymbol{\Gamma}_2 \\ \boldsymbol{\Gamma}_1\mathbf{N}^\top & \mathbf{I} \end{bmatrix} \quad (7.114)$$

with the unit matrix $\mathbf{I} \in \mathbb{R}^{k \times k}$. In Equation 7.113, complying with the definitions from Section 7.2.1.3, $\ddot{\mathbf{s}}$, $\dot{\mathbf{s}}$ and $\mathbf{s}$ are the acceleration, velocity and displacement vectors of the structure. $\ddot{\mathbf{d}}$, $\dot{\mathbf{d}}$ and $\mathbf{d}$ are corresponding motion vectors of the O-TLCDs. Furthermore, $\mathbf{M}_S$, $\mathbf{C}_S$ and $\mathbf{K}_S$ as well as $\mathbf{M}_D$, $\mathbf{C}_D$

and $\mathbf{K}_D$ are the mass, damping and stiffness matrices of the structure and the O-TLCDs, respectively. Moreover, $\mathbf{\Gamma}_1$ and $\mathbf{\Gamma}_2$ contain the geometric factors of the O-TLCDs. On the structure, the ground acceleration $\ddot{s}_g$ and the dynamic loads $\mathbf{f}$ are applied. Finally, in both EoMs, $\boldsymbol{\nu}$ and $\mathbf{N}$ represent the distribution of the earthquake load and the dampers on the structure, respectively.

The state vector of the system $\mathbf{z} \in \mathbb{R}^{2(2m+k)\times 1}$ and its time-derivative $\dot{\mathbf{z}} \in \mathbb{R}^{2(2m+k)\times 1}$ are defined as

$$\mathbf{z} = [\mathbf{s}\ \mathbf{d}\ \dot{\mathbf{s}}\ \dot{\mathbf{d}}]^\top, \quad \dot{\mathbf{z}} = [\dot{\mathbf{s}}\ \dot{\mathbf{d}}\ \ddot{\mathbf{s}}\ \ddot{\mathbf{d}}]^\top. \tag{7.115}$$

The state-space representation of the system reads

$$\dot{\mathbf{z}} = \mathcal{A}\mathbf{z} + \mathcal{B}\mathbf{u}, \tag{7.116}$$

where $\mathcal{A} \in \mathbb{R}^{2(2m+k)\times 2(2m+k)}$ and $\mathcal{B} \in \mathbb{R}^{2(2m+k)\times(2m+1)}$ are the transition and input matrices, which are described as

$$\mathcal{A} = \begin{bmatrix} \mathbf{0} & \mathbf{I} \\ -\mathbf{M}_M^{-1}\begin{bmatrix} \mathbf{K}_S & \mathbf{0} \\ \mathbf{0} & \mathbf{K}_D \end{bmatrix} & -\mathbf{M}_M^{-1}\begin{bmatrix} \mathbf{C}_S & \mathbf{0} \\ \mathbf{0} & \mathbf{C}_D \end{bmatrix} \end{bmatrix}, \tag{7.117}$$

$$\mathcal{B} = \begin{bmatrix} \mathbf{0} \\ -\mathbf{M}_M^{-1}\begin{bmatrix} \mathbf{M}_S\boldsymbol{\nu} + \tilde{\mathbf{N}}\mathbf{M}_D\tilde{\mathbf{N}}^\top\boldsymbol{\nu} & \mathbf{I} \\ \mathbf{\Gamma}_1\mathbf{N}^\top\boldsymbol{\nu} & \mathbf{0} \end{bmatrix} \end{bmatrix}. \tag{7.118}$$

The corresponding input vector $\mathbf{u} \in \mathbb{R}^{(2m+1)\times 1}$ reads

$$\mathbf{u} = [\ddot{s}_g\ \mathbf{f}]^\top, \tag{7.119}$$

where $\ddot{s}_g$ and $\mathbf{f}$ denote the ground acceleration and the dynamic loads, respectively.

7.2.2 Experimental Studies

The omnidirectional operation capability of the O-TLCD is investigated on a prototype by shaking table tests. The setup is shown in Figure 7.22. The prototype has four L-arms, which are connected to each other symmetrically around an origin assembling a cross-form in plan. PVC-U pipes with an identical inner diameter of 55 mm are used. The liquid level in the vertical columns is $V = 121$ mm. The horizontal length of each L-arm is $H/2 = 270$ mm. Furthermore, a prototype of a classical TLCD is prepared with the same inner diameter and the liquid level V. The total horizontal length of the TLCD is H. The dimensions of both prototypes are illustrated in Figure 7.23.

The prototypes are installed on a 0.50×0.50 m uniaxial shaking table, which is operated by a step motor in 0.1–50 Hz frequency range and ± 50 mm stroke. The tested prototype and the shaking table are interconnected by a rotatable platform, which allows to conduct multidirectional vibrations on the prototype within an inclination range of $\alpha =$

(a) (b)

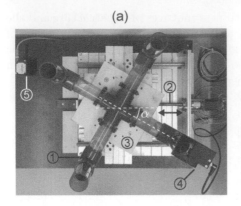 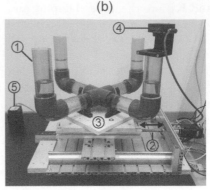

FIGURE 7.22
Plan (a) and front (b) views of the experimental setup utilized for the investigation of the omnidirectional operation capability of the O-TLCD: Prototype with four L-arms (1), uniaxial shaking table (2), rotatable platform (3), ultrasonic liquid level sensor (4), laser position sensor (5) of the shaking table. Excitation direction $\alpha = \{0, \pi/12, \pi/6, \pi/4, \pi/3, 5\pi/12, \pi/2\}$. Courtesy of O. Altay.

$\{0, \pi/12, \pi/6, \pi/4, \pi/3, 5\pi/12, \pi/2\}$. A Simulink model of the shaking table is compiled by a host-PC and uploaded to a target-PC, which then generates the control signal for the shaking table.

An ultrasonic sensor is used to measure dynamically the change of liquid level. A laser sensor is used simultaneously to measure the position of the shaking table. Both measurement signals are processed by an analog-to-digital (A/D) converter and recorded with 100 Hz sampling rate in Matlab/Simulink environment.

In first part of the study, frequency sweep tests are performed on the O-TLCD and TLCD prototypes to investigate their natural frequency. For this purpose, the harmonic signal function

$$s = s_0 \sin(2\pi f t) \qquad (7.120)$$

is used, where the shaking table amplitude is $s_0 = 4.5$ mm and the excitation frequency is $f = \{0.65, \ldots, 1.05\}$ Hz with 0.005 Hz increment in the expected resonance range and otherwise 0.02 Hz. The response curves are determined from the measured liquid maximum level changes as shown in Figure 7.24 (a). In this study, the excitation direction is $\alpha = 0$. The natural frequencies of the O-TLCD and TLCD prototypes are determined as $f_D = 0.830$ Hz and $f_D = 0.825$ Hz, respectively.

In Table 7.6, the parameters of both prototypes and their natural frequencies are given. Here, the natural frequencies, which are determined from experiments, are compared with the natural frequencies, which are calculated

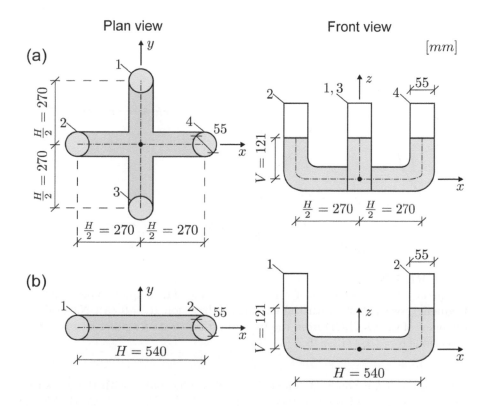

FIGURE 7.23
Plan and front views of damper prototypes used for the investigation of the omnidirectional operation capability of the O-TLCD: An O-TLCD with four L-arms (a) and a classical TLCD (b).

TABLE 7.6

Parameters of the O-TLCD and TLCD prototypes as well as their calculated and measured natural frequencies.

	V	$H/2$	$A_V = A_H$	L_1	$f_{D,cal}$	$f_{D,exp}$
O-TLCD	121 mm	270 mm	2376 mm^2	782 mm	0.797 Hz	0.830 Hz
TLCD	121 mm	270 mm	2376 mm^2	782 mm	0.797 Hz	0.825 Hz

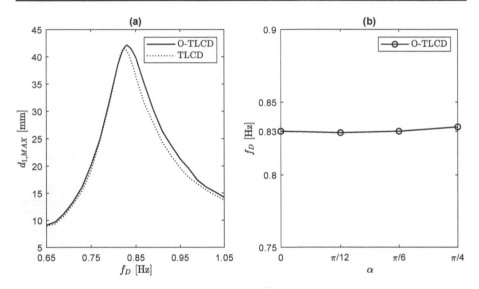

FIGURE 7.24

Frequency response curves of the O-TLCD and TLCD prototypes from the frequency sweep tests with the excitation direction $\alpha = 0$ (a). Natural frequencies of the O-TLCD from the free vibrations tests for different excitation directions $\alpha = \{0, \pi/12, \pi/6, \pi/4\}$ (b).

according to Equation 7.83. The calculated results match with the measured frequencies. The difference is 0.033 Hz for the O-TLCD and 0.025 Hz for the TLCD prototype and can be justified by the uncertainties included in the test rig and the measurement setup.

Furthermore, in the second part of the study, free vibration tests are conducted on the O-TLCD by applying an impulse signal, which is induced by the shaking table in excitation directions $\alpha = \{0, \pi/12, \pi/6, \pi/4\}$. Due to symmetry, the damper is expected to repeat its response for the excitation directions $\alpha = \{\pi/3, 5\pi/12, \pi/3\}$. From the FFT of the time histories of the measured liquid level, the natural frequencies are determined and depicted in Figure 7.24 (b). Accordingly, the natural frequency of the O-TLCD shows only a marginal change for the tested excitation directions, which is reasonable considering inaccuracies involved in the measurement signals and their FFTs.

Finally, the third part of the study investigates the amplitudes of the O-TLCD response and their prediction with the proposed modeling approach as introduced in Section 7.2.1.1. For this purpose, harmonic excitation is conducted on the prototype of the O-TLCD by the shaking table according to Equation 7.120 on alternating excitation directions of $\alpha = \{0, \pi/12, \pi/6, \pi/4, \pi/3, 5\pi/12, \pi/2\}$. The resonance frequency of the O-TLCD with $f = 0.830$ Hz is applied as excitation frequency. The excitation amplitude is $s_0 = 8.4$ mm. The recorded time histories of the liquid deflection are shown in Figure 7.25. The results are compared with the simulated responses. For the simulations, the head-loss coefficient of the O-TLCD is tuned as $\delta = 0.98 \cdot 10^{-3}$ mm^{-1} according to the recorded peak response at $\alpha = 0$. The response of the O-TLCD d_1 is calculated by solving the EoM from Equation 7.82 for d and then applying Equation 7.70.

As shown in Figure 7.25, the predicted responses match with the measured liquid deflection. As expected, with increasing α the liquid deflection decreases. At $\alpha = \pi/2$, the excitation direction is perpendicular to the L-arm, in which the liquid deflection is measured. Theoretically, in this case, the liquid deflection is expected to be zero. However, as shown in Figure 7.25 (c), for $\alpha = \pi/2$, a liquid motion is observed in the prototype due to sloshing effects, which occur in the columns and cause waves on the liquid surface. This effect begins to be noticeable approximately at $\alpha = \pi/4$ and increases with the increasing α. Still, the liquid deflection at $\alpha = \pi/2$ remains marginal. Therefore, the sloshing effect can be neglected and is expected to have only an insignificant influence on the vibration performance of the O-TLCD. In real applications, this effect can be prevented by installing baffles in the columns of the O-TLCD.

7.2.3 Numerical Studies

7.2.3.1 Study 1: Omnidirectional Control Capability

We consider a system consisting of a torsional rigid structure with an O-TLCD as shown in Figure 7.26. The structure is subjected to the ground acceleration $\ddot{s}_g$. Its resulting vibration direction is represented by the DoF s, which is inclined with $\alpha = \pi/6$ from the y direction. The corresponding modal mass, natural frequency and damping ratio of the structure are given in Table 7.7.

To reduce the seismic vibrations of the structure an O-TLCD with 3 L-arms and identical rectangular cross sections is attached to the structure. The dimensions of the damper can be determined using optimization schemes, such as [173], and listed in Table 7.7. Further parameters are depicted in Figure 7.26.

For an efficient vibration control, the natural frequency and inherent damping of the O-TLCD must be tuned to the parameters of the structure. To calculate the optimum natural frequency $\omega_{D,opt}$ and the optimum damping ratio $D_{D,opt}$ the tuning criteria of WARBURTON are employed [55]. These

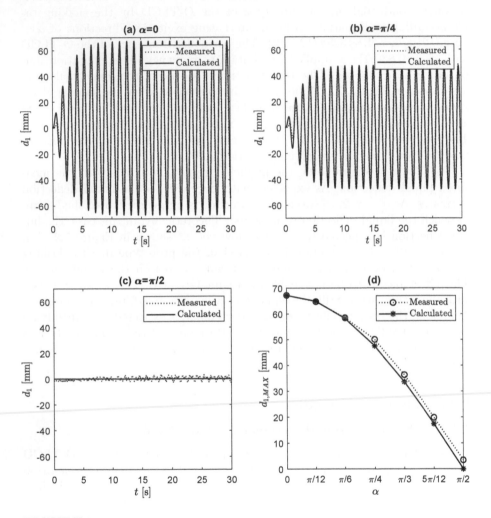

FIGURE 7.25

Measured and calculated liquid deflection response d_1 of the O-TLCD, which is subjected to a harmonic excitation at its resonance frequency with the excitation direction $\alpha = 0$ (a), $\alpha = \pi/4$ (b) and $\alpha = \pi/2$ (c). The measured and calculated peak responses $d_{1,\max}$ for different excitation directions $\alpha = \{0, \pi/12, \pi/6, \pi/4, \pi/3, 5\pi/12, \pi/2\}$ (d).

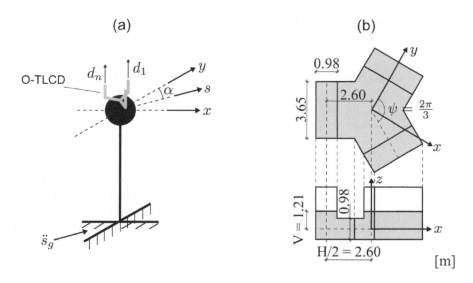

FIGURE 7.26
Structure with an O-TLCD subjected to ground acceleration $\ddot{s}_g$ (a). Resulting vibration response of the structure is in s direction. Liquid deflection in O-TLCD is represented by d_i. Plan and front views of the O-TLCD attached to the structure (b).

TABLE 7.7
Parameters of the O-TLCD and the structure.

Parameter	Value	Unit	Parameter	Value	Unit
H	5.20	m	V	1.21	m
$A_H = A_V$	3.58	m^2			
$L_1 = L_2$	7.62	m	$\gamma_1 = \gamma_2$	0.68	-
m_D	40.9	t	m_S	400.0	t
μ	10.2	%	μ^*	4.5	%
ω_S	1.76	rad s^{-1}	D_S	1.0	%
ω_D	1.60	rad s^{-1}	D_D	17.7	%
$\omega_{D,opt}$	1.62	rad s^{-1}	$D_{D,opt}$	10.5	%

equations are valid for low damped SDoF structures, which are equipped with the TMDs. For the O-TLCD, the WARBURTON criteria can be applied by substituting $d^* = d/\gamma_1$ in the EoMs according to the TMD analogy theory proposed by Hochrainer and Ziegler [110]. The corresponding optimum tuning frequency of the O-TLCD reads

$$\omega_{D,opt} = \omega_S^* \frac{\sqrt{1 - \mu^*/2}}{1 + \mu^*} \quad \text{with} \quad \omega_S^* = \omega_S \sqrt{m_S/m_S^*}, \tag{7.121}$$

where ω_S^* is the natural angular frequency of the structure. This is calculated using the increased structural mass

$$m_S^* = m_S \left(1 + \mu \left(1 - \gamma_1 \gamma_2\right)\right) \tag{7.122}$$

considering the passive portion of the liquid mass of the O-TLCD, which does not contribute to the vibration control. For the investigated O-TLCD, the increased structural mass is equivalent to 421.9 t. The corresponding natural angular frequency is $\omega_S^* = 1.71$ rad s^{-1}. Accordingly, the updated mass ratio reads

$$\mu^* = \frac{\mu \gamma_1 \gamma_2}{1 + \mu(1 - \gamma_1 \gamma_2)}. \tag{7.123}$$

For the investigated O-TLCD, the corresponding optimum tuning frequency $\omega_{D,opt}$ and updated mass ratio μ^* are provided in Table 7.7. The table shows also the natural angular frequency ω_D of the O-TLCD, which is calculated by Equation 7.83 and ω_D matches with $\omega_{D,opt}$. Accordingly, the natural frequency of the damper is tuned optimally.

The optimum damping ratio is calculated again using the updated mass ratio and reads

$$D_{D,opt} = \sqrt{\frac{\mu^*(1 - \mu^*/4)}{4(1 + \mu^*)(1 - \mu^*/2)}}. \tag{7.124}$$

For the investigated O-TLCD, the corresponding optimum damping ratio $D_{D,opt}$ is reported in Table 7.7. To calculate the corresponding head-loss coefficient $\delta_{D,opt}$, Equation 7.87 is applied for the resonance state with $\Omega = \omega_D$ as

$$\delta_{D,opt} = D_{D,opt} \frac{3\pi}{4d_{\max}}, \tag{7.125}$$

which gives $\delta_{D,opt} = 0.34$ m^{-1} for the geometrically maximum possible liquid displacement $d_{\max} = 0.72$ m. However, according to the results of a parametric study using seismic excitation, to prevent any exceedance of the liquid displacement, the head-loss coefficient is chosen as $\delta = 0.58$ m^{-1}. This is higher than $\delta_{D,opt}$ and corresponds to the damping ratio D_D given in Table 7.7. It is assumed that the damper together with its orifice is designed in such a way that it can produce the required damping.

The numerical study is conducted in Simulink/Matlab by implementing the corresponding EoMs of the structure and the O-TLCD. The DORMAND-PRINCE explicit time integration method is applied to solve the EoMs [178].

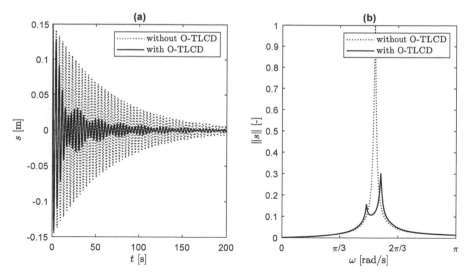

FIGURE 7.27
Free displacement response of the structure without and with O-TLCD (a).
Corresponding frequency content (b).

In the first part of the study, the free vibration response of the structure is investigated without and with the O-TLCD to check the tuning effect of its parameters. The time increment used in this investigation is 0.01 s. The free vibration is induced by applying an initial displacement of 0.15 m to the structure. The displacement time history and the corresponding frequency content are shown in Figure 7.27 (a) and (b).

With the O-TLCD, the displacement response dissipates quickly with a strong decay indicating the high control performance of the O-TLCD. This effect can be observed also from responses in the frequency-domain. The reduction of the norm value of the displacement is calculated as $\|x\| = x/x_{max}$, where x_{max} is the maximum displacement of the structure without O-TLCD. The two neighboring frequency response peaks arise due to the fact that O-TLCD with the structure ensembles a 2-DoF system. These peaks reach different displacement levels due to the slight difference between the optimum natural frequency $\omega_{D,opt} = 1.62$ rad s^{-1} and the real natural frequency $\omega_D = 1.60$ rad s^{-1} of the O-TLCD. The effects of the difference between the real damping ratio $D_D = 17.7$ % of the O-TLCD and its optimum damping ratio $D_{D,opt} = 10.5$ % can be observed from the low reduction of the response at resonance frequency region compared to the two neighboring frequency response peaks. For $D_D = D_{D,opt}$, the peaks would be narrower. However, as previously mentioned, high inherent damping keeps the liquid motion within the vertical column.

In the second part of the study, the seismic response of the structure is

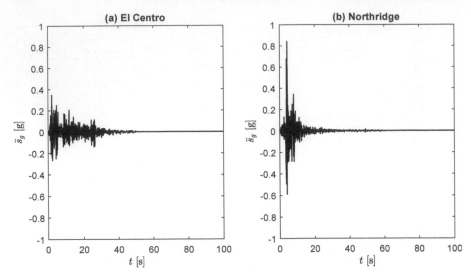

FIGURE 7.28
Time histories of the historic El Centro (a) and Northridge (b) earthquakes used in the numerical studies.

investigated without and with O-TLCD. Two historic California earthquakes are applied to the structure: The far-field earthquake El Centro and the near-field earthquake Northridge. The time histories of the earthquakes are shown in Figure 7.28. The properties of the earthquake records are listed below [179].

- *El Centro*: N-S component recorded at the Imperial Valley Irrigation District substation in El Centro, during the Imperial Valley, California Earthquake of 1940. RMS (Root mean square) = 0.05 g.

- *Northridge*: N-S component recorded at the County Hospital parking lot in Sylmar, during the Northridge, California Earthquake of 1994. RMS = 0.07 g.

The time increment used in this investigation is $\Delta t = 0.02$ s. The calculated time histories and the frequency content of the displacement response of the structure are depicted in Figure 7.29 (a) and (b). The corresponding liquid deflections in the columns of the O-TLCD are shown in Figure 7.29 (c).

From the time histories of the displacement response, the vibration control effect of the O-TLCD can be observed clearly both for the El Centro and for the Northridge earthquakes. The reduction values of the RMS and MAX (maximum) of the displacement response are calculated as

$$R_{\text{RMS}} = 1 - \frac{\text{RMS}_{s,\text{w O-TLCD}}}{\text{RMS}_{s,\text{wo O-TLCD}}} \quad \text{and} \quad R_{\text{MAX}} = 1 - \frac{\text{MAX}_{s,\text{w O-TLCD}}}{\text{MAX}_{s,\text{wo O-TLCD}}}, \quad (7.126)$$

and reported in Table 7.8. Accordingly, the RMS value is reduced during

TABLE 7.8

RMS and MAX values of the displacement response of the structure without and with O-TLCD during the El Centro and Northridge earthquakes.

	RMS	MAX	Unit
El Centro:			
without O-TLCD	0.061	0.248	m
with O-TLCD	0.036	0.215	m
Reduction	41.1	13.5	%
Northridge:			
without O-TLCD	0.193	0.718	m
with O-TLCD	0.106	0.640	m
Reduction	45.5	10.8	%

the El Centro earthquake by 41.4 % and during the Northridge earthquake by 45.5 %. The MAX value is reduced during the El Centro earthquake by 13.5 % and during the Northridge earthquake by 10.8 %. We remember that the mass ratio of the O-TLCD is $\mu = 10.2$ % with an active mass portion of $\mu^* = 4.5$ %. By increasing the mass ratio, a higher reduction for both the RMS and MAX values can be reached. The control effect is also shown by the frequency content.

From the time histories of the liquid deflection, it is observed that the maximum possible liquid deflection $d_{max} = 0.62 < 0.72$ m is not exceeded. It is noteworthy to mention that depending on the vibration direction of the structure the liquid deflection changes. This effect is clearly observed for d_2. For $\alpha = \pi/6$, the vibration direction is $\pi/2$ inclined from the second L-arm. Consequently, the liquid deflection in the corresponding column is $d_2 = 0$. For $\alpha = 0$, the maximum liquid deflection occurs in the first column with $\theta_1 = 0$ as $d_{max} = 0.62/\cos(\pi/6) = 0.71$ m (cf. Equation 7.70), which is still lower than the maximum possible liquid deflection.

7.2.3.2 Study 2: Semi-Active Control Capability

In Study 2, similar to Study 1, we consider again a torsional rigid structure. However, in contrast to Study 1, 20 semi-active O-TLCDs are installed on the structure as shown in Figure 7.30. Each of three columns of the dampers features closable cells allowing the damper to regulate the cross-sectional areas of its columns. Accordingly, the damper can adapt its natural frequency. Furthermore, the semi-active O-TLCD utilizes closable orifices to modulate its inherent damping. Analogous to Study 1, the structure is subjected to the free vibration and the ground acceleration $\ddot{s}_g$ in $\alpha = \pi/6$ direction.

The relevant parameters of the O-TLCD are listed in Table 7.9 together with the parameters of the structure. Further parameters are shown in

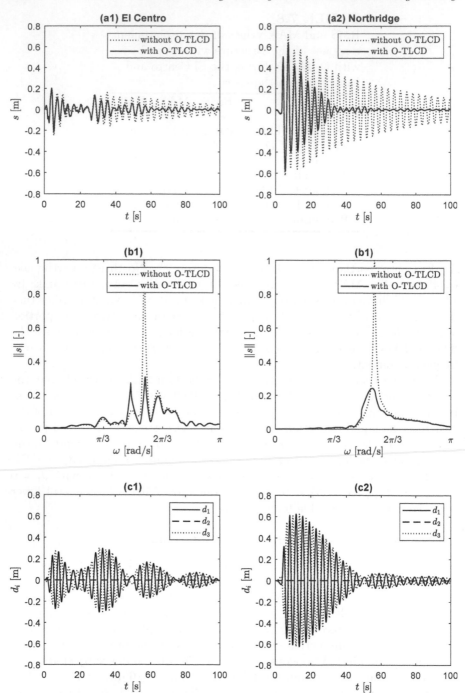

FIGURE 7.29

Seismic displacement response of the structure without and with O-TLCD during the El Centro (a1) and Northridge (a2) earthquakes. Corresponding frequency content (b1), (b2) and the liquid displacement in the columns of the O-TLCD (c1), (c2).

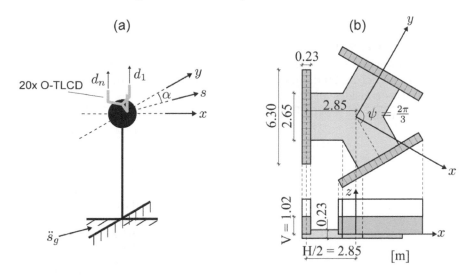

FIGURE 7.30
Structure with semi-active O-TLCDs subjected to ground acceleration $\ddot{s}_g$ (a).
Resulting vibration response of the structure is in s direction. Liquid deflection
in O-TLCD is represented by d_i. Layout of the semi-active O-TLCD (b).

Figure 7.30. The initial layout of the O-TLCD can be determined again by
optimization schemes, such as [173].

As it can be observed from Table 7.9, the semi-active O-TLCD can change
the cross-sectional area of its columns by sequentially closing the cells within
the range of an initial area of $A_{V,i} = 0.23$ m^2 and a final area of $A_{V,f} =
1.45$ m^2. This allows the damper to change its natural angular frequency
within the range of $\omega_D = 1.13$–2.17 rad s^{-1} as shown in Figure 7.31. With
this property, the damper can adapt its natural frequency and avoid any off-
tuning effects.

However, the frequency adaptation capability brings also some challenges
for the control, which are noteworthy and should be considered during the
design of the damper. In closed cells, the liquid mass kept in the cell cannot contribute to vibration control as previously shown in Figure 7.17. This
causes a reduction of the damper mass m_D, which is participating in vibration
control, and an increase of the structural mass m_S, which is to be controlled.
Consequently, both the initial natural angular frequency ω_S of the structure
and the mass ratio μ change by changing the cross-section area A_V as shown
in Figure 7.32 (a). Furthermore, by changing ratio of A_V/A_{II}, the geometric
factors γ_1 and γ_2 change as shown in Figure 7.32 (b), cf. Equation 7.83. Both
the change in μ and in $\gamma_{1,2}$ affect the updated mass ratio μ^* as shown in Figure 7.32 (a), cf. Equation 7.123. This may cause a variation of the restoring

TABLE 7.9

Parameters of the semi-active O-TLCDs and the structure. On the structure 20 dampers are installed. The mass and mass ratios consider all dampers.

Parameter	Value	Unit	Parameter	Value	Unit
H	5.70	m	V	1.02	m
A_H	0.62	m^2			
$A_{V,i}$	0.23	m^2	$A_{V,f}$	1.45	m^2
$L_{1,i}$	4.17	m	$L_{1,f}$	15.43	m
$L_{2,i}$	17.33	m	$L_{2,f}$	4.47	m
$\gamma_{1,i}$	1.37	-	$\gamma_{1,f}$	0.37	-
$\gamma_{2,i}$	0.33	-	$\gamma_{2,f}$	1.28	-
$m_{D,i}$	119.6	t	$m_{D,f}$	194.2	t
$m_{S,i}$	978.1	t	$m_{S,f}$	903.5	t
μ_i	12.2	%	μ_f	21.5	%
μ_i^*	5.2	%	μ_f^*	9.1	%
ω_S	1.86	rad s^{-1}	D_s	1.00	%
$\omega_{D,i}$	2.17	rad s^{-1}	$\omega_{D,f}$	1.13	rad s^{-1}
λ	13.61				

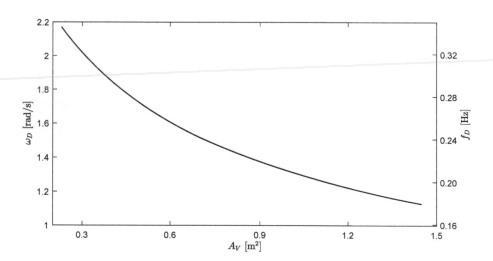

FIGURE 7.31

The frequency (ω_D, f_D) of the semi-active O-TLCD over the cross-sectional area A_V.

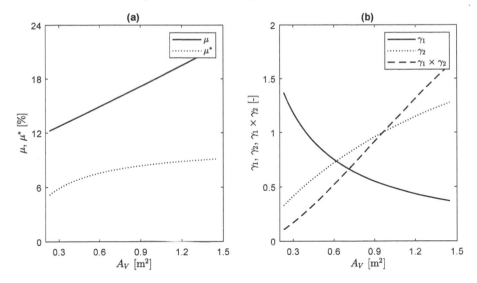

FIGURE 7.32
The mass ratio μ and the updated mass ratio μ^* (a) as well as the geometric factors γ_1 and γ_2 (b) over the cross-sectional area A_V of the columns.

forces as well as the liquid deflection of the damper as previously observed with the uniaxial S-TLCD [122], cf. Equations 7.82 and 7.91 as well as Section 7.1.

Both the change in updated mass ratio μ^* and the geometric factors $\gamma_{1,2}$ influence the optimum damping ratio $D_{D,opt}$ as shown in Figure 7.33 (a), cf. Equation 7.124. Accordingly, the inherent damping may require a compensation of the change via orifices. Finally μ^* and $\gamma_{1,2}$ influence also the optimum natural angular frequency $\omega_{D,opt}$ as shown in Figure 7.33 (b) for $\omega_S = 1.86$ rad s^{-1}, cf. Equation 7.121.

Finally, the natural angular frequency ω_S of the structure can also change by external and internal effects, such as caused by degradation effects, and affect $\omega_{D,opt}$, apart from the effects of the changing cross-sectional area A_V.

In Figure 7.34 (a), the cumulative effects on $\omega_{D,opt}$ and ω_D are plotted over the changing structural frequency ω_S and the cross-sectional area A_V. We observe that the depicted surfaces belonging to ω_D and $\omega_{D,opt}$ are intersecting. Accordingly, an appropriate choice of A_V can allow the damper to realize a natural frequency ω_D, which matches with the optimum frequency $\omega_{D,opt}$. As shown in Figure 7.34 (b), there is a PARETO curve, on which $\omega_{D,opt} = \omega_D$. Accordingly, in case of a change in ω_S, the semi-active O-TLCD can re-tune its natural frequency by regulating its A_V.

The closed-form solution of the best A_V configuration for an optimum frequency tuning can be derived for the WARBURTON tuning criteria [55] from

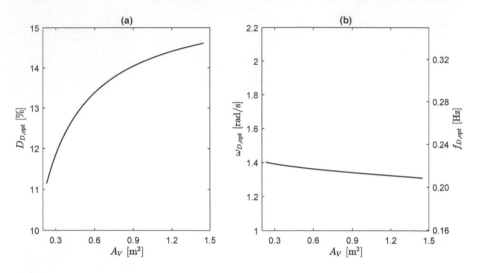

FIGURE 7.33

The optimum damping ratio $D_{D,opt}$ (a) and the optimum natural angular frequency $\omega_{D,opt}$ (b) over the cross-sectional area A_V of the columns for $\omega_S = 1.86$ rad s^{-1}.

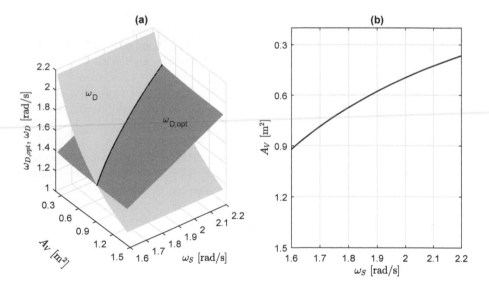

FIGURE 7.34

The optimum natural angular frequency $\omega_{D,opt}$ and the natural frequency of the damper ω_D over the cross-sectional area A_V and the natural angular frequency of the structure ω_S, which is changing due to external and internal effects, such as caused by degradation (a). PARETO curve, on which ω_D matches $\omega_{D,opt}$.

the equation

$$\omega_{D,opt} \overset{!}{=} \omega_D \tag{7.127}$$

$$\Leftrightarrow \omega_{D,opt} = \omega_S^* \frac{\sqrt{1 - \mu^*/2}}{1 + \mu^*} \overset{!}{=} \omega_D = \sqrt{\frac{2g}{L_1}}. \tag{7.128}$$

The updated natural angular frequency ω_S^* of the structure and the updated mass ratio μ^* consider the influence the mass portion of the damper, which is not participating in vibration control, as

$$\omega_S^* = \omega_S \sqrt{m_S/m_S^*} \quad , \quad \omega_S = \omega_{S,0} \sqrt{m_{S,0}/m_S}, \tag{7.129}$$

$$m_S = (m_{S,0} + m_{D,\max}) - m_D \quad , \quad m_S^* = m_S(1 + \mu)(1 - \gamma_1\gamma_2), \tag{7.130}$$

$$\mu = m_D/m_S \quad , \quad \mu^* = \frac{\mu\gamma_1\gamma_2}{1 + \mu(1 - \gamma_1\gamma_2)}. \tag{7.131}$$

Here, $\omega_{S,0}$ and $m_{S,0}$ are the natural angular frequency and the modal mass of the structure without the O-TLCDs and $m_{D,\max} = m_{D,f}$ is the maximum damper mass for $A_{V,f}$. The solution for the optimum A_V can be determined after manipulating Equation 7.128 and reads

$$\Leftrightarrow A_{V,opt} = \frac{8A_H m_{S+D}(-Vk_S + gm_{S+D})}{Hk_S(4m_{D+S} - 3A_H Hnk\rho)}, \tag{7.132}$$

where k, n and ρ are the number of dampers, the number of columns and the density of the damper fluid, respectively. Furthermore, m_{S+D} is the modal mass of the structure and the mass of the damper and k_S is the stiffness of the structure, which are given by

$$m_{S+D} = m_{S,0} + m_{D,\max}, \quad k_S = m_{S,0}\omega_{S,0}^2. \tag{7.133}$$

For the O-TLCDs and the structure of this numerical study, considering their parameters from Table 7.9, the $\omega_{S,0}$ dependent function of the optimum cross-sectional area reads

$$A_{V,opt}(\omega_{S,0}) = \frac{3.015}{\omega_{S,0}^2} - 0.258, \tag{7.134}$$

which corresponds to the PARETO curve shown in Figure 7.34 (b). In the equation, $A_{V,opt}$ and $\omega_{S,0}$ are in [m^2] and [rad s^{-1}], respectively.

Analogous to Study 1, the free vibration and the seismic responses of the structure are investigated. The free vibration is induced by an initial displacement of 0.30 m. For the seismic response analysis, the time history of the historic El Centro earthquake is used. Furthermore, the time history of the Kobe earthquake is used. The time histories of the earthquakes are shown in Figure 7.35. The properties of both records are listed below [179].

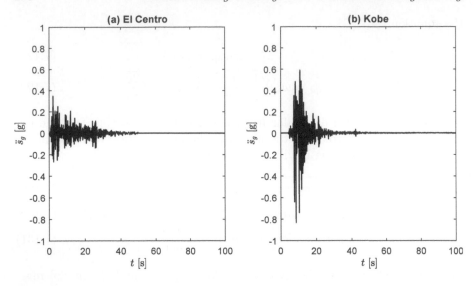

FIGURE 7.35
Time histories of the historic El Centro (a) and Kobe (b) earthquakes used in the numerical studies.

- *El Centro*: N-S component recorded at the Imperial Valley Irrigation District substation in El Centro, during the Imperial Valley, California Earthquake of 1940 with a PGA of 0.05 g.

- *Kobe*: N-S component recorded at the Kobe Japanese Meteorological Agency station, during the Hyogoken Nanbu Earthquake of 1995 with a PGA of 0.06 g.

The conducted case studies involve the uncontrolled structure, passively controlled structure with O-TLCDs and semi-actively controlled structure with O-TLCDs. For the semi-active control, two control approaches are investigated: (1) Frequency control (**O-TLCD 1**) and (2) Frequency+damping control (**O-TLCD 2**).

The natural frequencies of both the O-TLCD 1 and the O-TLCD 2 are tuned to the optimum frequency by regulating the cross-sectional area A_V according to the closed-form solution presented in Equation 7.134.

For the O-TLCD 1, the inherent damping is determined according to the same approach used in Study 1. For the initial A_V configuration, the head-loss coefficient is determined as $\delta_i = 1.63$ m^{-1}, which corresponds to $\lambda = 13.61$ with $L_{1,i} = 4.17$ m according to Equation 7.83, from a parametric study using free and seismic vibrations. Consequently, at final A_V configuration, the head-loss coefficient corresponds to $\delta_f = 0.44$ m^{-1} with $L_{1,f} = 15.43$ m. This head-loss coefficient limits the liquid motion matching with the maximum possible value of $d_{\max} = 0.90$ m. As shown in Figure 7.33, the required optimum damping ratio $D_{D,opt}$ varies for both A_V configurations between 11.5–14.6 %. These correspond to an optimum head-loss coefficient between $\delta_{D,opt} = 0.27$– 0.34 m^{-1}, which is lower than the chosen $\delta_{i,f}$ values. Accordingly, in the columns, any exceedance of the liquid displacement will be prevented. However, it is noteworthy that a minor performance loss is expected due to the deviation of the inherent damping from the optimum damping.

For the O-TLCD 2, the inherent damping is controlled by the displacement based *on-off groundhook controller*, cf. Section 3.3.2. Accordingly, corresponding to Equations 3.43 and 3.44, the loss factor of the O-TLCD is regulated as

$$s\dot{d} \leq 0 : \quad \lambda = \lambda_{\max}, \tag{7.135}$$

$$s\dot{d} > 0 : \quad \lambda = \lambda_{\min}, \tag{7.136}$$

where s and d are the DoFs of the structure and the O-TLCD, respectively. To reach an on-off behavior, $\lambda_{\min} = 2$ and $\lambda_{\max} = 200$ are chosen from a conducted parametric study on the O-TLCD response with different loss factors.

The response of the structure is computed, as in Study 1, using the DORMAND-PRINCE explicit time integration method in the Matlab/Simulink environment. The time increment is 0.01 s for the free vibration and 0.02 s for the seismic analyses.

In all studies, the natural frequency of the structure varies during the excitation event corresponding to Figure 7.36 (a). The frequency change is assumed to be induced by external effects. The frequency change can be identified by observers. An adaptive joint-state parameter observer is presented in Chapter 9.

Both the O-TLCD 1 and the O-TLCD 2 re-tune themselves by adapting their A_V as shown in Figure 7.36 (a,b). The corresponding head-loss coefficient of the O-TLCD 1 is shown in Figure 7.36 (c). The change of the head-loss coefficient of the O-TLCD 2 is shown in Figure 7.36 (d) for the Kobe earthquake.

The displacement time histories of the structure are shown in Figure 7.37. The responses in frequency-domain are depicted in Figure 7.38. The corresponding liquid motion is shown in Figure 7.39. The reduction of the RMS and MAX of the displacement response of the structure are reported in Table 7.10.

The results show that the displacement response of the structure can be controlled with the semi-active O-TLCD in a significantly more robust manner. A clear example is the El Centro earthquake, during which the vibration response of the structure is even increased by the passive O-TLCD due to the frequency de-tuning effects. The deterioration of the RMS and MAX are computed as -49.7 and -10.4 %, respectively.

With the semi-active O-TLCDs, the highest RMS reduction is observed during the Kobe earthquake by 71.1 % with the O-TLCD 2. The highest

TABLE 7.10

Reduction of RMS and MAX of the displacement responses of the
structure with the passive O-TLCDs as well as with the
semi-active O-TLCDs using the frequency control (O-TLCD 1)
and using the frequency+damping control (O-TLCD 2).

		Passive	Semi-active	
		O-TLCD	O-TLCD 1	O-TLCD 2
Free vibration	RMS [%]	37.6	55.1	63.2
Kobe	RMS [%]	50.0	63.6	71.1
	MAX [%]	36.3	39.4	39.4
El Centro	RMS [%]	−49.7	11.6	24.6
	MAX [%]	−10.4	17.7	18.6

MAX reduction is observed also during the Kobe earthquake by 39.4 % again
with the O-TLCD 2. Both the RMS and MAX reduction of the O-TLCD 1
is lower than the O-TLCD 2. This comparison shows that, besides frequency
tuning, the control effect of the semi-active O-TLCD can be further increased
by regulating its inherent damping.

The superior performance of the semi-active O-TLCDs can be observed
also from the frequency spectra in Figure 7.38. Particularly, the O-TLCD 2
shows a consistent control effect in a large frequency range.

From the comparison of the liquid motion, we observe in Figure 7.39 a
higher activation of the liquid in case of the O-TLCD 2. In all cases, the
maximum liquid deflection is lower than the maximum possible value of
$d_{\mathrm{max}} = 0.90$ m.

7.3   Conclusion

In this section, two semi-active TLCDs are introduced, which can adapt their
natural frequency as well as inherent damping. The first damper operates
in uniaxial direction and referred to as S-TLCD. The S-TLCD comprises a
U-shape tank, which is partially filled with a Newtonian fluid. The S-TLCD
provides mechanisms for the real-time continuous tuning of both natural frequency and damping parameters. The governing equations of the S-TLCD are
derived based on the non-stationary incompressible liquid flow. The EoM is
formulated first for an SDoF system with single S-TLCD. The natural frequency corresponds to a scaled liquid length. The scaling factors are derived
from the tank geometry. Further geometric parameters, which influence the
efficiency of the S-TLCD, are formulated. The interaction force between the
S-TLCD and the structure is derived from the equation of momentum, which
is induced by the liquid motion. For MDoF systems with several S-TLCDs, the

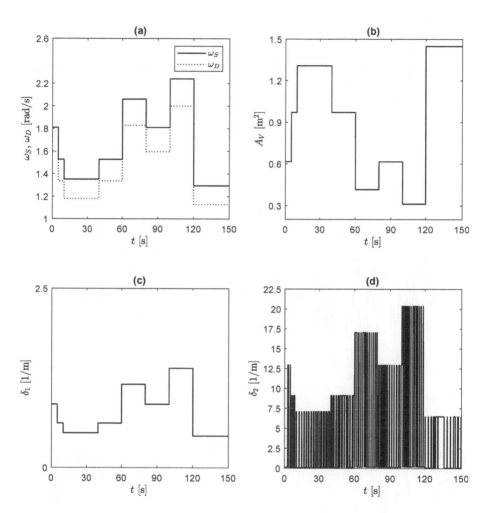

FIGURE 7.36
The change of the angular natural frequency ω_S of the structure and re-tuning of the angular natural frequency ω_D of the O-TLCD 1 and 2 (a). The change of the cross-sectional area A_V for the re-tuning of the damper to the corresponding optimum frequency (b). The head-loss coefficient δ of the O-TLCD 1 due to the change of the cross-sectional area A_V (c). The head-loss coefficient of the O-TLCD 2 during Kobe earthquake regulated by the displacement based on-off groundhook controller (d).

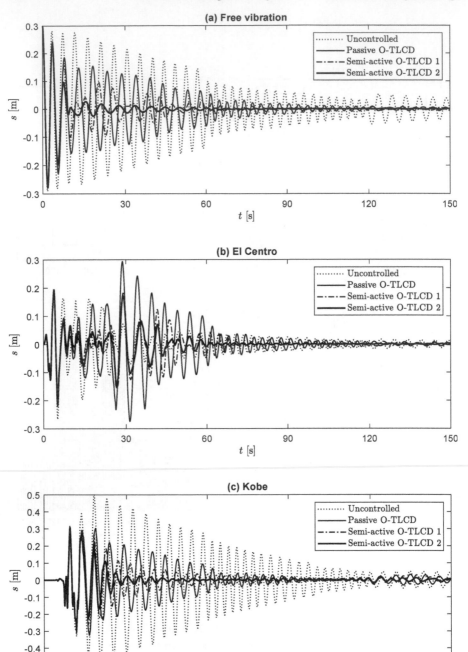

FIGURE 7.37
Time histories of the free and seismic displacement responses of the uncontrolled structure, structure with the passive O-TLCDs as well as with the semi-active O-TLCDs using frequency control (O-TLCD 1) and using the frequency+damping control (O-TLCD 2).

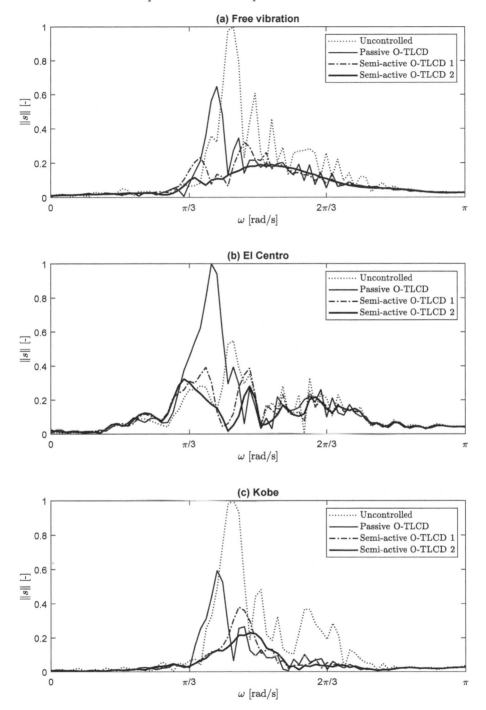

FIGURE 7.38
Frequency spectra of the free and seismic displacement responses of the uncontrolled structure, structure with the passive O-TLCDs as well as with the semi-active O-TLCDs using frequency control (O-TLCD 1) and using the frequency+damping control (O-TLCD 2).

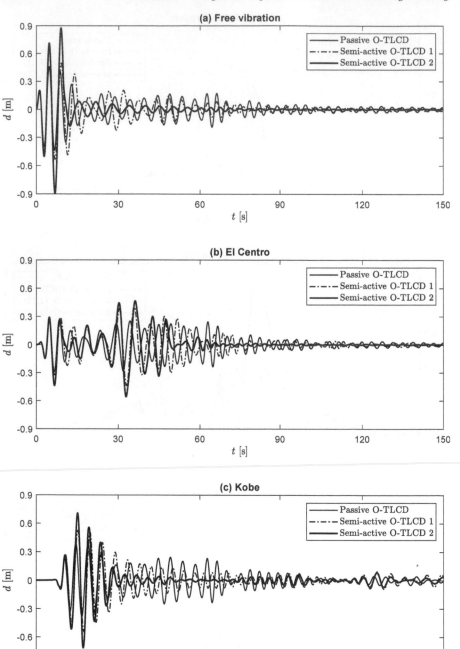

FIGURE 7.39

Time histories of the liquid motion of the passive O-TLCDs as well as the semi-active O-TLCDs using the frequency control (O-TLCD 1) and using the frequency+damping control (O-TLCD 2) during the free and seismic vibration of the structure.

state-space representation is formulated allowing the simulation of S-TLCDs with MDoF systems including sensors and control algorithm.

The natural frequency equation of the S-TLCD is validated by experimental studies using a prototype of the damper, which is attached on a shaking table. Further experiments are conducted to investigate the control performance of the damper. The results show that the natural frequency of the S-TLCD can be controlled in the intended manner. Effects on the inherent damping of the S-TLCD are investigated. The results show that the damping can also be controlled in the expected manner. Further experimental investigations are performed to study the effects influencing the control performance of the S-TLCD. For this investigation, the S-TLCD is attached to a model structure, which is excited by the shaking table. The results show that the tank geometry influences directly the efficiency of the damper. Hereby the most relevant aspect is turns out to be the natural frequency tuning. During the experimental tests, the natural frequency adaptation enabled the S-TLCD a performance improvement of over 77%.

The second semi-active TLCD operates omnidirectionally and referred to as O-TLCD. This damper consists of $n \geq 3$ columns, which are point symmetrically distributed around an origin and connected with each other via horizontal passages. The layout allows the damper to control translational vibrations of structures in a multidirectional manner. The natural frequency and the damping behavior of the damper can be adapted in real-time corresponding to the requirements of the structure. The mathematical modeling approaches are derived for the SDoF as well as the MDoF structures incorporating semi-active O-TLCDs. The EoMs are determined and formulated in state-space representation. Experimental studies are presented validating the operational capabilities of the damper. The investigations include frequency and response analyses by shaking table tests. Furthermore, the efficiency of the damper is investigated by numerical studies concerning various load cases including free, harmonic and seismic loading. The responses of the controlled structure is analyzed both in frequency and time-domain. The results are compared both to conventional TLCDs and passive control strategies showing the superior performance of the proposed damper.

General conclusions of the studies regarding the semi-active TLCDs are:

- General EoM formulations are derived for both semi-active TLCDs, which cover also the classical TLCDs.

- A reduction method is introduced to represent the O-TLCD with n-columns as an SDoF system.

- Conventional TLCDs operate primarily in one direction and lose their efficiency with changing excitation direction. However, the O-TLCD controls vibrations in every translational direction equally.

- The natural frequency of the O-TLCD is independent of the excitation direction.

- Semi-active adaptation capabilities of both dampers increase their control performance significantly compared to passive strategies. The dampers can control structures efficiently even after abrupt changes without any off-tuning effects.

8

Damping Systems Using Shape Memory Alloys

8.1 Introduction

Smart materials exhibit unique characteristics, which can also be controlled via both mechanical deformation and environmental conditions, such as temperature, magnetic and electric fields as well as stress. Because of this unique ability, smart materials play an important role in the development of new types of structural control approaches. In this regard, piezoelectric ceramics, MR and ER liquids have been studied so far. Furthermore, particularly for structural engineering, SMAs attract attention due to their superior mechanical qualities.

SMAs are metallic two-phase polycrystal alloys, such as NiTi (Nickel-titanium, Nitinol) and CuAlZn (Copper-aluminum-zinc), which show pseudoelastic material behavior due to their distinctive memory and superelasticity abilities. Below the so-called martensitic finish temperature M_f, the memory effect can be triggered by external heating, which enables deformed SMAs to remember and recover their original shape. On the other hand, above the austenite finish temperature A_f, the superelasticity enables SMAs to recover deformations just upon unloading without any external heating. During cyclic loading superelastic SMAs create closed hysteresis loops and convert vibration energy to heat energy by phase transformation.

Conventional steel and lead metallic dampers achieve energy dissipation through inelastic deformation, cf. Section 4.2. The deformation path of these devices ends up with a non-recoverable plastic deformation. Therefore, after each strong vibration, such as seismic events, these devices need to be replaced. However, the superelastic behavior allows SMAs to recover large deformations with even over 8 % strain and, therefore, damping systems incorporating SMAs count as an attractive alternative to the existing vibration mitigation systems. Apart from their capability to recover large deformations, SMAs exhibit also excellent fatigue and corrosion resistance.

Numerous application examples of SMA-based passive dampers have been already accomplished. Due to their unique mechanical properties, SMAs are expected to play a decisive role also in the development of new types of maintenance-free dampers and semi-active systems. In particular, the uniaxially loaded SMA wires seem to be promising due to their superior damping performance resulting from the efficient temperature release. However, the computation of the SMA response evokes some challenges and impairs the design process. In particular, the dynamic effects governing the damping behavior are not fully covered yet. Such effects occur not only during cyclic but also during random loading, such as earthquakes, which involves highly alternating loading characteristics and poses a challenge for the numerical modeling. This chapter introduces the relevant dynamic effects for SMA based vibration control systems and proposes improvements for existing modeling approaches to cover these effects both under cyclic and random loading.

Section 8.2 presents examples of damping systems incorporating SMAs. Section 8.3 introduces the superelastic material behavior. Section 8.4 derives the required mathematical formulations for the improvement of the modeling approaches. Sections 8.6, 8.5 and 8.7 present results of the numerical, experimental and hybrid investigations of the respective improvements. Section 8.8 concludes the chapter.

8.2   Application Examples

After the first large-scale SMA applications in aerodynamics in 1970s, such as in aircrafts, followed by further applications in medical sector, such as in orthodontia, SMAs count nowadays, at least in these fields, as a state-of-the-art technology. Most of these applications use the superelasticity and some are based also on the shape memory effect. In structural engineering, a significant part of the research focuses on the development of dampers using the superelastic deformation induced hysteretic damping.

Pioneering studies on the structural application of SMAs have been accomplished in 1990s by Aiken et al. at the Pacific Earthquake Engineering Research Center (PEER) headquartered at the University of California in Berkeley. These were based on a model structure, which was equipped with SMA based dampers [180]. Thereafter, in the framework of the research project MANSIDE (Memory Alloys for New Seismic Isolation and Energy Dissipation Devices), a damping system for buildings and bridges has been developed by Dolce et al. [181] using SMA wires. A further research project ISTECH (Shape Memory Alloy Devices for Seismic Protection of Cultural Heritage Structures) focused on the development of SMA dampers for historical structures as reported by Indirli and Castellano [182] and involved full-scale applications on real structures, such as earthquake resistant connection of the historic gable and the main structure of the Basilica San Francesco and the bell tower of the Church of San Giorgio in Italy. For the horizontal connections, the developed approach combines steel ties with devices incorporating several thin Nitinol wires. For the retrofitting of slender tower structures, vertical steel bars are

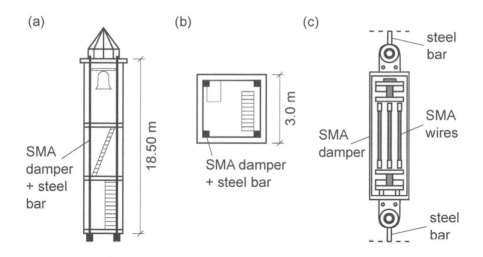

FIGURE 8.1
Application of SMA wires on slender structures. Approach developed during the ISTECH research project. Section (a) and plan (b) view of a bell tower are shown. Steel bars are combined with devices (c) incorporating SMA wires.

combined with SMA wires. The steel bars are prestressed and anchored to the building top and foundation as shown in Figure 8.1.

Zhang and Zhu developed another novel damper by combining SMA wires with a friction damper, which has also a semi-active adaptation character [183]. Li et al. conducted shaking table tests on a reduced-scale five-story building model to investigate the efficiency of two further SMA wire-based dampers [184]. Furthermore, Boroschek et al. performed shaking table tests on a three-story steel frame model retrofitted by copper-based SMA wires [185]. Ozbulut et al. tested the performance of SMA bracing elements on a steel frame model structure with three floors and four columns [186]. Dolce et al. applied reduced-scale shaking table tests to evaluate a SMA wire damping system consisting of energy dissipating steel braces combined with re-centering SMA braces [187]. Moreover, Dolce et al. tested a reduced-scale RC frame structure with a base isolation system incorporating SMAs [188]. Johnson et al. conducted further shaking table tests on SMA restrainers on reduced-scale bridges [189].

Besides these pioneering studies, more recent investigations have been performed by numerous researchers, such as Li et al. [190] on dampers with deformation amplification of SMA wires, Qiu and Zhu [191] on braces using SMA wires, Fang et al. [192] on dampers with annealed SMA cables as well as Issa and Alam [193] on SMA bars for frame structures. Furthermore, Torra et al. [194] and Casciati et al. [195] conducted experimental investigations on bridges using SMA wires.

A detailed review of application examples can be found in the literature, such as by Ozbulut et al. [196] as well as Fang and Wang [197].

8.3 Superelastic Material Behavior

Above the austenite finish temperature A_f, the parent phase of the SMAs is austenite, which has a body-centered cubic crystal structure (B2) and is crystallographically more ordered. The less-ordered low temperature phase martensite is the product phase and has a monoclinic crystal structure (B19'). A mechanical stress-induced deformation initiates a phase transformation, during which the martensite lattice is formed. This new structure is more stable for high stress values. Due to unloading, SMA returns back to the austenite phase, which is more stable for low stress values above the A_f temperature.

The effects of the phase transformation on the material behavior of the SMAs can be observed from the stress σ and strain ε responses as shown in Figure 8.2 (a). Here, σ_s^{AM} and σ_f^{AM} are the stress levels, at which the forward austenite-martensite transformation starts and finishes. Correspondingly, σ_s^{MA} and σ_f^{MA} are the stress levels, at which the reverse martensite-austenite transformation starts and finishes. At (1), the SMA is in its parent phase — austenite. At (2), the phase transformation is completed and the SMA is in its martensite phase. During the forward phase transformation, martensite bands nucleate and spread in the SMA. The material deforms under constant stress. Accordingly, the stress-strain curve forms a stress plateau. During the reverse transformation, martensite bands revert to austenite. The SMA recovers its deformation without changing its stress and a lower stress plateau is formed in the stress-strain curve.

In Figure 8.2 (b), the change of the crystal structure is shown corresponding to the austenite and martensite phases. At (1), corresponding to the austenite phase, the more-ordered body centered cubic crystal structure is shown. At (2), corresponding to the martensite phase, the detwinned monoclinic crystal structure is shown.

Figure 8.3 illustrates the phase transformation zones for SMAs. M_f, M_s, A_s and A_f represent the start and finish temperatures of the transformation between austenite and martensite phases. Austenite is the stable phase at high temperatures. Conversely, martensite is the stable phase at low temperatures. Furthermore, the loading and unloading patterns are depicted for a superelastic SMA above the austenite finish temperature A_f. The figure shows also the phase transformation start and finish stress levels, which correspond to the intersection of corresponding zone boundary with the material temperature. Accordingly, the critical stress levels depend on the material temperature and if the material temperature changes, the critical stress levels change as well. This aspect describes the thermomechanically coupled material behavior

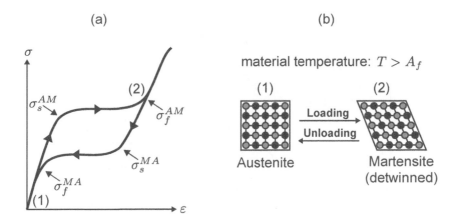

FIGURE 8.2
Stress-strain response of a superelastic SMA to a mechanical stress-induced deformation (a). Critical stress levels of the phase transformations σ_s^{AM}, σ_f^{AM}, σ_s^{MA} and σ_f^{MA} are shown. Change of the crystal lattice structure of the superelastic SMA (b). In both (a) and (b), (1) and (2) correspond to austenite and martensite phases, respectively.

of SMAs and is characteristic particularly for their response under dynamic loading.

During the forward austenite-martensite phase transformation, SMAs generate heat, which is proportional to the volume fraction of martensite and must be released to the environment. Accordingly, the forward transformation is an exothermic process. During the reverse martensite-austenite transformation, SMAs absorb heat. Accordingly, the reverse transformation is an endothermic process.

At high strain rates, during the forward transformation, the available time span for heat release shortens. SMAs start to keep most of the generated heat, which increases their material temperature. Consequently, the high temperature phase (austenite) becomes energetically more stable. This development delays the forward austenite-martensite transformation. The critical stress level σ_f^{AM} increases and causes an increase in the forward transformation slope of the stress-strain response.

On the other hand, during the reverse martensite-austenite transformation, SMAs try to return to the stable austenite phase as quickly as possible due to the high strain rates and the associated high latent heat. The critical stress level σ_s^{MA} increases and causes an increase in the reverse transformation slope. As a result, hysteresis loop reduces, which corresponds to a reduction of the hysteretic damping and, consequently, the energy dissipation performance of the SMA deteriorates.

For SMA applications under dynamic loading, particularly for SMAs

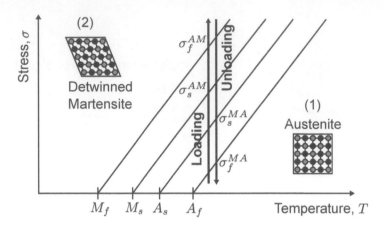

FIGURE 8.3
The temperature dependency of the critical phase transformation stress levels. The shown loading path of a superelastic SMA above the austenite finish temperature A_f.

incorporated in vibration control systems, the strain rate dependency of the material behavior must be considered in modeling. Accordingly, mathematical models used for the design of such superelastic SMA applications must be able to reflect these effects accurately.

8.4   Mathematical Modeling

For the computation of the superelastic SMA responses micro, micro-macro and macro models have been proposed. Microscopic approaches describe the material behavior at microstructural levels. These models are based on the Ginzburg-Landau theory or molecular dynamics. Studies proposed by Falk [198] as well as Foiles and Daw [199] are the earliest examples of microscopic approaches.

Micro-macro approaches describe the SMA behavior at micro and meso levels by using micromechanical and micro-plane/micro-sphere models. The macroscopic response is then computed by a scale transition. Some of the earliest examples in this category are the models proposed by Delaey et al. [200], Govindjee and Hall [201] and Brocca et al. [202].

Both microscopic and mesoscopic models can reflect the SMA behavior with high accuracy. However, SMAs used in structural engineering and, particularly, in structural control involve large volumes. Accordingly, the modeling of such applications with microscopic and mesoscopic approaches require high computational effort, which may be critical particularly for the real-time applications, such as with semi-active dampers. Therefore, due to their superior computing efficiency, macroscopic models are preferred for the simulation of these systems. Macroscopic models comprise generally phenomenological approaches based on thermodynamic principles and experimental data.

In their seminal work, Tanaka and Nagaki propose a continuum mechanical modeling framework based on internal variables, which can calculate the response of uniaxially loaded SMAs [203]. Inspired by this approach, Liang and Rogers developed a 3D model for the calculation of the superelastic response and the shape memory effect [204]. They introduced evolution equations for the calculation of the martensite volume fraction. Other pioneering 3D constitutive models have been proposed by Raniecki et al. [205–207], Boyd and Lagoudas [208], Leclercq and Lexcellent [209], Aurichhio et al. [210] and Souza et al. [211], which established the fundamentals for the further developments, such as by Reese and Christ [212], Auricchio and Bonetti [213], Kohlhaas and Klinkel [214] as well as Praster et al. [215]. An extensive review of these models is reported by Cisse et al. [216].

In structural control, for the computation, particularly 1D models are preferred. Compared to the 3D models, this approach saves valuable computational resources. Moreover, uniaxial SMA wires seem to exhibit most suitable properties for the vibration control applications as previously mentioned in Section 8.1.

1D macroscopic models have been proposed by Graesser and Cozzarelli [217], Brinson [218] as well as Brinson and Lammering [219], which represent both superelastic and shape memory effects. However, these models do not include strain rate-dependent effects, which are, as previously mentioned in Section 8.3, decisive for the damping performance. Wilde et al. proposed an extended version of the Graesser-Cozzarelli model, which incorporates also the strain hardening effects [220]. Ren et al. reported a further development of the same model, which considers some of the strain rate dependent effects using the least square method [221].

In particular, the macroscopic 1D models proposed by Auricchio et al. [222] as well as Zhu and Zhang [223] seem to be quite promising for structural control applications. They are both based on thermodynamic frameworks, which already include some of the dynamic effects and allow further developments. Therefore, this chapter focuses on these models and presents them in subsequent sections with the proposed improvements.

8.4.1 Modeling of the Strain Rate Dependent Entropy Effect

The thermomechanically coupled material behavior governs the damping performance of SMAs. As introduced in Section 8.3, SMA wires exhibit high latent heat under high strain rates, due to which the austenite phase becomes energetically more stable. Consequently, the increased material heat affects the evolution of martensite bands.

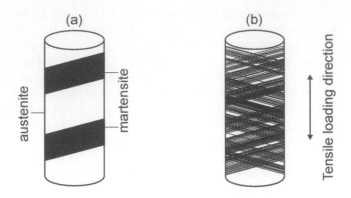

FIGURE 8.4
Strain rate and associated latent heat dependent evolution of the martensite bands of an SMA wire under uniaxial quasi-static (a) and dynamic (b) tensile loading.

Figure 8.4 illustrates schematically the strain rate-dependent latent heat effect for an SMA wire under uniaxial tensile loading. The evolution of martensite bands are shown for a quasi-static loading pattern with low strain rates (a) and a dynamic loading pattern with high strain rates (b). The quantity, position and propagation of martensite bands in the wire change with changing strain rate. In quasi-static case, the martensite bands are sparse but wide due to low latent heat evolution. However, in dynamic case, the martensite bands are spread in the wire and narrow.

The strain rate-dependent evolution and propagation of the martensite bands shown in Figure 8.4 were also experimentally observed by Shaw and Kyriakides [224] by photographically recording the surface changes of coated SMA dog bone test specimens under different strain rates. Simultaneously, temperature changes were monitored by an infrared imaging radiometer. The findings of this study have been confirmed by numerous studies, such as by Xiao et al. [225] by using a charge-coupled device camera and tracking the material temperature with an infrared camera.

The evolution of transition fronts causes inclusions or dislocations in the lattice structure of material. Accordingly, the stability of the material structure decreases with increasing number of coexistent transformation fronts, which is a characteristic case under dynamic loading patterns. At micro level, with increasing strain rate and higher phase transformation rate, the degree of atomic disorder and the motion of the atoms increases, which corresponds to an increase in material entropy, cf. Tatar and Yildirim [226]. Consequently, during the reverse transformation, SMA strives for the more stable austenite grid structure and tries to return to the stable austenite phase as quickly as possible.

This material behavior affects the strain-stress response of the material and deteriorates its damping performance, as will be shown in Section 8.5 also by experimental results.

To reflect this effect mathematically, the entropy change of the material needs to be calculated in a strain rate dependent manner. A possible way for this purpose is derived in this section using the constitutive SMA modeling framework proposed by Auricchio et al. [222]. This modeling framework uses the theory of irreversible thermodynamics introduced by Auricchio and Sacco [227] and a general internal variable framework proposed by Lubliner and Auricchio [228]. As a phenomenological uniaxial 1D constitutive model, it aims to represent the thermomechanically coupled superelastic behavior of SMAs. The model is developed in the small deformation regime, which is an adequate assumption for the so far developed SMA-based damping devices.

In the model, the martensite volume fraction is represented by the internal variable $\xi \in [0, 1]$. Three different evolutionary equations of ξ (linear, power, exponential) are proposed in rate form. The input variable of the model is the total strain ε.

For the derivation of the governing equations, we consider an SMA wire under tensile loading with the uniaxial mechanical stress $\sigma \geq 0$.

8.4.1.1 Elastic Modulus

Based on ξ, the REUSS scheme is adapted for the calculation of the YOUNG's modulus of the SMA according to the homogenization theory presented by Auricchio and Sacco [229] as well as Ikeda et al. [230]. The YOUNG's modulus reads

$$E = \frac{E_A E_M}{E_M + (E_A - E_M)\xi},$$ (8.1)

which yields at the beginning of the phase transformation $E(\xi = 0) = E_A$ and at the end $E(\xi = 1) = E_M$ corresponding to the austenite and martensite phases and their YOUNG's moduli E_A and E_M.

8.4.1.2 Strain Decomposition

Within the small deformation regime, the strain is decomposed into two parts as

$$\varepsilon = \varepsilon_{el} + \varepsilon_{in},$$ (8.2)

where ε_{el} and ε_{in} are the elastic and inelastic strain portions, respectively. ε_{in} occurs during the phase transformation of the tensile loaded SMA wire, depends also on ξ and is given by

$$\varepsilon_{in} = \varepsilon_L \xi,$$ (8.3)

where ε_L is the maximum deformation at $\xi = 1$ after the martensite lattice alignment is fully completed and referred as the maximum residual strain, cf. Funakubo [231]. Accordingly, the elastic portion of the total strain ε of the wire reads

$$\varepsilon_{el} = \varepsilon - \varepsilon_{in}$$ (8.4)

and before the beginning of the phase transformation at $\xi = 0$, the elastic strain and the total strain are equivalent. During the phase transformation ($\xi > 0$), the elastic strain is calculated with Equation 8.4 using the input variable, the total strain ε, and the internal variable, the martensite volume fraction ξ.

8.4.1.3 Free Energy Formulation

The constitutive model uses the following formulation for the calculation of the HELMHOLTZ free energy potential, which is based on the formulation proposed by Raniecki and Bruhns [232] for single-phase materials:

$$\psi = [(u_A - T\eta_A) - \xi(\Delta u - T\Delta\eta)] + C\left[(T - T_0) - T\ln\frac{T}{T_0}\right]$$
$$+ \frac{1}{2}E\varepsilon_{el}^2 - (T - T_0)E\alpha\varepsilon_{el}. \tag{8.5}$$

Here, u_A and η_A represent the initial internal energy and entropy levels at the parent austenite phase, respectively. Δu and $\Delta\eta$ are ≥ 0 and represent the difference between the internal energy and the entropy levels of the austenite and martensite phases. T and T_0 are the absolute and reference material temperatures, respectively. C is the heat capacity of the material and α is the thermal expansion factor.

8.4.1.4 Stress Definition

The mechanical stress of the material is calculated from the free energy formulation as

$$\sigma = \frac{\partial\psi}{\partial\varepsilon} = E\left[\varepsilon_{el} - (T - T_0)\alpha\right], \tag{8.6}$$

where the effect of the thermal expansion on the stress can be neglected for SMA wires with small cross-section due to $\alpha \approx 0$.

8.4.1.5 Heat Equation

The heat equation is formulated using the free energy ψ according to the first law of thermodynamics as shown by Lemaitre and Chaboche [233] from

$$C\dot{T} = -\frac{\partial\psi}{\partial\xi}\dot{\xi} + T\frac{\partial^2\psi}{\partial T\partial\varepsilon}\dot{\varepsilon} + T\frac{\partial^2\psi}{\partial T\partial\xi}\dot{\xi} - \gamma(T - T_0) \tag{8.7}$$

and in its written out formulation as

$$C\dot{T} = b - \gamma(T - T_0), \tag{8.8}$$

where b is the heat source, γ the heat convection coefficient of the material and T_0 the initial temperature of the SMA, which can be assumed to be equal to the ambient temperature.

Based on Leclercq and Lexcellent [209], the constitutive model calculates the heat source as

$$b = H_{tmc} + D_{mec},\tag{8.9}$$

where H_{tmc} represents the heat production initiated by the thermomechanically coupled behavior of the material and D_{mec} represents the mechanical dissipation. According to the second law of thermodynamics, the CLAUSIUS-DUHEM inequality $D_{mec} \geq 0$ applies. Both components are calculated using the free energy formulation as

$$H_{tmc} = T\left(\frac{\partial^2 \psi}{\partial T \partial \varepsilon}\dot{\varepsilon} + \frac{\partial^2 \psi}{\partial T \partial \xi}\dot{\xi}\right) = T\left[-E\alpha\dot{\varepsilon} + (\varepsilon_L E\alpha + \Delta\eta)\dot{\xi}\right],\tag{8.10}$$

$$D_{mec} = -\frac{\partial \psi}{\partial \xi}\dot{\xi} = \dot{\xi}\left(\varepsilon_L \sigma - T\Delta\eta + \Delta u\right),\tag{8.11}$$

where $\dot{\varepsilon}$ and $\dot{\xi}$ are the time derivatives of the total strain ε and the martensite volume fraction ξ, respectively.

For $\alpha \approx 0$, after neglecting the thermal expansion parts, the heat equation becomes

$$C\dot{T} = \underbrace{\dot{\xi}\varepsilon_L}_{\dot{\varepsilon}_{in}}\sigma + \Delta u\dot{\xi} - \gamma(T - T_0).\tag{8.12}$$

8.4.1.6 Kinetic Rules

As shown also experimentally, such as by Shaw [234], the phase transformation of SMAs can be initiated both by stress and temperature. Accordingly, in the constitutive model, the driving force of the phase transformation is represented by

$$F = \sigma + TA,\tag{8.13}$$

which corresponds to the content of the mechanical dissipation given in Equation 8.11. In Equation 8.13, A is

$$A = \frac{\Delta\eta}{\varepsilon_L}\tag{8.14}$$

and approximates the slope of the phase transformation zones of the stress-temperature diagram, which were previously depicted in Figure 8.3.

For the determination of the martensite volume fraction, the model introduces kinetic rules in rate form as first-order differential equations. The most sophisticated version is given for the forward and reverse transformations as

$$\dot{\xi} = \begin{cases} \beta^{AM}\dfrac{\dot{F}}{(F - R_f^{AM})^2}H(1 - \zeta) & \text{for } A \to M \\[2ex] \beta^{MA}\dfrac{\dot{F}}{(F - R_f^{MA})^2}H\xi & \text{for } M \to A \end{cases},\tag{8.15}$$

where the coefficients β^{AM} and β^{MA} are so-called speed parameters, which are used to cover strain rate effects on the stress response of the material. H is an activation factor, which is calculated as

$$H = \begin{cases} 1 & \text{when} \begin{cases} \dot{F} > 0 & \text{and} & R_s^{AM} < F < R_f^{AM} & \text{for} & A \to M \\ \dot{F} < 0 & \text{and} & R_f^{MA} < F < R_s^{MA} & \text{for} & M \to A \end{cases} \\ 0 & \text{otherwise} \end{cases},$$

$$(8.16)$$

where

$$R_s^{AM} = \sigma_s^{AM} + T_0 A, \qquad\qquad R_f^{AM} = \sigma_f^{AM} + T_0 A, \qquad (8.17a)$$
$$R_s^{MA} = \sigma_s^{MA} + T_0 A, \qquad\qquad R_f^{MA} = \sigma_f^{MA} + T_0 A. \qquad (8.17b)$$

As previously introduced, $\sigma_s^{AM}, \sigma_f^{AM}, \sigma_s^{MA}$ and σ_f^{MA} are the critical stress levels, at which the forward and reverse transformations start and finish. Correspondingly, T_0 represents the critical temperature levels, at which the transformations start and finish.

8.4.1.7 Integration and Solution Algorithms

For the computation of the internal variable ξ, the corresponding stress σ as well as the absolute material temperature T from the strain ε, the constitutive model is formulated in a time discrete manner by integrating the continuous heat equation (Equation 8.8) and the kinetic rules (Equation 8.15), which were previously introduced in rate form. The integration occurs over the time interval $[t_n, t]$. In the following representations, quantities with the subscript n correspond to the time point t_n. Accordingly, quantities without the subscript correspond to the time point t.

The integration of the heat equation through backward EULER integration scheme reads

$$C \frac{T - T_n}{t - t_n} = b_I + \gamma(T - T_0), \qquad (8.18)$$

where

$$b_I = H_{I,tmc} + D_{I,mec} \qquad (8.19)$$

with the integrated heat production due to the thermomechanical coupling and the mechanical dissipation

$$H_{I,tmc} = T \left[-E\alpha \frac{\varepsilon - \varepsilon_n}{t - t_n} + (\varepsilon_L E\alpha + \Delta\eta) \frac{\xi - \xi_n}{t - t_n} \right], \qquad (8.20)$$

$$D_{I,mec} = \frac{\xi - \xi_n}{t - t_n} (\varepsilon_L \sigma - T\Delta\eta + \Delta u). \qquad (8.21)$$

Accordingly, the integrated kinetic rules read in residual form

$$R = (\xi - \xi_n)(F - R_f^{AM})^2 - \beta^{AM}(F - F_n)H(1 - \xi) = 0 \text{ for } A \to M \quad (8.22a)$$
$$R = (\xi - \xi_n)(F - R_f^{MA})^2 - \beta^{MA}(F - F_n)H\xi = 0 \qquad \text{for } M \to A. \quad (8.22b)$$

At each time step, the martensite volume fraction is determined in an iterative fashion using the NEWTON-RAPHSON method starting with $\xi^1 = \xi_n$. Further iteration steps are calculated by

$$\xi^{i+1} = \xi^i - \frac{R(\xi^i)}{\frac{\partial R(\xi^i)}{\partial \xi}}. \tag{8.23}$$

The SMA stress σ and the absolute material temperature T are calculated subsequently by solving the Equations 8.6 and 8.18.

8.4.1.8 Rate Dependent Formulation of Entropy Change

In the free energy formulation shown in Equation 8.5, the difference between the entropy levels of the austenite and martensite phases $\Delta\eta$ is a time invariant scalar. However, as previously discussed, with increasing strain rate the instability of the martensitic state and accordingly the entropy increases. Therefore, the entropy difference must be defined as a variable, which dependents on the strain rate. To include this entropy effect, as proposed in [235], in the free energy formulation of the constitutive model, the scalar $\Delta\eta$ is replaced by the time variant variable $\Delta\eta(t)$. Equation 8.5 is rewritten as

$$\psi = [(u_A - T\eta_A) - \xi(\Delta u - T\Delta\eta(t))] + C\left[(T - T_0) - T\ln\frac{T}{T_0}\right] + \frac{1}{2}E\varepsilon_{el}^2 - (T - T_0)E\alpha\varepsilon_{el}, \tag{8.24}$$

where η_A represents the entropy level of the SMA in its pure austenite state. For the formulation of the variable $\Delta\eta(t)$, the entropy is calculated from the CLAUSIUS-DUHEM inequality using the free energy potential as

$$\eta = -\frac{\partial\psi}{\partial T} = \eta_A + \xi\Delta\eta + C\ln\frac{T}{T_0} \tag{8.25}$$

by neglecting the thermal expansion terms. By replacing the constant $\Delta\eta$ with the time dependent function $\Delta\eta(t)$ for $\xi < 1$ as

$$\Delta\eta(t) = \eta(t) - \eta_A \tag{8.26}$$

the calculation of the entropy reads

$$\eta(t) = \eta_A + \frac{C}{1-\xi}\ln\frac{T}{T_0}. \tag{8.27}$$

Accordingly, also the entropy change becomes variable as

$$\Delta\eta = \frac{C}{1-\xi}\ln\frac{T}{T_0}. \tag{8.28}$$

The rate dependency of the entropy change arises directly from ξ, which depends on the phase transformation rate, and from T, which depends both on $\dot{\varepsilon}$ and $\dot{\xi}$ as formulated in the heat equation.

Accordingly, the start and finish stress levels of the reverse martensite-austenite transformation change depending on the entropy level and are reformulated as

$$R_{s,f}^{MA}(\eta) = \sigma_{s,f}^{MA} + T_0 \frac{\eta(t)}{\varepsilon_L}. \tag{8.29}$$

Furthermore, the speed parameter is written also in an entropy dependent form as

$$\beta^{MA}(\eta) = \frac{\eta(t)}{\eta_A}. \tag{8.30}$$

As discussed at the beginning of this section, during the reverse transformation, SMA tries to return to the stable austenite phase as quickly as possible due to increased material entropy at high strain rates. Equations 8.29 and 8.30 enable the constitutive model to represent this effect more accurately.

8.4.2   Modeling of the Strain Rate Dependent Latent Heat Evolution

In previous Section 8.4.1, we considered the strain rate effects that influence the SMA damping response by an adaptive calculation scheme for the entropy, which is expected to be efficient during the reverse martensite-austenite transformation. However, as mentioned earlier in Section 8.3, the trigger of the effects is the increase in latent heat, which causes the parent austenite phase to become more stable and directly influences the stress-strain response. Therefore, this section derives an alternative approach and proposes a phenomenological method, which includes strain rate effects directly by the latent heat. In general, this section proposes a strain rate sensitive calculation of the thermal energy.

Due to its continuum thermomechanical framework, the macroscopic model of Zhu and Zhang [223] is chosen for this approach. Similar to the previously introduced constitutive model of Auricchio et al. [222], the macroscopic 1D model is based on the first and second law of thermodynamics and uses the energy balance and the CLAUSIUS-DUHEM inequality. The internal variable of the model is again the martensite volume fraction ξ and the input variable is the total strain ε. Analogous to the model of Auricchio et al. [222], the model is developed in the small deformation regime.

For the derivation of the governing equations, we consider also in this section an SMA wire under tensile loading with the uniaxial stress $\sigma \geq 0$.

8.4.2.1   Elastic Modulus

The YOUNG's modulus considers both austenite and martensite phases and is formulated based on ξ according to Liang and Rogers [204] as well as Sato and Tanaka [236], which is different from the previously adapted REUSS scheme.

The YOUNG's modulus reads

$$E = E_A + \xi(E_M - E_A), \tag{8.31}$$

where E_A and E_M are the YOUNG's moduli corresponding to the austenite and martensite phases, respectively.

8.4.2.2 Strain Decomposition

The total and inelastic strain portions are calculated according to Equations 8.3 and 8.4 as

$$\varepsilon = \varepsilon_{el} + \varepsilon_{in} = \varepsilon_{el} + \varepsilon_L \xi, \tag{8.32}$$

where ε_L is the maximum residual strain and small deformations are assumed.

8.4.2.3 Free Energy Formulation

The free energy formulation used in the constitutive model is similar to free energy equation of Sadjadpour and Bhattacharya [237] and given for unit mass as

$$\psi = \frac{L}{T_{cr}}(T - T_{cr})\xi - C_S T \ln\frac{T}{T_0} + \frac{1}{2\rho}E\varepsilon_{el}^2, \tag{8.33}$$

where the thermal expansion term is neglected. In Equation 8.33, T, T_{cr} and T_0 are the absolute, transformation and reference temperatures, respectively. Furthermore, C_S and ρ are respectively the specific heat capacity and the density of the SMA material. The parameter L represents the latent heat associated with the phase transition.

8.4.2.4 Stress Definition

The mechanical stress is calculated from the free energy formulation and is equivalent to Equation 8.6 of the previous constitutive model. Without the thermal expansion term it reads

$$\sigma = \rho\frac{\partial\psi}{\partial\varepsilon} = E\varepsilon_{el}. \tag{8.34}$$

8.4.2.5 Heat Equation

According to the first law of thermodynamics applied to Equation 8.7 and using the free energy formulation presented in Equation 8.33, the heat energy equation of the constitutive model yields

$$C_S \dot{T}\rho = \dot{\varepsilon}_{in}\sigma + L\rho\dot{\xi} - \gamma(T - T_0) \tag{8.35}$$

for $\alpha \approx 0$, cf. Equation 8.12. γ is, as previously introduced, the heat convection coefficient and can be calculated for unit volume of V from

$$\gamma = \frac{k}{V}, \tag{8.36}$$

where k is the heat transfer coefficient of the SMA.

8.4.2.6 Kinetic Rules

The model uses kinetic rules for the determination of the martensite volume fraction based on the sigmoid function. As an improved version of the kinetic rules of Liang and Rogers [204], they are able to describe inner loops, which can occur during the hysteretic response and are more stable compared to the exponential functions, as used by other models, such as by Tanaka [238]. Furthermore, the sigmoid function enables the kinetic rules a more smooth transition at the critical stress levels. The kinetic rules are described as

$$\xi = \begin{cases} \xi_0 + (1 - \xi_0)\mathrm{sig}\left[-a_M\left(T - T_M - \frac{\sigma}{c_M}\right)\right] & \text{for} \quad A \to M \\ \xi_0\mathrm{sig}\left[-a_M\left(T - T_A - \frac{\sigma}{c_A}\right)\right] & \text{for} \quad M \to A \end{cases}, \quad (8.37)$$

where ξ_0 is the initial value of the martensite fraction and T_M, T_A are critical temperatures given as

$$T_M = \frac{M_s + M_f}{2}, \quad T_A = \frac{A_s + A_f}{2}. \quad (8.38)$$

Here, M_s, M_f, A_s and A_f are the start and finish temperatures of the martensite and austenite phases as shown in Figure 8.3.

In Equation 8.37, a_M and a_A are material constants, which are defined as

$$a_M = \frac{\ln(10000)}{M_s - M_f}, \quad a_A = \frac{\ln(10000)}{A_f - A_s} \quad (8.39)$$

and c_M and c_A correspond to the slopes of the phase transformation zones of the stress-temperature diagram shown in Figure 8.3. The formulation provides with the sigmoid functions 0.01 and 0.99 for the martensite fraction corresponding to the transformation begin and end, respectively.

8.4.2.7 Integration and Solution Algorithms

For the computation of the internal variable ξ and the material response σ and T the presented mechanical stress and heat equations as well as the kinetic rules are reformulated in rate form as

$$\dot{\xi} = f_1(\sigma, T, \xi, \varepsilon, \dot{\varepsilon}), \quad (8.40a)$$
$$\dot{T} = f_2(\sigma, T, \xi, \varepsilon, \dot{\varepsilon}), \quad (8.40b)$$
$$\dot{\sigma} = f_3(\sigma, T, \xi, \varepsilon, \dot{\varepsilon}), \quad (8.40c)$$

where ε is the input variable, which is known and, accordingly, $\dot{\varepsilon}$ is also known. The differential equations are solved in time-domain simultaneously without iteration using the fourth-order RUNGE-KUTTA method.

8.4.2.8 Rate Dependent Formulation of Latent Heat Evolution

As previously introduced in the free energy formulation (Equation 8.33), the latent heat energy of the phase transition is covered by the parameter L, which is constant. However, as this parameter is directly related to the strain rate-dependent entropy evolution, from a mathematical point of view a variable is required, which can reflect the strain rate dependent response.

For this purpose, as proposed in [239], the latent heat constant L is replaced with the variable L_p in the free energy formulation (Equation 8.33) and accordingly in the further governing equations of mechanical stress (Equation 8.34) and heat energy (Equation 8.35) as well as in the kinetic rules (Equation 8.37). The variable L_p is determined phenomenologically as

$$L_p = c_1 e^{c_2 \varepsilon} + L_0, \tag{8.41}$$

where L_0 is the constant initial latent heat corresponding to the quasi-static loading case and $c_{1,2}$ are material parameters. With the time-dependent input variable ε, this equation enables the constitutive model to consider strain rate-dependent effects in the calculation of the latent heat. Therefore, the model is now expected to calculate the dynamic load cases more accurately, as will be shown later in Sections 8.5 and 8.6. In Section 8.6.2, the phenomenological determination of Equation 8.41 and $c_{1,2}$ will be introduced together with case studies showing the strain rate dependency of the latent heat calculation.

8.5 Experimental Studies

This section presents the results of the experimental studies performed on SMA wires and on a frame structure incorporating SMA wires. Section 8.5.1 is concerned with the effects of entropy on the SMA response. The results presented in this section will be used in Section 8.6.1 to validate the proposed modeling approach for the entropy effect. Section 8.5.2 is concerned with the phenomenological latent heat formulation. The results presented in this section will be used in Section 8.6.2 to validate the proposed modeling approach of the latent heat evolution. Section 8.5.3 investigates the responses of a frame structure with SMA wires by shaking table tests. The results of this section will be used in Section 8.6.3 to derive and validate a modeling approach for SMA controlled frame structures.

8.5.1 Entropy Effects

The first part of the experimental studies is concerned with cyclic tensile tests, which were performed on SMA wires with the setup shown in Figure 8.5. A uniaxial 0.50×0.50 m shaking table (1) generated the load patterns. The shaking table was controlled in a closed-loop by measuring the position of the table with a laser sensor. A fixed bearing (2) was mounted on one side of the shaking table as a reference point relative to the motion of the table. Trained Nitinol (Ni-55.8%-Ti-44.05%-C-0.05%-O-0.05%-Fe-0.05%) wires (3)

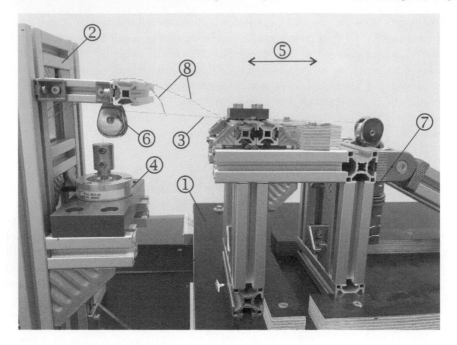

FIGURE 8.5
Side-view of the test setup of the tensile tests applied to SMA wires. (1)
Shaking table, (2) fixed bearing, (3) SMA wire, (4) load cell, (5) tensile direc-
tion, (6) pulley, (7) weights for pre-stressing, (8) 2x Type T thermocouples.
Courtesy of O. Altay.

of 150 mm length and 0.20 mm diameter were attached on one end to a load
cell and oriented parallel to the motion direction (5) of the table. A pulley (6)
was used to ensure a centric loading on the load cell and to avoid any buckling
effects on the wire. Weights (7) were attached on the opposite wire end to
apply a pre-stress of $\sigma_0 = 1.40$ MPa and to initiate an earlier martensitic
transformation.

The ambient room temperature was $T = 22.5°C$ (~ 295.7 K), which is
above the austenite finish temperature of the tested wires ($A_f = 263.2$ K).
The material and thermodynamic parameters of the tested wires are presented
in Table 8.1. These parameters will be used also in Section 8.6.1 for the numerical modeling of the wire response using the constitutive model introduced
in Section 8.4.1.

As shown in Figure 8.6, cyclic tensile tests were applied to the trained
SMA wires at frequencies 0.05, 0.10, 0.50, 1.00 and 2.00 Hz with a maximum
strain amplitude of 4 %. The measured stress-strain responses are shown
in Figure 8.7 With increasing strain rate, the shape of the stress-strain response changes significantly. In particular, the slope of both the forward and

TABLE 8.1

Material and thermodynamic parameters of the SMA wires used to investigate the entropy effects.

Parameter		Value	Unit
YOUNG's moduli	E_A, E_M	32350, 18550	MPa
Maximum residual strain	ε_l	3.34	%
Transformation stress levels	$\sigma_s^{AM}, \sigma_f^{AM}$	85, 305	MPa
	$\sigma_s^{MA}, \sigma_f^{MA}$	225, 200	MPa
Internal energy difference	Δu	1320	MPa
Austenitic entropy level	η_A	$9.3 \cdot 10^{-3}$	MPa K^{-1}
Heat convection coefficient	γ	0.1	-
Heat capacity	C	4.0	MPa K^{-1}
Density	ρ	6.5	g cm^3
Reference material temperature	T_0	295.7	K
Thermal expansion factor	α	0	K^{-1}

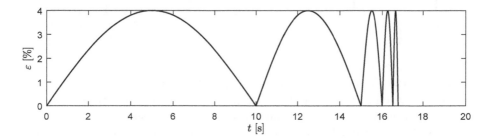

FIGURE 8.6

Cyclic tensile loading pattern applied to the SMA wires. Excitation frequencies are 0.05, 0.10, 0.50, 1.00 and 2.00 Hz. Maximum strain amplitude is 4 %.

the reverse transformation plateaus increases. In the figure, the solid lines show the response of the martensitic SMA. The dashed lines show the reverse martensite-austenite transformation. Due to the increased loading rate, the material temperature increases as shown in Figure 8.8. The high temperature phase (austenite) becomes energetically more stable and the SMA tries to return to the austenite phase as quickly as possible. Consequently, the time frame for martensitic response shown by the solid line reduces. After about 0.5 Hz, it has almost disappeared. As mentioned previously in Section 8.4.1, this effect can be represented mathematically by the entropy change.

Figure 8.8 shows the temperature evolution of the SMA measured by thermocouples during the cyclic tensile tests with 0.05, 0.10 and 0.50 Hz frequencies. The temperature increases rapidly at high strain rate and reaches with the excitation frequency 0.50 Hz approximately 25°C, which is higher than 23.5°C reached with the excitation frequency 0.05 Hz.

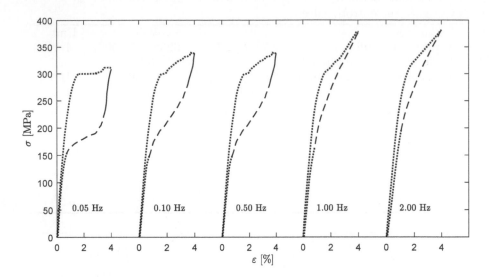

FIGURE 8.7

Results of the cyclic tensile tests applied to SMA wires to investigate the strain rate dependent entropy effect. Dashed lines show the reverse martensite-austenite transformation. Solid lines show the martensitic response of the SMA. With increasing strain rate the time frame of martensitic response reduces.

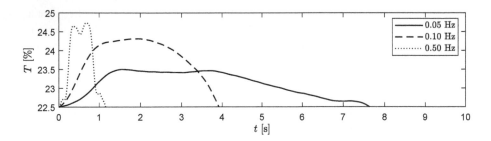

FIGURE 8.8

Temperature evolution measured during the cyclic tensile tests with 0.05, 0.10 and 0.50 Hz frequencies and a maximum strain amplitude of 4 %.

TABLE 8.2

Material and thermodynamic parameters of the SMA wires used for the latent heat formulation.

Parameter		Value	Unit
YOUNG's moduli	E_A, E_M	29000, 14100	MPa
Maximum residual strain	ε_l	4.34	%
Transformation temperatures	A_s, A_f	263.2, 285.2	K
	M_f, M_s	231.4, 247.7	K
Critical stress-temperature-slope	c_A, c_M	6.57, 6.13	$N(mm^2K)^{-1}$
Specific heat capacity	C_S	800	$J(kgK)^{-1}$
Heat transfer coefficient	k	0.021	$W\ K^{-1}$
Thermal expansion factor	α	0	K^{-1}

8.5.2 Phenomenological Latent Heat Formulation

The second part of the experimental studies is concerned with phenomenological formulation of the latent heat. Cyclic tensile tests were conducted using the previously introduced setup on trained Nitinol (Ni-55.90%-Ti-43.95%-C-0.02%-O-0.04%-Fe-0.02%) wires.

The same wire set was installed in the third part of the experimental study on a frame structure, to which shaking table tests were applied, cf. Section 8.5.3.

Material and thermodynamic parameters of the tested wires are presented in Table 8.2. These parameters will be used in Sections 8.6.2 and 8.6.3 also for the numerical modeling of the wire response using the constitutive model introduced in Section 8.4.2. Again the ambient room temperature was $T = 22.5°C$ (~ 295.7 K), which is above the austenite finish temperature of the tested wires ($A_f = 285.2$ K).

Figure 8.9 shows the measured stress-strain responses of the SMA wires for the excitation frequencies 0.05 and 1.00 Hz with a maximum amplitude of 6 %. Similar to the previous experiments (cf. Figure 8.7), with increasing strain rate the slopes of the transformation plateaus increases and consequently the shape of the hysteretic area changes.

To determine the constant initial latent heat L_0 as well as the material parameters $c_{1,2}$ of the latent heat evolution equation

$$L_p(t) = c_1 e^{c_2 \varepsilon(t)} + L_0 \tag{8.42}$$

the cyclic tensile tests were repeated with the maximum strain amplitudes 2, 4 and 6 % at 1 Hz excitation frequency. The stress-strain responses were calculated for the same loading pattern using the approach presented in Section 8.4.2. From a comparison of the measured and calculated wire responses, the best matching L_0 and $c_{1,2}$ parameters are chosen. The determined final L_p function and its strain rate dependency will be presented in Section 8.6.2.

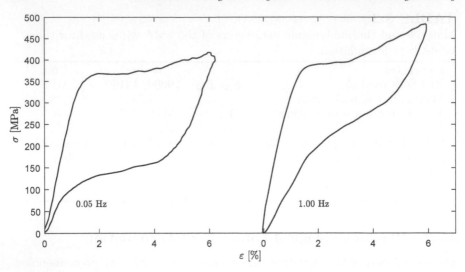

FIGURE 8.9
Results of the cyclic tensile tests applied to SMA wires to determine the latent heat formulation.

8.5.3 Shaking Table Tests on a Frame Structure Incorporating SMA Wires

The third part of the experimental studies is concerned with the shaking table tests, which were performed on a three-story steel frame structure. The structure incorporated trained Nitinol wires. The test setup is shown in Figure 8.10. The structural parameters of the frame are given in Table 8.3. The fundamental frequency and the inherent damping of the structure without SMAs were determined as $f_1 = 1.9$ Hz and $D_1 = 1.0$ %, respectively. With SMAs, the fundamental frequency of the structure was increased to $f_1 = 2.3$ Hz.

The properties of the SMA wires were already introduced in previous

TABLE 8.3
Parameters of the three-story steel frame structure.

Parameter		Value	Unit
Total height	h	2.11	m
Total mass[1]	m_T	72.1	kg
Floor mass	$m_1 \approx m_2 \approx m_3$	14.3	kg
Natural frequencies[2]	f_1, f_2, f_3	1.9, 6.0, 9.2	Hz
Natural frequencies[1]	f_1, f_2, f_3	2.3, 8.0, 12.7	Hz
Damping ratio of 1., 2. mode[2]	D_1, D_2	1.0, 0.5	%

[1] with SMAs
[2] without SMAs

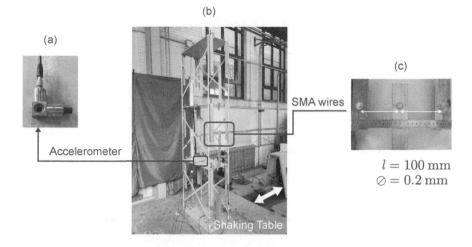

FIGURE 8.10
Uniaxial shaking table tests on a three-story steel frame structure with SMA wires. Accelerometers (a) are measuring the response of the structure (b). The wires are attached to the structure via steel angle bars (c). Shaking table tests were conducted at the seismic testing facility at *Tsinghua University*.

Section 8.5.2. Material and thermodynamic parameters of the wires were given in Table 8.2. The SMAs have the same diameter of 0.2 mm. The wires were attached to the structure via steel angle bars. The floors of the structure were restrained by four 100 mm SMAs wires, which were pre-stressed with $\sigma_0 = 1.40$ N mm^{-2}. Depending on the motion direction of the structure, tensile loads were applied to two out of four wires.

The tests have been carried out on a shaking table of the seismic testing facility at *Tsinghua University*. Using the shaking table, base excitation was applied to the structure. The structure was mounted on the table by anchorage screws. Perpendicular to the excitation direction the structure was reinforced by cross bracing. To prevent the SMA wires from tearing, the base excitation amplitude was limited by keeping the maximum story drift of the bottom floor under 6 mm, which corresponds to a maximum 6 % wire strain. For the measurement of the structural response, accelerometers were attached on each floor and on the shaking table. The properties of the shaking table and the accelerometers are given in Table 8.4.

Historic far-field earthquake El Centro and the near-field earthquakes Kobe and Taft are scaled and applied to the structure. The time histories of the earthquakes are shown in Figure 8.11. The properties of the earthquake records are listed below [179, 240].

- *El Centro*: N-S component recorded at the Imperial Valley Irrigation District

TABLE 8.4

Properties of the shaking table and the parameters of the accelerometers.

Parameter	Value	Unit
Shaking table:		
Size	1.5×1.5	m
Payload capacity	2	t
Max. acceleration with max. payload	1.2	g
Stroke	± 0.2	m
Excitation frequency range	0.1–50	Hz
Accelerometers:		
Measurement range	50	g
Resolution	$0.2 \cdot 10^{-3}$	g
Frequency range	0.5–6000	Hz

substation in El Centro, during the Imperial Valley, California Earthquake of 1940 with a scaled PGA of 0.07 g.

- *Kobe*: N-S component recorded at the Kobe Japanese Meteorological Agency station, during the Hyogoken Nanbu Earthquake of 1995 with a scaled PGA of 0.05 g.

- *Taft*: N-S component recorded at Taft, California, on a U.S. Coast and Geodetic Survey instrument, during the Arvin and Tehachapi Earthquake of 1952 with a scaled PGA of 0.05 g.

The acceleration and displacement responses of the structure without and with SMAs due to the earthquakes are shown in Figures 8.12, 8.13 and 8.14. An SMA induced significant reduction of the vibrations is observed during all earthquakes. In Figure 8.15, the RMS and MAX values of the responses are depicted. Also here the control effect of SMA wires can be clearly observed. However, during the Taft earthquake, the maximum acceleration responses of the second and third floors are increased slightly. In Table 8.5, the reduction amount of RMS and MAX values are calculated for acceleration and displacements responses by

$$R_{\ddot{x}} = 1 - \frac{\ddot{x}_{\text{w SMA}}}{\ddot{x}_{\text{wo SMA}}} \quad \text{and} \quad R_x = 1 - \frac{x_{\text{w SMA}}}{x_{\text{wo SMA}}}, \qquad (8.43)$$

where the subscripts $\cdot_{\text{w SMA}}$ and $\cdot_{\text{wo SMA}}$ correspond to the responses with and without SMAs respectively. Accordingly, the highest reduction of the acceleration is recorded during the Kobe earthquake with 59.3 % for the RMS and 36.3 % for the MAX values. The lowest reduction occurs during the Taft earthquake with 15.5 % for the RMS and −8.0 % for the MAX values. For displacement the best performance is recorded during the El Centro earthquake with 52.3 % for the RMS and 25.6 % for the MAX values. The lowest

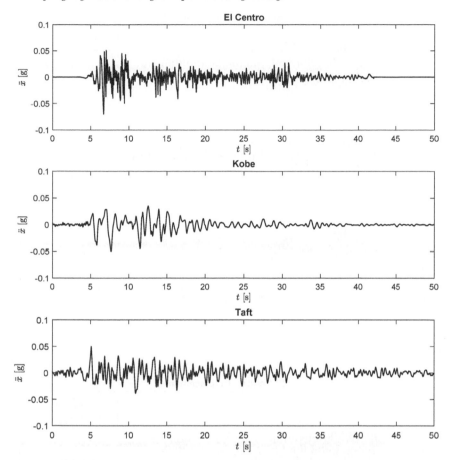

FIGURE 8.11
Time histories of the historic El Centro (a), Kobe (b) and Taft (b) earthquakes
used in the experimental studies.

reduction is recorded also for the displacement during the Taft earthquake
with 30.0 % for the RMS and 22.9 % for the MAX values.

In general, the experiments revealed that the SMA wires can control vibrations of the frame structure efficiently. However, as shown particularly by
the results of the Taft earthquake the performance may deviate depending
on the response characteristics of the structure. Accordingly, these must be
taken into account for the design of the SMA wires. This aspect will be further
detailed in Sections 8.6.3 and 8.7.

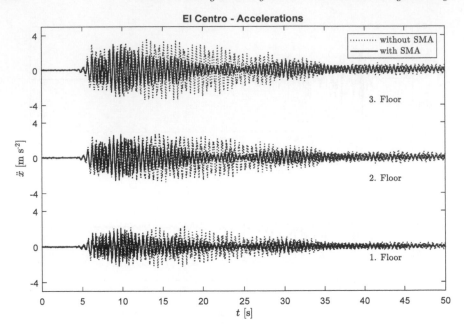

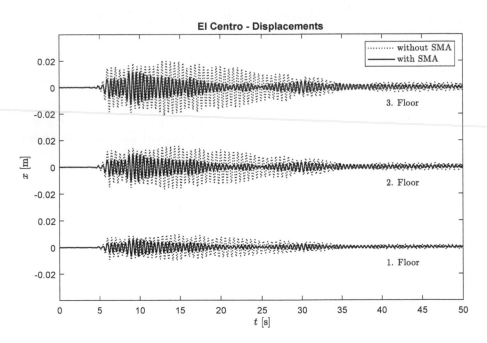

FIGURE 8.12

Acceleration and displacement responses of the frame structure without and with SMAs due to the El Centro earthquake.

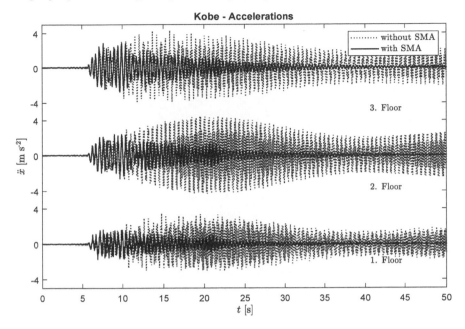

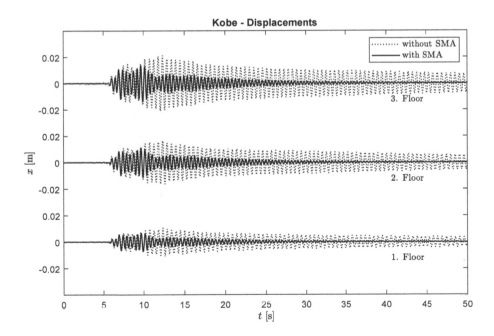

FIGURE 8.13
Acceleration and displacement responses of the frame structure without and with SMAs due to the Kobe earthquake.

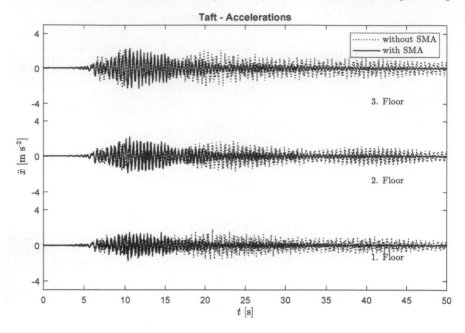

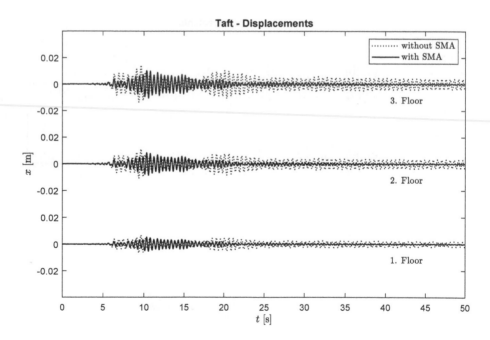

FIGURE 8.14
Acceleration and displacement responses of the frame structure without and
with SMAs due to the Taft earthquake.

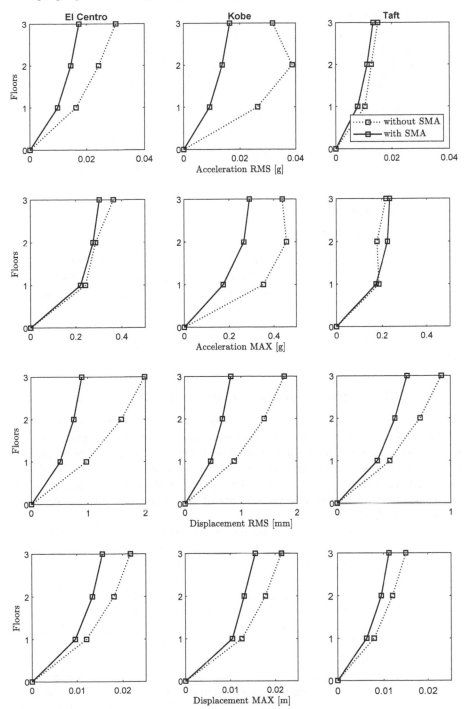

FIGURE 8.15

The RMS and MAX values of acceleration and displacement responses of the frame structure without and with SMAs due to the El Centro, Kobe and Taft earthquakes.

TABLE 8.5

Reduction of RMS and MAX values of acceleration and displacement responses of the frame structure without and with SMAs due to the El Centro, Kobe and Taft earthquakes.

Parameter	El Centro [%]	Kobe [%]	Taft [%]
RMS reduction of acceleration	41.0	59.3	15.5
MAX reduction of acceleration	17.0	36.3	−8.0
RMS reduction of displacement	52.3	52.0	30.0
MAX reduction of displacement	25.6	24.6	22.9

8.6 Numerical Studies

This section presents the results of the numerical computation of the SMA response using the constitutive model improvements presented in Section 8.4. Section 8.6.1 is concerned with the simulation of the strain rate-dependent entropy effect. Section 8.6.2 is concerned with the phenomenological formulation of the latent heat evolution. Finally, Section 8.6.3 applies the latent heat formulation to the response simulation of the three-story frame incorporating SMA wires.

8.6.1 SMA Wire Response Considering Entropy Effect

The SMA response is simulated using the constitutive model presented in Section 8.4.1. The cyclic tensile loading patterns of excitation frequencies 0.05, 0.10, 0.50, 1.00 and 2.00 Hz are applied with the maximum strain amplitude of 4 %. The same loading patterns were previously applied experimentally to the SMA wires in Section 8.5.1, cf. Figure 8.6. For the calculations, the same material and thermodynamic parameters of the experimentally tested wires are used, cf. Table 8.1.

The numerical and experimental results are compared in Figure 8.16. Numerical results with both excluding and including the strain rate-dependent entropy effect are shown. At the low excitation frequency of 0.05 Hz, i.e., under quasi-static loading, the response of the SMA is modeled with both numerical approaches quite accurately. However, as soon the strain rate increases, the modeling requires the inclusion of the entropy effects. In particular, at high frequencies, starting with 0.50 Hz, the improved modeling approach including the entropy effect shows its merits and can track the dynamic material behavior accurately.

The reason for the superior performance of the improved SMA model can be observed from the calculated entropy time histories, which are shown in Figure 8.17 for the excitation frequencies 0.05, 0.10 and 0.50 Hz. The evolution of the entropy is able to track the measured temperature development.

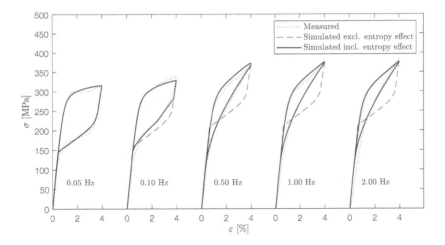

FIGURE 8.16
Measured and calculated stress-strain responses of the SMA wires due to the cyclic tensile loading. The calculation is performed both excluding and including the strain rate-dependent entropy effect.

The measured temperature time histories were shown previously in Figure 8.8 and repeated here for the convenience of the reader. Accordingly, the highest entropy change is calculated for the dynamic loading case with 0.50 Hz excitation frequency, which corresponds also to the highest measured temperature. Furthermore, as intended in the constitutive model, the lowest entropy change occurs during the quasi-static loading with 0.05 Hz, which corresponds also to the lowest measured temperature.

As the modeling approach can simulate the material response by considering the entropy change more accurately, also the estimated amount of dissipated vibration energy is expected to approximate the reality better. This can be indeed observed in Figure 8.18 from the comparison of the dissipated energies calculated using the force-displacement responses of the SMA wires. At quasi-static excitation, the experimental results can be replicated by both including and excluding the entropy dependent effects accurately. However, with increasing excitation frequency, the model excluding the entropy effects struggles to track the real energy dissipation. In particular, in high frequency range, when the entropy change is not considered, the energy dissipation is always overestimated. The SMA model including the entropy change effects reflects the real energy dissipation more accurately and corresponds to the experimental findings.

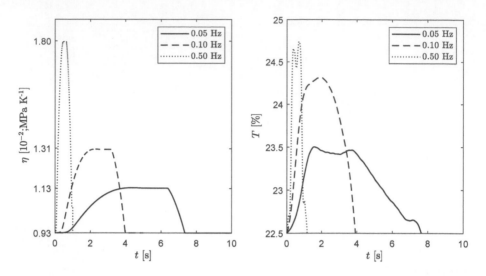

FIGURE 8.17
Calculated entropy evolution (a) and measured temperature (b) during the cyclic tensile tests with 0.05, 0.10 and 0.50 Hz frequencies and a maximum strain amplitude of 4 %.

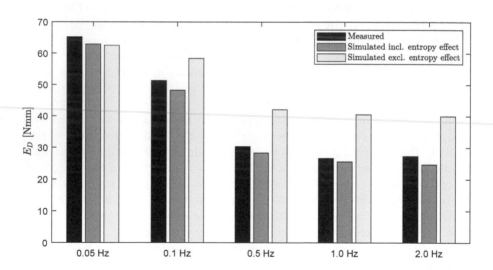

FIGURE 8.18
Dissipated vibration energy calculated from the force-displacement responses of the SMA wires during the cyclic tensile loading with 0.05, 0.10 and 0.50 Hz frequencies and a maximum strain amplitude of 4 %. Experimental results are compared with simulations including and excluding entropy change-dependent effects.

8.6.2 SMA Wire Response Considering Latent Heat Evolution

As previously introduced in Sections 8.4.2 and 8.5.2, for the computation of the latent heat the phenomenological formulation

$$L_p(t) = c_1 e^{c_2 \varepsilon(t)} + L_0 \tag{8.44}$$

is proposed. A parametric study is conducted using the formulation. The results are compared with the experiments. The study revealed that for the tested SMA wires, the material responses can be simulated most accurately with the following parameters:

$$L_0 = 4400 \ [\text{J kg}^{-1}], \quad c_1 = 711 \ [\text{J kg}^{-1}], \quad c_2 = 81 \ [\text{-}]. \tag{8.45}$$

The corresponding exponential curve of the latent heat evolution L_p is shown in Figure 8.19. As also shown in Equation 8.45, L_p is a variable, which directly depends on the strain amplitude and, consequently, on the strain rate of the wire. To clarify this, we consider two case studies. In the first case study, as shown in Figure 8.20, the latent heat evolution is calculated for a cyclic tensile loading pattern using 1 Hz excitation frequency with 2, 4 and 6 % maximum strain amplitudes. In the second case study, as shown in Figure 8.21, the latent heat evolution is calculated for a cyclic tensile loading pattern with 0.5 and 1.0 Hz excitation frequencies and 6 % maximum strain amplitude. In the first case study, the maximum latent heat increases for increasing strain amplitudes as expected. In the second case study, although the maximum strain amplitudes are the same, the latent heat reaches its maximum earlier for the excitation with higher frequency. This effect is directly caused by the strain rate applied to the wire and shows the strain rate dependency of the latent heat formulation. Furthermore, the strain rate dependency of the formulation is confirmed by the time derivatives of the latent heat $\dot{L}_p$, which are also shown in both figures. In both cases, the maximum value of the time derivative increases with the increasing strain rate.

Figure 8.22 compares the measured and calculated stress-strain responses of the SMA wires for the quasi-static and dynamic loading scenarios with 0.05 and 1.00 Hz excitation frequencies, respectively, and 6 % maximum strain amplitude. The measurements were previously introduced in Section 8.5.2. The calculations are performed both with a constant L_p of 10 kJ kg^{-1} corresponding to the approach of Zhu and Zhang [223] and with the variable improved formulation from Equation 8.45. With the variable L_p, the maximum values of the latent heat for both excitation patterns with 0.05 and 1.00 Hz frequencies are almost identical due to the same maximum strain amplitude of 6 %. The calculated L_P values reach 96.1 and 95.7 kJ kg^{-1}. The corresponding maximum values of the time derivatives of the latent heat $\dot{L}_p$ are 83.5 kJ(kgs)$^{-1}$ for the quasi-static case and 747.1 kJ(kgs)$^{-1}$ for the dynamic case.

The simulation with the constant L_p can reflect the quasi-static response. However, with this original formulation, the constitutive model struggles to

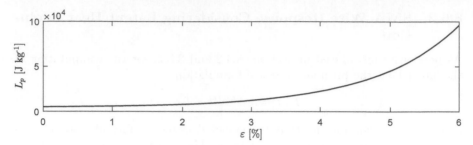

FIGURE 8.19
Latent heat evolution L_p phenomenologically determined from cyclic tensile tests. ε is the total strain applied to the wires.

estimate the SMA response in the dynamic case. On the other hand, the simulation with the proposed variable L_p formulation can track the SMA response in both quasi-static and dynamic loading cases. In particular, in dynamic case, the proposed approach can reflect the real response of the wires more accurately.

8.6.3   Response of a Frame Structure Incorporating SMA Wires

This section is concerned with a 3-DoF frame structure incorporating SMA wires. The structure has been previously investigated experimentally by shaking table tests as reported in Section 8.5.3. In this section, the responses of the structure are numerically computed by using the improved SMA modeling approach with the phenomenological latent heat formulation as discussed in Sections 8.5.2 and 8.6.2.

The frame structure is depicted in Figure 8.23. To model the structure we consider the following EoM

$$\mathbf{M}\ddot{\mathbf{x}} + \mathbf{C}\dot{\mathbf{x}} + \mathbf{K}\mathbf{x} + \mathbf{f_S}(x) = \mathbf{f}(t), \qquad (8.46)$$

where $\mathbf{M}$, $\mathbf{C}$ and $\mathbf{K}$ are the mass, damping and stiffness matrices, and $\ddot{\mathbf{x}}$, $\dot{\mathbf{x}}$ and $\mathbf{x}$ are the acceleration, velocity and displacement responses of the structure. $\mathbf{f_S}(x)$ represents the restoring forces generated by the SMA wires. The structure is excited by a shaking table. The corresponding dynamic forces applied to the structure are written in a force vector as

$$\mathbf{f}(t) = -\mathbf{M}[\ddot{x}_g, \ddot{x}_g, \ddot{x}_g]^\top, \qquad (8.47)$$

where $\ddot{x}_g$ represents the ground acceleration, which is induced by the shaking table.

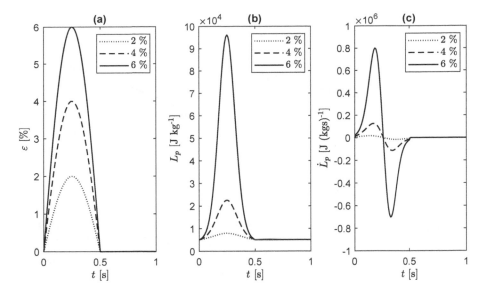

FIGURE 8.20
Time histories of the applied total strain (a), the latent heat evolution L_p (b) as well as its time derivative $\dot{L}_p$ (c) for cyclic tensile loading with 1 Hz excitation frequency and 2, 4 and 6 % maximum strain amplitudes.

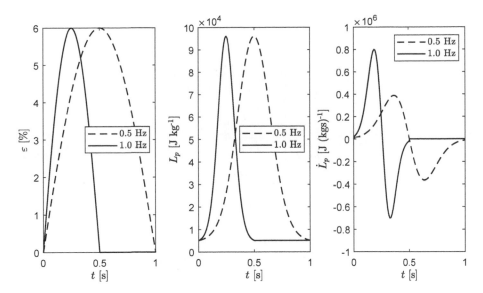

FIGURE 8.21
Time histories of the latent heat evolution L_p and its time derivative $\dot{L}_p$ for cyclic tensile loading with 0.5 and 1.0 Hz excitation frequencies and 6 % maximum strain amplitude.

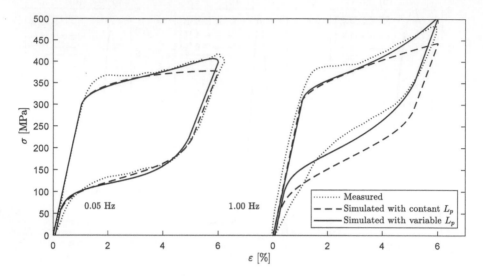

FIGURE 8.22
Measured and calculated stress-strain responses of the SMA wires due to the cyclic tensile loading. The calculation performed both with constant and variable latent heat formulation L_p.

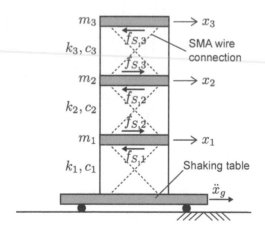

FIGURE 8.23
3-DoF frame structure numerically modeled with the SMA wires mounted between the floors of the structure and the shaking table.

TABLE 8.6
Natural frequencies of the numerical model of
the frame structure and their comparison
with the measured frequencies.

	f_1	f_2	f_3	Unit
Numerical model	1.9	5.2	7.3	Hz
Experimental model	1.9	6.0	9.2	Hz
Difference	0	0.8	1.9	%

The mass and stiffness matrices of the structure read

$$\mathbf{M} = \text{diag}[m_1, m_2, m_3] \tag{8.48}$$
$$= \text{diag}[2.55, 2.55, 2.11] \cdot 10^{-2} \text{ t}$$

$$\mathbf{K} = \begin{bmatrix} k_1 + k_2 & -k_2 & 0 \\ -k_2 & k_2 + k_3 & -k_3 \\ 0 & -k_3 & k_3 \end{bmatrix} \tag{8.49}$$
$$= \begin{bmatrix} 3242 & -1621 & 0 \\ -1621 & 3242 & -1621 \\ 0 & -1621 & 1621 \end{bmatrix} \cdot 10^{-2} \text{ kN m}^{-1},$$

where m_i and k_i are the mass and stiffness parameters of the structure corresponding to the DoF i. The natural frequencies of the numerical model of the frame structure are calculated using the matrices $\mathbf{M}$ and $\mathbf{K}$. As reported in Table 8.6, the calculated frequencies match with the frequencies determined from the measurements, cf. Section 8.5.3.

The damping matrix of the structure is computed by the RAYLEIGH damping as

$$\mathbf{C} = \alpha\mathbf{M} + \beta\mathbf{K} \tag{8.50}$$
$$= \begin{bmatrix} 0.92 & -0.18 & 0 \\ -0.18 & 0.92 & -0.18 \\ 0 & -0.18 & 0.65 \end{bmatrix} \cdot 10^{-2} \text{ kNs m}^{-1},$$

where the RAYLEIGH damping coefficients α and β are calculated from

$$\begin{bmatrix} \alpha \\ \beta \end{bmatrix} = \frac{2\omega_1\omega_2}{\omega_2^2 - \omega_1^2} \begin{bmatrix} \omega_2 & -\omega_1 \\ -1/\omega_2 & 1/\omega_1 \end{bmatrix} \begin{bmatrix} D_1 \\ D_2 \end{bmatrix} \tag{8.51}$$
$$= \begin{bmatrix} 0.2234 \text{ s}^{-1} \\ 0.1081 \cdot 10^{-3} \text{ s} \end{bmatrix},$$

The natural angular frequencies $\omega_1 = 1.9 \cdot 2\pi$ rad s^{-1} and $\omega_2 = 6.0 \cdot 2\pi$ rad s^{-1} with the corresponding damping ratios of the first two modes $D_1 = 1.0$ % and $D_2 = 0.5$ %, which are used for the calculation of the RAYLEIGH damping coefficients, were determined from experiments in Section 8.5.3 and presented previously in Tables 8.3 and 8.6.

The vector of the restoring forces of the SMA wires reads

$$\mathbf{f_S}(x) = [f_{S,1} - f_{S,2}, \; f_{S,2} - f_{S,3}, \; f_{S,3}]^\top, \tag{8.52}$$

$$f_{S,i} = \begin{cases} A\sigma_i & \text{for} \quad \sigma_i \geq 0 \\ 0 & \text{for} \quad \sigma_i < 0 \end{cases}, \tag{8.53}$$

where the force $f_{S,i}$ corresponds to the restoring force of the SMA wire set i, cf. Figure 8.23. Each wire strain ε_i is calculated from the interstory drift between the DoFs. The wire stress σ_i is computed from the strain by the constitutive model according to the approach proposed in Section 8.4.2. $A = \pi \cdot 10^{-2}$ mm^2 is the cross-sectional area of each SMA wire. The restoring force $f_{S,i}$ is activated, corresponding to the operational configuration of the wires, only in case of a tensile wire strain.

GUI-λ time-domain integration method is used with $\lambda = 4$ to solve the equation of motion. For further information on this method, the reader is referred to Section 8.7 and [241].

8.6.3.1 Study 1: Harmonic Response

The first study is concerned with the computation of the responses of the structure without and with SMAs due to harmonic excitation induced by the shaking table. The excitation frequency is 1 Hz with an acceleration amplitude of $\ddot{x}_g = 1.263$ m s^{-2}. At the beginning and at the end of the excitation signal, ramp functions are applied to smooth the shaking table motion. The measured and calculated acceleration and displacement responses of the structure without and with SMAs are shown in Figures 8.24 and 8.25. To investigate the accuracy of the calculation we compare the RMS values by

$$\Delta\text{RMS} = \frac{\text{RMS}_{\text{exp}}}{\text{RMS}_{\text{num}}}, \tag{8.54}$$

where the subscript $\cdot_{\text{exp}}$ indicates the measured and $\cdot_{\text{num}}$ the calculated values of the acceleration and displacement responses of the structure. Furthermore, to determine the accuracy of the used SMA constitutive model, the modeling error is calculated in % with

$$\epsilon = |\Delta\text{RMS}_{\text{wSMA}} - \Delta\text{RMS}_{\text{woSMA}}| \cdot 100, \tag{8.55}$$

where the subscripts $\cdot_{\text{wSMA}}$ and $\cdot_{\text{woSMA}}$ indicate the cases with and without SMA, respectively. This approach allows us to exclude any possible inaccuracies of the numerical model of the frame structure.

The comparison of the results is reported in Table 8.7. The modeling error ϵ for the harmonic excitation is at most 14.4 %, which is associated with the acceleration response of the third floor. Due to the reduced vibration

TABLE 8.7

Comparison of the measured and calculated responses
of the frame structure under harmonic excitation.

	1. Floor	2. Floor	3. Floor	Unit
Acceleration:				
$\Delta \text{RMS}_{\text{woSMA}}$	1.014	1.114	1.284	m s^{-2}
$\Delta \text{RMS}_{\text{wSMA}}$	1.336	1.218	1.158	m s^{-2}
ϵ	5.2	10.4	14.4	%
Displacement:				
$\Delta \text{RMS}_{\text{woSMA}}$	1.006	0.935	1.016	m
$\Delta \text{RMS}_{\text{wSMA}}$	0.927	0.823	0.891	m
ϵ	7.9	11.2	12.5	%

amplitude, the amount of error is generally low for the first floor responses.
The mean value of the modeling error for all three floors is 10.0 % for the
acceleration and 10.5 % for the displacement.

As it can be observed both from the depicted time histories of the structural response and the calculated error values, the proposed improvement
allows the constitutive modeling approach to track the restoring forces generated by the SMA wires accurately.

8.6.3.2 Study 2: Seismic Response

The second study is concerned with the computation of the responses of the
structure without and with SMAs due to seismic excitation induced by the
shaking table. The historic far-field Kobe and the near-field El Centro earthquakes are investigated. Section 8.5.3 introduced previously properties of the
earthquakes and depicted their time histories in Figure 8.11. For the sake of
brevity, we consider only the measured and calculated acceleration responses
of the structure without and with SMAs as shown in Figures 8.26 and 8.27.
The comparison of the RMS values and the modeling error ϵ are presented in
Table 8.8. The modeling error ϵ for the seismic excitation is at most 10.9 %,
which is comparable with the previously presented results of harmonic excitation. The mean value of modeling error of three floors is 7.2 % for Kobe
earthquake and 5.7 % for El Centro earthquake, which are also comparable
with the harmonic excitation case. Accordingly, the investigation of the seismic response concludes that the proposed approach is able to track accurately
the restoring forces of the SMAs also in case of seismic excitation.

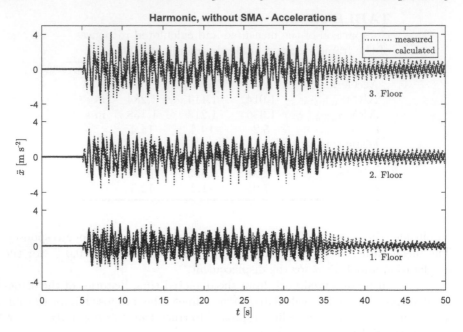

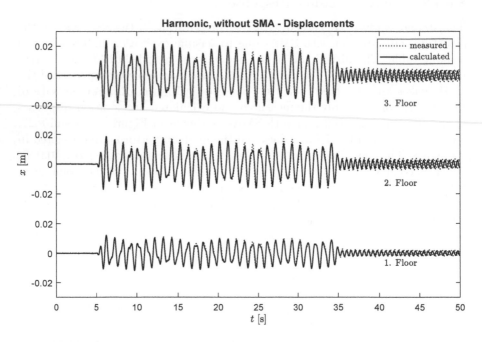

FIGURE 8.24

Measured and calculated acceleration and displacement responses of the frame structure without SMAs due to harmonic excitation.

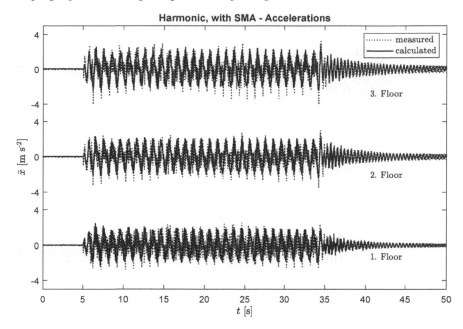

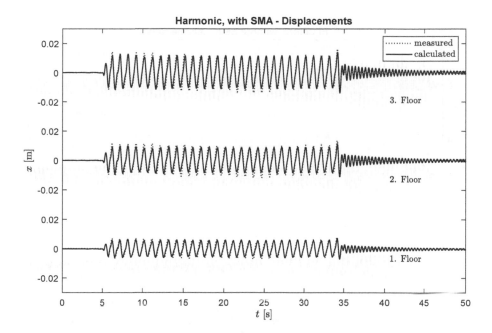

FIGURE 8.25

Measured and calculated acceleration and displacement responses of the frame structure with SMAs due to harmonic excitation.

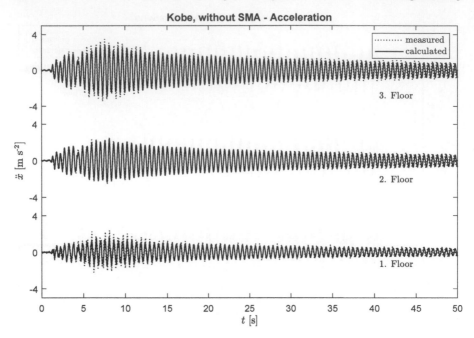

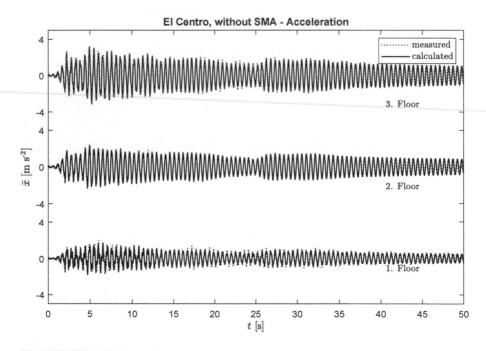

FIGURE 8.26

Measured and calculated acceleration and displacement responses of the frame structure without SMAs due to seismic excitation.

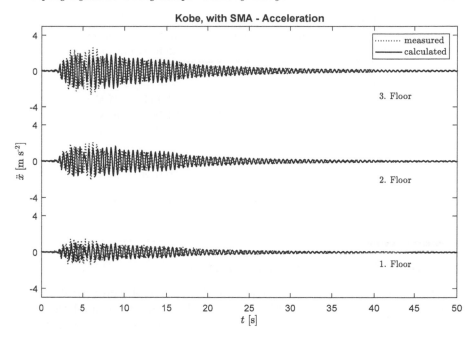

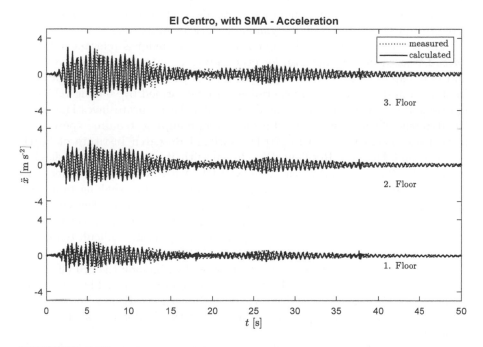

FIGURE 8.27

Measured and calculated acceleration and displacement responses of the frame structure with SMAs due to seismic excitation.

TABLE 8.8

Comparison of the measured and calculated responses
of the frame structure due to seismic excitation.

	1. Floor	2. Floor	3. Floor	Unit
Kobe:				
$\Delta \text{RMS}_{\text{woSMA}}$	1.133	1.071	1.117	m s^{-2}
$\Delta \text{RMS}_{\text{wSMA}}$	1.149	0.980	1.008	m s^{-2}
ϵ	1.6	9.1	10.9	%
El Centro:				
$\Delta \text{RMS}_{\text{woSMA}}$	1.062	0.970	1.002	m s^{-2}
$\Delta \text{RMS}_{\text{wSMA}}$	1.034	0.909	0.918	m s^{-2}
ϵ	2.5	6.1	8.4	%

8.7 Real-Time Hybrid Simulations

The *Real-time hybrid simulation* (RTHS) technique couples numerical simulations with dynamic physical testing methods. The numerical simulation part computes the responses of a substructure of the investigated structural system in real-time while the physical testing is applied to the rest of the structure by a shaking table or actuators. The approach considers dynamic effects and accordingly allows to investigate large-size complex systems more accurately with less effort than other simulation methods, such as conventional shaking table test methods and pseudo-dynamic (hybrid) simulations.

The analysis of the rate–dependent material behavior is relevant not only for SMA-based devices but also for other auxiliary vibration control systems and may require experiments supplementary to the numerical investigations. In structural dynamics, such experiments are performed by shaking tables or multi-axial actuators as shown in Figure 8.28. An accurate investigation of the dynamic effects, such as inertia forces and damping, require experiments in real-size. Facilities, such as NIED (National Research Institute of Earth Sciences and Disaster Resilience) in Japan with its 20 m × 15 m shaking table and 1200 t payload capacity or NHERI (Natural Hazards Engineering Research Infrastructure) in USA with its 7.6 m × 12.2 m shaking table and 2200 t payload capacity, can perform such tests in real-size. However, the number of large test facilities, which are capable of real-size investigations, is limited. Furthermore, such tests are expensive and generally difficult to repeat. Moreover, for conventional shaking table tests the consideration of SSI effects is difficult. Some approaches have been proposed to include the SSI effects in experiments, such as by using shear-boxes as a rigid container filled with soil proposed by Mizuno et al. [242]. However, such methods struggle to represent boundary effects accurately and are limited by the payload capacity of the shaking tables.

To reduce the experimental costs of real-size tests *pseudo-dynamic*

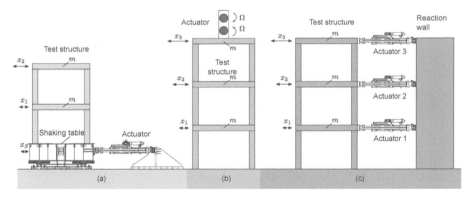

FIGURE 8.28
Conventional experimental methods used to investigate the dynamic responses
of structures. Shaking table tests (a), multi-axial actuators (b,c).

(hybrid) simulations have been proposed, such as by Hakuno et al. [243] and
Takanashi et al. [244]. This method applies quasi-static tests to structures or
to their substructures as shown in Figure 8.29. The test setup involves static
actuators to induce incremental displacements and load cells to measure the
restoring forces. The measured force signals are transmitted to a computer, in
which inertia and damping forces are added manually to simulate the entire
structure. In this approach, only the numerical simulation part is performed
dynamically, whereas the physical testing part, in contrast to the RTHS, is
performed quasi-statically. Therefore, the accuracy of this approach is insuf-
ficient particularly for systems with rate-dependent material behavior.

The RTHS is an improved version of the hybrid simulations. The first
RTHS application was performed in 1990s by Nakashima et al. on a structure
with a viscous damper using a dynamic actuator [245, 246]. The RTHS can
reflect accurately rate-dependent effects of materials in real-time. By using
substructuring techniques it is capable of simulating large systems, including
soil-foundation systems to consider SSI.

Section 8.7.1 describes the theory behind the RTHS technique and presents
its governing equations. Section 8.7.2 elaborates important aspects related
to the numerical part of the simulation method. Section 8.7.3 introduces
the hardware setup required for the application of the method. Finally, Section 8.7.4 is concerned with the previously introduced frame structure and
investigates the control performance of the installed SMA wires considering
SSI effects.

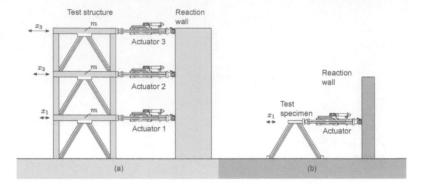

FIGURE 8.29
Pseudo-dynamic (hybrid) simulation method applied to a frame structure to investigate its dynamic behavior (a). The method is applied to a substructure of the frame structure (b).

8.7.1   Governing Equations

8.7.1.1   Equations of Motion

We consider a seismically excited $n+p$ DoF frame structure and divide it into two substructures. The first substructure involves the lower floors of the structure and the second substructure involves the upper floors of the structure. The first n DoF substructure is numerically modeled with a computer by a linear discrete model and the second p DoF substructure is physically tested on a shaking table. Figure 8.30 shows this approach by a $1+1$ DoF frame structure. The numerical part simulates the earthquake excitation on the first substructure and calculates the seismic responses of its top DoF. In real-time, these responses are forwarded to the shaking table in the form of excitation signals and applied to the second substructure. Interaction forces between the second substructure and the shaking table are measured simultaneously by sensors and applied to the first substructure in the numerical part.

Mathematically we describe the two substructures by their EoM as

$$\mathbf{M_N}\ddot{\mathbf{x}}_\mathbf{N} + \mathbf{C_N}\dot{\mathbf{x}}_\mathbf{N} + \mathbf{K_N}\mathbf{x_N} = \mathbf{F_E} + \mathbf{F_G}, \tag{8.56}$$

$$\mathbf{M_P}\ddot{\mathbf{x}}_\mathbf{P} + \mathbf{C_P}\dot{\mathbf{x}}_\mathbf{P} + \mathbf{K_P}\mathbf{x_P} = \mathbf{F_S}, \tag{8.57}$$

where the mass, damping and stiffness matrices are represented by $\mathbf{M_{N,P}}$, $\mathbf{C_{N,P}}$, $\mathbf{K_{N,P}} \in \mathbb{R}^{n \times n}$ as well as $\in \mathbb{R}^{p \times p}$ corresponding to the numerical and physical substructures. The acceleration, velocity and displacement vectors are represented by $\ddot{\mathbf{x}}_\mathbf{N,P}$, $\dot{\mathbf{x}}_\mathbf{N,P}$ and $\mathbf{x_{N,P}} \in \mathbb{R}^{n \times 1}$ as well as $\in \mathbb{R}^{p \times 1}$ corresponding to the numerical and physical substructures.

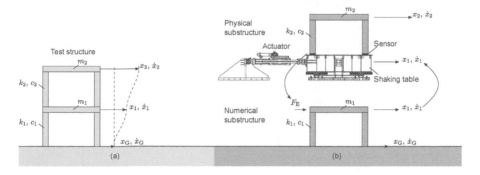

FIGURE 8.30
Seismically excited 2-DoF frame structure (a). Substructuring applied to the frame according to RTHS technique (b).

The mass matrices of both substructures are defined as

$$
\mathbf{M_N} =
\begin{bmatrix}
m_1 & 0 & \cdots & 0 & 0 \\
0 & m_2 & \cdots & 0 & 0 \\
\vdots & \vdots & \ddots & \vdots & \vdots \\
0 & 0 & \cdots & m_{n-1} & 0 \\
0 & 0 & \cdots & 0 & m_n
\end{bmatrix},
\tag{8.58}
$$

$$
\mathbf{M_P} =
\begin{bmatrix}
m_{n+1} & 0 & \cdots & 0 & 0 \\
0 & m_{n+2} & \cdots & 0 & 0 \\
\vdots & \vdots & \ddots & \vdots & \vdots \\
0 & 0 & \cdots & m_{n+p-1} & 0 \\
0 & 0 & \cdots & 0 & m_{n+p}
\end{bmatrix}.
\tag{8.59}
$$

The damping matrices of both substructures are defined as

$$
\mathbf{C_N} =
\begin{bmatrix}
c_1 + c_2 & -c_2 & \cdots & 0 & 0 \\
-c_2 & c_2 + c_3 & \cdots & 0 & 0 \\
\vdots & \vdots & \ddots & \vdots & \vdots \\
0 & 0 & \cdots & c_{n-1} + c_n & -c_n \\
0 & 0 & \cdots & -c_n & c_n
\end{bmatrix},
\tag{8.60}
$$

$$
\mathbf{C_P} =
\begin{bmatrix}
c_{n+1} + c_{n+2} & -c_{n+2} & \cdots & 0 & 0 \\
-c_{n+2} & c_{n+2} + c_{n+3} & \cdots & 0 & 0 \\
\vdots & \vdots & \ddots & \vdots & \vdots \\
0 & 0 & \cdots & c_{n+p-1} + c_{n+p} & -c_{n+p} \\
0 & 0 & \cdots & -c_{n+p} & c_{n+p}
\end{bmatrix}.
\tag{8.61}
$$

Similarly, the stiffness matrices of both substructures are defined as

$$\mathbf{K_N} = \begin{bmatrix} k_1 + k_2 & -k_2 & \cdots & 0 & 0 \\ -k_2 & k_2 + k_3 & \cdots & 0 & 0 \\ \vdots & \vdots & \ddots & \vdots & \vdots \\ 0 & 0 & \cdots & k_{n-1} + k_n & -k_n \\ 0 & 0 & \cdots & -k_n & k_n \end{bmatrix}, \tag{8.62}$$

$$\mathbf{K_P} = \begin{bmatrix} k_{n+1} + k_{n+2} & -k_{n+2} & \cdots & 0 & 0 \\ -k_{n+2} & k_2 + k_3 & \cdots & 0 & 0 \\ \vdots & \vdots & \ddots & \vdots & \vdots \\ 0 & 0 & \cdots & k_{n+p-1} + k_{n+p} & -k_{n+p} \\ 0 & 0 & \cdots & -k_{n+p} & k_{n+p} \end{bmatrix}. \tag{8.63}$$

The response vectors of the system read

$$\ddot{\mathbf{x}}_\mathbf{N} = [\ddot{x}_1 \ldots \ddot{x}_n]^\top, \tag{8.64}$$

$$\dot{\mathbf{x}}_\mathbf{N} = [\dot{x}_1 \ldots \dot{x}_n]^\top, \tag{8.65}$$

$$\mathbf{x}_\mathbf{N} = [x_1 \ldots x_n]^\top, \tag{8.66}$$

$$\ddot{\mathbf{x}}_\mathbf{P} = [\ddot{x}_{n+1} \ldots x_{n+p}]^\top, \tag{8.67}$$

$$\dot{\mathbf{x}}_\mathbf{P} = [\dot{x}_{n+1} \ldots \dot{x}_{n+p}]^\top, \tag{8.68}$$

$$\mathbf{x}_\mathbf{P} = [x_{n+1} \ldots x_{n+p}]^\top. \tag{8.69}$$

In Equation 8.56, the interaction force of the physical substructure with the shaking table is defined as

$$\mathbf{F_E} = \begin{bmatrix} 0 \\ \vdots \\ F_E \end{bmatrix}_{n \times 1} \quad \text{with } F_E = c_{n+1}(\dot{x}_{n+1} - \dot{x}_n) + k_{n+1}(x_{n+1} - x_n). \tag{8.70}$$

In the same EoM, the earthquake force is defined as

$$\mathbf{F_G} = \begin{bmatrix} F_G \\ 0 \\ \vdots \end{bmatrix}_{n \times 1} \quad \text{with } F_G = c_1 \dot{x}_G + k_1 x_G. \tag{8.71}$$

In Equation 8.57, the force applied by the shaking table to the physical substructure is defined as

$$\mathbf{F_S} = \begin{bmatrix} F_S \\ 0 \\ \vdots \end{bmatrix}_{p \times 1} \quad \text{with } F_S = c_{n+1} \dot{x}_n + k_{n+1} x_n. \tag{8.72}$$

8.7.1.2 State-space Representation

For the control of the shaking table and actuators, it is preferable to rewrite the presented EoMs of the $n + p$-DoF frame structure in the state-space representation, cf. Section 3.2. This approach allows to consider the measured sensor signals and the applied control forces in a *cyber-physical framework*. The fundamental equations of the system read

$$\dot{\mathbf{z}}_i = \mathcal{A}_i \mathbf{z}_i + \mathcal{B}_i \mathbf{u}_i, \tag{8.73}$$

$$\mathbf{y}_i = \mathcal{C}_i \mathbf{z}_i + \mathcal{D}_i \mathbf{u}_i, \tag{8.74}$$

where $i = \mathbf{N}$ or $\mathbf{P}$ corresponding to the numerical or physical substructure. The transition, input, output and feedthrough matrices are defined as

$$\mathcal{A}_i = \begin{bmatrix} \mathbf{0} & \mathbf{I} \\ -\mathbf{M}_i^{-1}\mathbf{K}_i & -\mathbf{M}_i^{-1}\mathbf{C}_i \end{bmatrix}, \qquad \mathcal{B}_i = \begin{bmatrix} \mathbf{0} & \mathbf{0} \\ -\mathbf{M}_i^{-1} & -\mathbf{M}_i^{-1} \end{bmatrix}, \tag{8.75}$$

$$\mathcal{C}_i = \begin{bmatrix} \mathbf{I} & \mathbf{0} \\ \mathbf{0} & \mathbf{I} \\ -\mathbf{M}_i^{-1}\mathbf{K}_i & -\mathbf{M}_i^{-1}\mathbf{C}_i \end{bmatrix}, \qquad \mathcal{D}_i = \begin{bmatrix} \mathbf{0} & \mathbf{0} \\ \mathbf{0} & \mathbf{0} \\ -\mathbf{M}_i^{-1} & -\mathbf{M}_i^{-1} \end{bmatrix}. \tag{8.76}$$

Here, we assume that all displacement, velocity and acceleration responses of the structure are observed. $\mathbf{0}$ and $\mathbf{I}$ are zero and unit matrices, respectively $\in \mathbb{R}^{n \times n}$ as well as $\in \mathbb{R}^{p \times p}$ corresponding to the DoFs of the numerical and physical substructures. The state, output and input vectors of the state-space representation read

$$\mathbf{z}_i = [x_1 \dots x_{n/p} \ \dot{x}_1 \dots \dot{x}_{n/p}]^\top, \tag{8.77}$$

$$\dot{\mathbf{z}}_i = [\dot{x}_1 \dots \dot{x}_{n/p} \ \ddot{x}_1 \dots \ddot{x}_{n/p}]^\top, \tag{8.78}$$

$$\mathbf{u}_{\mathbf{N}} = (F_{\mathbf{E}} + F_{\mathbf{G}})\Gamma_{\mathbf{N}}, \tag{8.79}$$

$$\mathbf{u}_{\mathbf{P}} = F_{\mathbf{S}}\Gamma_{\mathbf{P}}, \tag{8.80}$$

where $\Gamma_{\mathbf{N}} \in \mathbb{R}^{n \times 1}$ and $\Gamma_{\mathbf{P}} \in \mathbb{R}^{p \times 1}$ are incidence vectors corresponding to the DoFs, at which the interaction between the substructures occurs and the ground motion is induced on the system. $F_{\mathbf{E}}$, $F_{\mathbf{G}}$ and $F_{\mathbf{S}}$ are interaction, earthquake and shaking table forces as previously introduced in Equations 8.70, 8.71 and 8.72, respectively.

8.7.2 Numerical Simulation Part

8.7.2.1 Numerical Modeling

In the RTHS, the modeling approach used in the numerical simulation part influences directly the performance of the analysis. As the entire simulation is conducted in real-time, computationally efficient models are required. For this purpose, the first RTHS applications used discrete models with lumped masses and reduced DoFs. A pioneering example is the RTHS-based investigation of

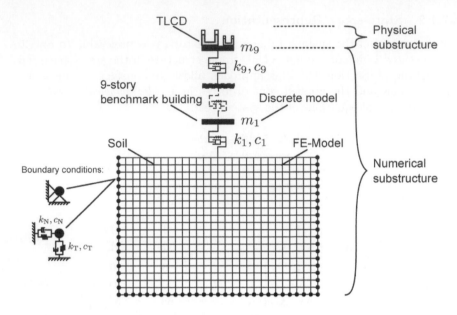

FIGURE 8.31

The RTHS framework applied by Zhu et al. to a 9-story benchmark building with TLCDs considering soil-foundation model with fixed and semi-fixed boundary conditions as numerical substructure [1].

Nakashima and Masaoka on a 5-DoF frame structure with base isolation [247]. Karavasilis et al. developed the nonlinear finite-element program *HybridFEM* for the 2D real-time simulation of frame structures [248]. Chen and Ricles [249] as well as Chae et al. [250] applied this program for the simulation of frame structures with 122 and 514-DoFs retrofitted by magnetorheological dampers.

For larger structural models, Zhou et al. developed a Simulink-environment using FE, which can simulate a maximum of 132-DoFs [251]. Zhu et al. enhanced the computing ability of this simulation framework to 1240 DoFs [252]. Figure 8.31 shows the application of this RTHS framework by Zhu et al. [1] on a 9-story benchmark building [179] with TLCDs using an FE-based soil-foundation model. This model will be used in Section 8.7.4 for the RTHS of the frame structure incorporating SMA wires.

8.7.2.2 Time Integration Methods

With increasing model size and DoFs, the solution of the numerical substructure requires more efficient numerical time integration methods. In this regard, due to required iteration, implicit algorithms are usually not preferred [253]. Explicit algorithms are far more commonly used in the RTHS, such as the *central difference method* [245] as well as the explicit formulation of the NEWMARK-*Beta method* [254]. In the NEWMARK-Beta method, the displacement and velocity vectors and the corresponding EoM read

$$\mathbf{x_{N_{k+1}}} = \mathbf{x_{N_k}} + \Delta t_N \dot{\mathbf{x}}_{\mathbf{N}_k} + \frac{\Delta t_N^2}{2} \ddot{\mathbf{x}}_{\mathbf{N}_k}, \tag{8.81}$$

$$\dot{\mathbf{x}}_{\mathbf{N}_{k+1}} = \dot{\mathbf{x}}_{\mathbf{N}_k} + \frac{\Delta t_N}{2} \left(\ddot{\mathbf{x}}_{\mathbf{N}_k} + \ddot{\mathbf{x}}_{\mathbf{N}_{k+1}} \right), \tag{8.82}$$

$$\mathbf{M_N} \ddot{\mathbf{x}}_{\mathbf{N}_{k+1}} + \mathbf{C_N} \dot{\mathbf{x}}_{\mathbf{N}_{k+1}} + \mathbf{K_N} \mathbf{x}_{\mathbf{N}_{k+1}} = \mathbf{F_{E_{k+1}}} + \mathbf{F_{G_{k+1}}}, \tag{8.83}$$

where $\ddot{\mathbf{x}}_{\mathbf{N}_k}$, $\dot{\mathbf{x}}_{\mathbf{N}_{k+1}}$ and $\mathbf{x}_{\mathbf{N}_k}$ are acceleration, velocity and displacement vectors of the numerical substructure at time step t_{k+1} and Δt_N is the time increment.

Chang [255] modified the NEWMARK formulation to ensure the unconditional stability for linear systems by adding two weighing matrices $\boldsymbol{\beta}_1$ and $\boldsymbol{\beta}_2$ to Equation 8.81 as

$$\mathbf{x_{N_{k+1}}} = \mathbf{x_{N_k}} + \boldsymbol{\beta}_1 \Delta t_N \dot{\mathbf{x}}_{\mathbf{N}_k} + \boldsymbol{\beta}_2 \frac{\Delta t_N^2}{2} \ddot{\mathbf{x}}_{\mathbf{N}_k}, \tag{8.84}$$

which consider the effects of the mass, damping and stiffness matrices and read

$$\boldsymbol{\beta}_1 = \boldsymbol{\beta}_2 \left(\mathbf{I} + \frac{1}{2} \Delta t_N \mathbf{M_N^{-1}} \mathbf{C_N} \right), \tag{8.85}$$

$$\boldsymbol{\beta}_2 = \left(\mathbf{I} + \frac{1}{2} \Delta t_N \mathbf{M_N^{-1}} \mathbf{C_N} + \frac{1}{4} \Delta t_N^2 \mathbf{M_N^{-1}} \mathbf{K_N} \right)^{-1}. \tag{8.86}$$

Furthermore, Gui et al. [241] developed a family of integration algorithms (GUI-λ). A special case of this method is the *CR algorithm* of Chen and Ricles [256], which is commonly known in the context of RTHS applications. As a double explicit algorithm with second-order precision, the GUI-λ method can realize the accuracy level required for the RTHS. The displacement and velocity vectors of the succeeding time step $k + 1$ are formulated as

$$\mathbf{x_{N_{k+1}}} = \mathbf{x_{N_k}} + \Delta t_N \dot{\mathbf{x}}_{\mathbf{N}_k} + \boldsymbol{\alpha} \frac{\Delta t_N^2}{2} \ddot{\mathbf{x}}_{\mathbf{N}_k}, \tag{8.87}$$

$$\dot{\mathbf{x}}_{\mathbf{N}_{k+1}} = \dot{\mathbf{x}}_{\mathbf{N}_k} + \boldsymbol{\alpha} \frac{\Delta t_N}{2} \ddot{\mathbf{x}}_{\mathbf{N}_k}, \tag{8.88}$$

where the corresponding EoM is analogous to Equation 8.83. The matrix $\boldsymbol{\alpha}$ reads

$$\boldsymbol{\alpha} = \lambda (2\lambda \mathbf{M_N} + \lambda \Delta t_N \mathbf{C_N} + 2\Delta t_N^2 \mathbf{K_N})^{-1} \mathbf{M_N}, \tag{8.89}$$

where λ determines the numerical characteristics of the method.

8.7.2.3 Delay Compensation

The accuracy of the RTHS technique depends on the time integration method used for the solution of the numerical substructure. On the other hand, the stability of the method depends on the transmission rate of the numerical

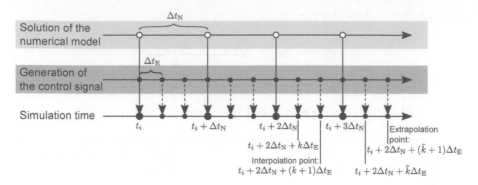

FIGURE 8.32
Delay compensation by interpolating and extrapolating missing sampling points in the numerical solution.

results to the physical substructure as reported by Zhu et al. [257]. The sampling rate of the numerical solution Δt_N is not equal to the sampling rate of the controller of the shaking table Δt_E. Generally, depending on the number of DoFs and nonlinearities of the numerical model, the sampling rate of the numerical solution is lower:

$$\Delta t = \Delta t_N - \Delta t_E \geq 0. \tag{8.90}$$

For one solution step of the numerical model, the shaking table requires already several response signals from the numerical substructure. The sampling rate of the controller of the shaking table depends on the hardware as well as the test setup, and can be higher than $f_E \geq 2048$ Hz. The corresponding time step is $\Delta t_E \leq 0.5 \cdot 10^{-3}$ s. For a numerical solution with the time step size $\Delta t_N = 0.01$ s, the delay between both substructures would be consequently $\Delta t \geq 9.5 \cdot 10^{-3}$ s.

For a continuous generation of the control signal of the shaking table actuators, the delay between the sampling rates must be compensated. For this purpose, interpolation and extrapolation methods can be used as shown in Figure 8.32. To prevent stability problems, the computational resources are reserved for the generation of the control signal by applying interpolation and extrapolation methods. In the meantime, the solution procedure of the numerical model is stopped temporarily, cf. [253].

The application of interpolation and extrapolation methods is shown in Figure 8.33. If the displacement response of the numerical model x_{N_n} and $x_{N_{n+1}}$ at two adjacent solution steps are known, the controller will still require also information about the displacement path between these responses. To compensate this missing information, Figure 8.33 (a) applies interpolation method. On the other, the solution of the next step $x_{N_{n+1}}$ might have a delay. To bridge the time gap required for the solution, Figure 8.33 (b) applies

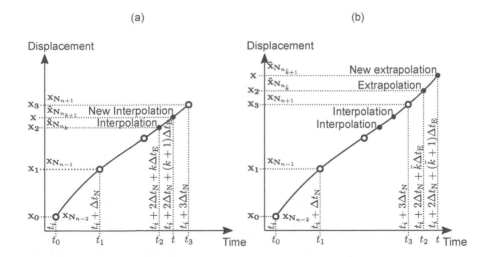

FIGURE 8.33
Application of the interpolation (a) and extrapolation (b) methods for the generation of the control signal and preventing delay between substructures of the RTHS.

extrapolation for the generation of the missing points of the control signal. For a higher accuracy, both during interpolation and extrapolation, not only the existing points, but also the displacement information from the previous interpolations and extrapolations are used as further control points as shown in Figure 8.33 and marked by $t_{1,2,3}$.

For the interpolation and extrapolation of control signal points, polynomial formulations can be used as introduced by Bonnet et al. [258]. An example is the LAGRANGE polynomial of third-order, which interpolates a fifth point $\mathbf{x}$ from four control points $\mathbf{x_{0,1,2,3}}$ as

$$t_0 = t_i \qquad\qquad \mathbf{x_0} = \mathbf{x_{N_{n-2}}}, \qquad (8.91)$$

$$t_1 = t_i + \Delta t_N \qquad\qquad \mathbf{x_1} = \mathbf{x_{N_{n-1}}}, \qquad (8.92)$$

$$t_2 = t_i + 2\Delta t_N + k\Delta t_E \qquad\qquad \mathbf{x_2} = \mathbf{\hat{x}_{N_{n_k}}}, \qquad (8.93)$$

$$t_3 = t_i + 3\Delta t_N \qquad\qquad \mathbf{x_3} = \mathbf{x_{N_{n+1}}}, \qquad (8.94)$$

$$t = t_i + 2\Delta t_N + (k+1)\,\Delta t_N \qquad\qquad \mathbf{x} = \mathbf{\hat{x}_{N_{n_{k+1}}}}, \qquad (8.95)$$

where

$$\hat{\mathbf{x}}_{\mathbf{N}_{n_{k+1}}} = \sum_{i=0}^{3} \mathbf{x_i} l_i(t), \tag{8.96}$$

$$l_i(t) = \prod_{\substack{j=0 \\ j \neq i}}^{3} \frac{t - t_j}{t_i - t_j}. \tag{8.97}$$

Here, $\mathbf{x}_{\mathbf{N}_{n-2}}$, $\mathbf{x}_{\mathbf{N}_{n-1}}$ and $\mathbf{x}_{\mathbf{N}_{n+1}}$ are vectors of displacement at relevant DoFs of the numerical substructure, which are simulated by the shaking table and actuators on the physical substructure. The discrete displacement responses correspond to the time points t_0, t_1 and t_3 of a time window, which begins with the initial time point t_i. To generate a continuous displacement on the physical substructure, as a 3th point, the last interpolation point $\hat{\mathbf{x}}_{\mathbf{N}_{n_k}}$ is used. k is the number of points required by the control signal. For the extrapolation, in the equations, k is replaced by $\tilde{k}$ as shown in Figures 8.32 and 8.33.

8.7.3 Physical Testing Part

The physical part of the RTHS consists of several hardware components. A possible configuration of the RTHS framework is shown in Figure 8.34. A servo-hydraulic shaking table or actuators induce the dynamic excitation of the physical system. The motion of the shaking table and actuators is driven by a controller, which calculates the required displacement and velocity from the results of the numerical part. The realized motion is measured by sensors, such as capacitive displacement transducers. A data acquisition (DAQ) system obtains the measured data and processes it together with further sensor signals coming from other transducers, such as load cells and strain gauges, which are measuring the interaction forces between the substructures. Furthermore, DAQs are also used to observe the dynamic response of the physical substructure, such as by accelerometers.

For the control of the servo-hydraulic system, the numerical substructure must be solved in real-time. For this purpose, generally, two computers are used: Host-PC and Target-PC, which are connected to each other via TCP/IP-network (Transmission control protocol/internet protocol). The Host-PC monitors the hybrid simulation process. The model of the numerical substructure is prepared in this computer usually in Matlab/Simulink- or Opensees-environment [259], and uploaded to the Target-PC. The Target-PC is a high performance computer, which can conduct real-time operations. The numerical model is solved by the Target-PC in real-time. Here, a control signal, which is required for the excitation of the physical substructure, is simultaneously generated and forwarded to the controller of the servo-hydraulic system. To minimize the delay of data transfer, fiber optical or serial high-speed cables are used, which are connected with the Target-PC and the controller by

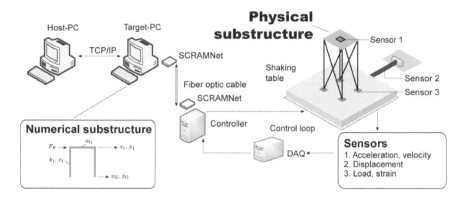

FIGURE 8.34
Hardware component configuration of the RTHS technique.

SCRAMNet-cards (Shared common RAM network). The solution of the numerical model and the generation of the control signal can also be distributed to two Target-PCs to optimize the simulation performance, cf. [252].

8.7.4 Response of a Frame Structure Incorporating SMA Wires Considering Soil-Structure Interaction

As previously described in Section 8.3, the vibration energy dissipated by SMA wires corresponds to the hysteresis surface enclosed by their stress-strain response. Accordingly, to achieve highest possible energy dissipation, the design goal is to maximize the enclosed hysteresis area without exceeding a maximum possible strain limit, which prevents any permanent deformation or tearing of the wires. On the other hand, below a certain strain level, SMA wires respond only linear elastically without any energy dissipation. Consequently, at least a certain strain level must be reached for the activation of the vibration control effect. These two strain levels depend on the chosen SMA properties and the loading pattern, which is expected to be applied to the wires.

Both the properties of the SMAs and the loading pattern must match to obtain the most efficient vibration control performance. The most relevant SMA wire properties are, besides their alloy composition, the wire geometry given by length and diameter. On the other hand, the loading pattern applied to the wires is characterized by the dynamic response of the structure. Accordingly, the design procedure of the vibration control systems incorporating SMA wires requires an accurate simulation and testing of the dynamic structural response.

As commonly known and also presented in the literature, such as by Kausel [260], Lou et al. [261], Mitropoulou et al. [262] as well as Anand and Satish Kumar [263], SSI can have a significant influence on the dynamic response of a

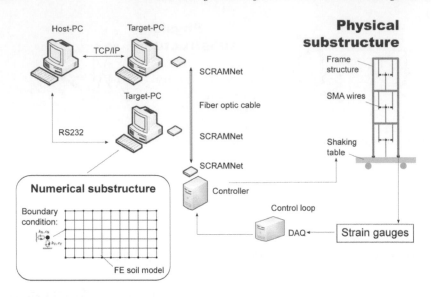

FIGURE 8.35
The RTHS framework used for the investigation of control performance of SMA wires considering the SSI effects.

structure. Therefore, the design procedure of SMA-based control devices must consider SSI. However, as discussed in Section 8.7, conventional test methods, such as by shaking tables, cannot consider SSI accurately.

This section presents the seismic performance investigation of a frame structure, which is incorporating SMA wires, considering SSI effects. The investigation of the same structure is presented previously in Sections 8.5.3 and 8.6.3 excluding SSI by conventional shaking table tests. In this section, the presented investigation uses the RTHS technique according to Ding et al. [264] to consider the SSI effects. Figure 8.35 shows the outline of the RTHS setup used in this investigation.

8.7.4.1 Numerical Simulation Part

The numerical substructure is a 20 m × 10 m FE soil model with 2 × 2 m mesh size. The model is discretized by 50 quadrilateral elements with 66 nodes and has 132 DoFs. The model is implemented using the Simulink environment developed by Zhou et al. [251]. The absorption of the energy produced by the scattering waves in the semi-infinite soil model is represented by an artificial viscous-spring boundary with pairs of dashpots and springs as proposed by Liu and Li [265].

The initial mass density of the soil is chosen as 2000 kg m^{-3}. Furthermore, for the YOUNG's modulus and POISSON's ratio 200 MPa and 0.2 are chosen respectively. Considering the dynamic properties of the tested frame structure, particularly its low mass, the parameters of the soil model are scaled-down according to the procedure presented by Wang et al. [266] and Lu et al. [267]. The chosen mass, damping and stiffness scales are $C_m = 400$, $C_c = 400$ and $C_k = 400$, respectively. Accordingly, the mass density, YOUNG's modulus and POISSON's ratio of the modeled soil are adapted as 5 kg m^{-3}, 0.5 MPa and 0.2, respectively.

For the solution of the numerical substructure the GUI-λ integration method is used, cf. Section 8.7.2. According to the previous investigations performed by Gui et al. $\lambda = 4$ enables an unconditionally stable integration scheme for nonlinear systems [241].

For the delay compensation, an adaptive forward prediction algorithm is used as proposed by Wallace et al. [268] based on the LAGRANGE polynomial method, cf. Section 8.7.2.

8.7.4.2 Physical Simulation Part

The physical substructure is the frame structure with SMA wires. On the frame structure, seismic excitation is simulated by the uniaxial shaking table. As for the conventional shaking table tests of Section 8.5.3, seismic testing facility at *Tsinghua University* is also used for this study together with the required RTHS hardware. The interaction forces between the shaking table and the frame structure are measured by strain gauges mounted on the legs of the structure. Furthermore, accelerometers are used to measure the structural response.

A distributed real-time calculation system is used to improve the performance of the simulation. In this system, the FE model of the soil is solved by a Target-PC. Accordingly, this Target-PC performs the so-called response analysis task (RAT) as presented by Zhu et al. [1]. A further Target-PC performs simultaneously the so-called signal generation task (SGT) and generates the control signal for the controller of the shaking table.

8.7.4.3 Results and Discussion

The scaled historic far-field earthquake El Centro as well as the near-field earthquakes Kobe and Taft are applied to the frame structure with the SMA wires corresponding to the previously presented conventional shaking table investigations. The configuration of the SMA wires corresponds to the previous study presented in Section 8.5.3.

The acceleration and displacement responses of the structure without and with SMAs due to the earthquakes are shown in Figures 8.36, 8.37 and 8.38. In Figure 8.39, the RMS and MAX of the displacement and acceleration responses are depicted. In Table 8.9, the reduction amount of RMS and MAX values are reported, which are calculated according to Equation 8.43, for the two cases excluding and including SSI effects.

From both time histories and the presented RMS, MAX results, we observe a significant change in performance picture of the SMA wires compared

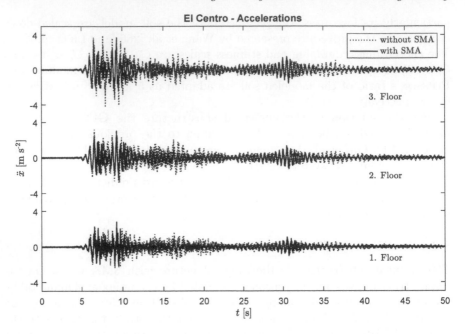

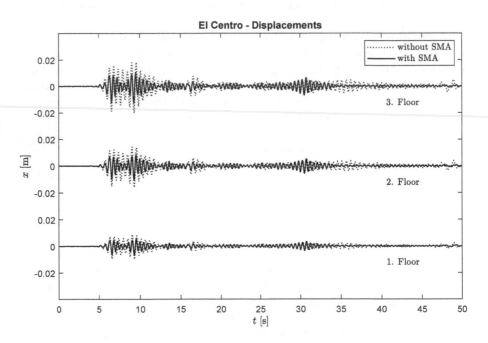

FIGURE 8.36

Acceleration and displacement responses of the frame structure without and with SMAs due to the El Centro earthquake considering SSI.

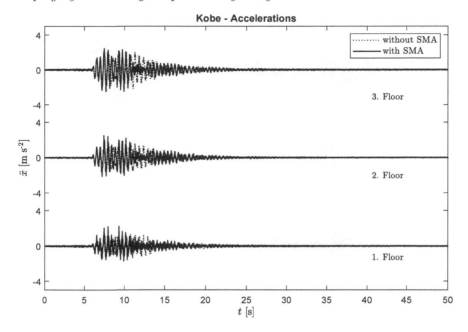

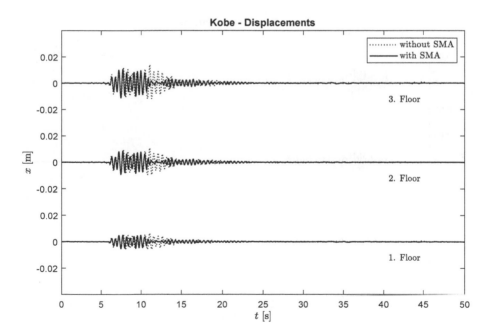

FIGURE 8.37

Acceleration and displacement responses of the frame structure without and with SMAs due to the Kobe earthquake considering SSI.

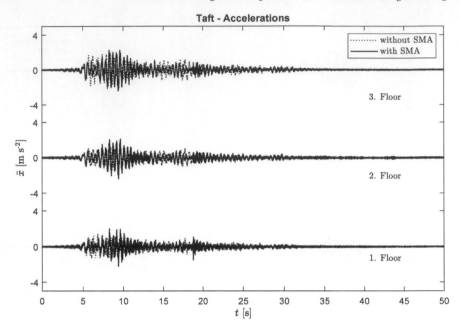

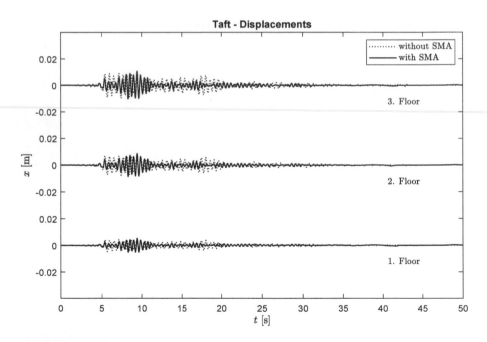

FIGURE 8.38
Acceleration and displacement responses of the frame structure without and with SMAs due to the Taft earthquake considering SSI.

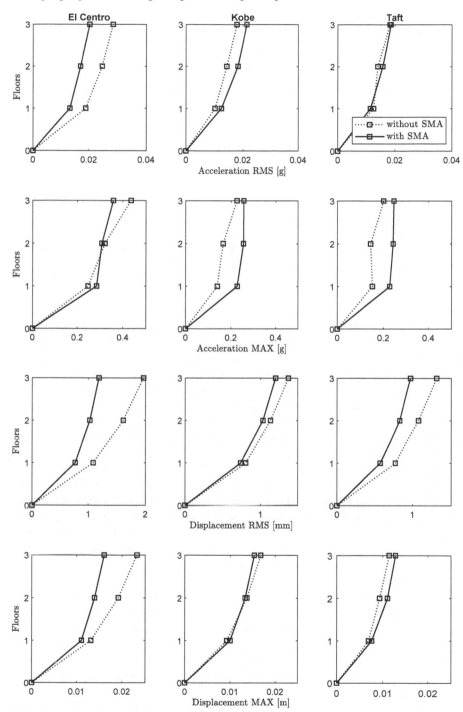

FIGURE 8.39
The RMS and MAX values of acceleration and displacement responses of the frame structure without and with SMAs due to the El Centro, Kobe and Taft earthquakes considering SSI.

TABLE 8.9

Reduction of RMS and MAX values of the acceleration and displacement responses of the frame structure without and with SMAs due to the El Centro, Kobe and Taft earthquakes excluding and including SSI.

Parameter	El Centro [%]	Kobe [%]	Taft [%]
Without SSI:			
RMS reduction of acceleration	41.0	59.3	15.5
MAX reduction of acceleration	17.0	36.3	−8.0
RMS reduction of displacement	52.3	52.0	30.0
MAX reduction of displacement	25.6	24.6	22.9
With SSI:			
RMS reduction of acceleration	29.8	−22.8	−2.5
MAX reduction of acceleration	17.8	−13.6	−23.0
RMS reduction of displacement	36.0	10.4	25.0
MAX reduction of displacement	26.0	2.8	−13.6

to the previous study excluding SSI (cf. Section 8.5.3). For the El Centro earthquake, the SMAs can control the vibration response of the structure efficiently. However, including SSI, the reduction of the RMS values is not as efficient as before excluding SSI. For both near field earthquakes, the vibration control deteriorates significantly. Both the RMS and MAX values of the acceleration response of the structure are increased by the SMAs. The RMS value of the displacement is reduced. However, the reduction amount is not as high as excluding SSI. An insignificant reduction of the MAX value of displacement is observed for the Kobe earthquake. For the Taft earthquake, this value is increased by the SMAs.

From the time histories, we observe that for the case including SSI effects even without SMA wires the vibration response amplitudes of the structure are quite low compared to the previous study results without SSI. There, the structural responses were mostly of resonance type with long transient vibration phases. Accordingly, for the case with SSI, it can be concluded that the SMA wires are not as highly activated as the case without SSI and the control performance of the SMAs is deteriorated.

To clarify this statement, we compare in Figure 8.40 the stress-strain responses of the SMA wires for the El Centro earthquake excluding and including SSI effects. In Figure 8.40 (a), the wire response is calculated using the approach presented in Section 8.6.3 from the interstory drift of the frame structure for the case excluding the SSI effects. In Figure 8.40 (b), the calculation is repeated for the case including SSI effects. In Figure 8.40 (c), the wire set configuration is optimized by combining two different wire lengths 100 mm and 70 mm.

From the stress-strain responses, we observe a reduction of the area and number of hysteresis loops, if SSI effects are considered. From Figure 8.40 (a) excluding SSI, we obtain the dissipated energy as 1299.0 J mm^{-3}. The dissipated energy reduces by considering the SSI in Figure 8.40 (b) to 573.1 J mm^{-3}. In Figure 8.40 (c), by combining the wire sets the energy dissipation value can be increased to 921.1 J mm^{-3}.

This noticeable change in the control performance of the SMA wires shows the importance of an accurate calculation and testing of the structural response. If the design of the wires is performed based on a wrong estimation of the structural response, the SMAs can even increase the vibrations. However, if the design matches with the requirements of the structure, a significant vibration control effect can be reached with SMA wires.

8.8   Conclusion

This chapter is concerned with the superelastic SMA–based damping systems, in particular with their dynamic response and mathematical modeling. Such dampers dissipate vibration energy by the phase transformation and the associated hysteretic damping. However, studies in this chapter and in other references show that the damping behavior depends strongly on the characteristics of the loading pattern.

As a two-phase material, SMAs undergo thermocoupled phase transformations between the austenite and martensite phases. Dynamic events with high strain rates limit the time required for the release and absorption of temperature and impair the phase transformation process. Consequently, the shape of the hysteretic loop surface enclosed by the stress-strain response reduces and the energy dissipation performance deteriorates. On the other hand, the material response, particularly the dynamic response, must be modeled accurately for an accurate design of SMAs in damping systems.

To consider the strain rate dependent effects, the chapter presents two different modeling approaches, which are based on the macroscopic constitutive SMA models of Auricchio et al. [222] and of Zhu and Zhang [223]. These models use the martensitic volume fraction as an internal variable and the strain, stress and temperature of the SMAs as external variables. In the first approach, the dynamic loading effects are considered by a time variant formulation of the change in entropy level of the SMAs in order to account for the increase in latent heat of the SMAs. The second approach is concerned with the variable strain rate-dependent formulation of the latent heat evolution of the SMAs.

The chapter derives the governing equations for both approaches and integrates them into the constitutive models. Experimental cyclic tensile tests are performed on the SMA wires for a phenomenological determination of the required modeling parameters. Furthermore, within these investigations, the dynamic SMA response is analyzed and computed by the proposed modeling

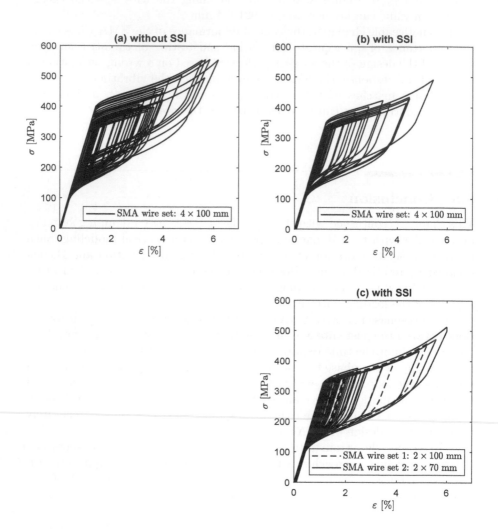

FIGURE 8.40

Stress-strain responses of the SMA wires during the El Centro earthquake. Responses are calculated for a wire set excluding SSI effects (a) and including SSI effects (b). The wire set configuration is optimized for the case including SSI in (c).

approaches. The measured and calculated results are compared for different load cases showing the accuracy of the mathematical modeling.

Shaking table tests are applied to a frame structure with superelastic SMA wires to investigate the vibration control performance of the SMA. The response of the structure is computed again using the SMA models and validated by the measured test results. On the same structure, the RTHS technique is applied to consider the SSI effects and particularly their influence on the vibration control of SMAs. For this purpose, a numerical soil model is solved and shaking table tests are conducted in real-time. The study showed that the SSI can significantly affect the dynamic response of the structure and, consequently, the SMA performance.

General conclusions of the studies regarding the dynamic response and the mathematical modeling of the superelastic SMAs are:

- Dynamic loading patterns with high strain rate deteriorate the release of temperature, which is generated by the evolution of martensite fraction in SMAs.

- Experimental studies show that the increase in material temperature causes a shift of the critical stress levels of the phase transformation and influences the SMA response as well as the amount of dissipated energy.

- This effect must be considered by mathematical models as it directly influences the performance of the SMA-based vibration control devices.

- An efficient and accurate way for this is the time-variant strain rate-dependent formulation of the latent heat evolution in the constitutive SMA models.

- Besides mathematical modeling of the dynamic effects, also the structural parameters must be determined accurately for a better SMA performance.

- An important structural aspect, which influences the vibration control level of SMAs, is the SSI effect on the structure.

- The SSI can be considered in performance assessment studies of SMA-based damping systems by the RTHS technique.

9

System Identification

9.1 Introduction

Control algorithms of active and semi-active systems require information about the state and parameters of structures. The state is described by their displacements, velocities and accelerations at the DoFs and the parameters are defined by dynamic properties, such as their natural frequencies, damping and mode shapes of structures. For the identification of the state and parameters of systems, numerous system identification methods have been proposed. However, particularly for semi-active vibration control systems, a real-time operation capability of the used identification method is decisive. Based on the KALMAN filter and its improved version, the unscented KALMAN filter, this chapter introduces an adaptive observer for the joint identification of the state and parameters of MDoF structures. The observer is particularly suitable for semi-active systems.

Section 9.2 presents a summary of existing identification methods. In Section 9.3, the KALMAN filter is introduced with its governing equations. In Section 9.4, the unscented KALMAN filter is introduced. The proposed adaptive observer is described in Section 9.5. In Section 9.6, numeric parametric studies are conducted to investigate the performance of the observer. The findings of the chapter are concluded in Section 9.7.

9.2 System Identification Methods

The existing system identification methods can be distinguished into non-recursive and recursive methods. The non-recursive methods operate in an offline manner and can detect the information after the measurement is completed. Some examples of these methods are the autoregressive moving average (ARMA) method, the stochastic subspace identification method [269–271] and the frequency-domain decomposition (FDD) method [270, 272].

Although the non-recursive methods can identify systems very accurately, for the semi-active vibration control systems it is necessary to use an observer based on a recursive identification method, which operates online and can identify the system in real-time. This approach allows the control system to react to the vibrations almost in real-time with a minor delay. A further requirement arises due to the fact that vibration control systems introduce generally nonlinearities into systems. Moreover, the response of the structure itself may have some nonlinearities. Therefore, the chosen system identification method must be able to predict these nonlinearities as well.

One of the most prominent examples of the recursive methods is the *least squares estimation* (LSE). As a time-domain method, LSE estimates system information by minimizing an error value calculated from the squared difference between the measured and expected values for the tracking of nonlinear MDoF systems [273–278]. However, one drawback of the LSE is that the method requires the measurement of the displacement and velocity of the structure. In particular, the displacement measurement is difficult to apply for civil engineering structures, as most of the measurement techniques require a fixed reference point. Furthermore, most of the dynamic measurement systems are based on accelerometers. Although displacement and velocity can be calculated from the acceleration signals by integration, inaccuracies arise due to the signal noise.

Another example of the recursive system identification is the *sequential Monte Carlo* (SMC) method, which is also referred to as the *particle filter* (PF) method. As a Bayesian approximation technique, the SMC uses time-domain samples (particles) to represent the posterior probability of a state or parameter of a system [279–289]. However, for complex systems, the required computational effort may be very high as the number of required samples increases with increasing complexity.

Besides the SMC method, another approach for recursive system identification is the KALMAN *filter* (KF), which is also a Bayesian approximation technique in time-domain [290]. The method identifies the behavior of linear systems by weighting the deviation between their measured and estimated responses with the so-called KALMAN *gain*, which is calculated from the covariances of the estimation error and of the noise comprised in sensors and model of the system. The method is described in Section 9.3.

For the identification of nonlinear systems, two modifications of the KF exist: the *extended* KALMAN *filter* (EKF) and the *unscented* KALMAN *filter* (UKF), which also operate in a recursive manner.

In the EKF, the system response is approximated by the first-order linearization of the system and measurement equations through TAYLOR series expansion. As one of the most common identification methods for nonlinear systems, the EKF has been used for numerous applications, such as [291–306]. However, the linearization procedure of the EKF involves the derivation of JACOBIAN matrices at each time step, which generally requires a considerable computational effort. Furthermore, the approximation is accurate only for systems, which are almost linear. Therefore, for higher nonlinearities, the UKF method is preferable, which is also computationally superior as shown in [307–309].

The UKF uses the *unscented transformation* (UT) method as proposed by Julier et al. [310–313] and extended by other researchers, such as Wan and Van Der Merwe [314], Van Zandt [315] and by Tenne and Singh [316]. A set of sample points are chosen deterministically, which capture the statistical representation of the state and parameters of the system. These points are propagated through the nonlinear system and the measurement functions of the system without linearization. After propagation, the chosen points still represent the statistics accurately as described in Section 9.4. Accordingly, the UKF can consider nonlinearities of systems more accurately than the EKF method. Furthermore, the UKF does not involve any computationally costly calculation and is, therefore, more efficient than the EKF method. The UKF has been applied for numerous cases, such as by Sitz et al. [317], Popercu and Wong [318], Voss et al. [319], Wu and Smyth [309] and by Chatzi and Smyth [289] for the tracking of nonlinear MDoF systems as well as by Miah et al. [320, 321] and by Dertimanis and Chatzi [322] for the tracking of semi-active and active systems and by Van Der Merwe et al. [323, 324], Chen et al. [325], Crassidis and Markley [326], Choi et al. [327, 328], Lee and Alfriend [329], Romanenko and Castro [330] and by Li et al. [331] for the navigation and tracking of processes as well as for the training of neural networks.

As previous observations show, particularly after strong dynamic events, such as earthquakes, structures may exhibit a significant stiffness degradation causing up to 20 % abrupt reduction in the natural frequency. As also reported by Naeim [332], after the Northridge Earthquake in California, the stiffness degradation was initiated both by the loss of non-structural elements and damages in the core structure. The time span between the main event and aftershocks may be short. Therefore, the re-tuning of structural control devices and particularly the adaptation of semi-active systems require a quick and if possible a real-time identification of structural changes by a joint state-parameter observer.

Although the UKF can identify nonlinear systems, such an abrupt change in parameters cannot be covered by conventional algorithms in real-time. Furthermore, the supplementary damping applied by the vibration control impedes the identification of the systems. Therefore, an adaptive definition scheme is required to enable the UKF method to be operable both before and after the abrupt change. In case of a substantially high structural damage, such an observer can also be useful for the determination of the remaining structural seismic capacity and if necessary for the initiation and coordination of rescue services. In Section 9.5, such an adaptive observer is introduced for the joint identification of the state and parameters of MDoF structures incorporating vibration control devices.

9.3 KALMAN Filter

For the derivation of the governing equations of the KF, we consider the discrete state-space model (cf. Section 3.2) of a linear system over the sampling interval Δt from t_k to t_{k+1}. The model reads

$$\mathbf{z}_{k+1} = \mathcal{A}_d \mathbf{z}_k + \mathcal{B}_d \mathbf{u}_k + \mathbf{w}_k, \qquad (9.1)$$

$$\mathbf{y}_k = \mathcal{C} \mathbf{z}_k + \mathcal{D} \mathbf{u}_k + \mathbf{v}_k, \qquad (9.2)$$

where $\mathcal{A}_d \in \mathbb{R}^{2n \times 2n}$ is the discrete transition matrix, $\mathcal{B}_d \in \mathbb{R}^{2n \times n}$ the discrete input matrix, $\mathcal{C} \in \mathbb{R}^{m \times 2n}$ the output matrix and $\mathcal{D} \in \mathbb{R}^{m \times n}$ the feedthrough matrix as previously introduced in Section 3.2 for n DoFs and m sensors. Furthermore, in Equations 9.1 and 9.2, $\mathbf{z}_k \in \mathbb{R}^{2n \times 1}$, $\mathbf{u}_k \in \mathbb{R}^{n \times 1}$ and $\mathbf{y}_k \in \mathbb{R}^{m \times 1}$ are the state, input and output vectors, respectively at time t_k.

In Equations 9.1 and 9.2, $\mathbf{w}_k \in \mathbb{R}^{2n \times 1}$ and $\mathbf{v}_k \in \mathbb{R}^{m \times 1}$ are process and measurement noise vectors, respectively, which are assumed to be of zero-mean, distributed according to a GAUSSIAN distribution and are independent from each other. Accordingly, their covariance matrices read

$$E[\mathbf{v}_i \mathbf{v}_j^\top] = \mathbf{Q}_i \qquad \text{for } i = j, \text{ otherwise } \mathbf{0}, \qquad (9.3)$$

$$E[\mathbf{w}_i \mathbf{w}_j^\top] = \mathbf{R}_i \qquad \text{for } i = j, \text{ otherwise } \mathbf{0}, \qquad (9.4)$$

$$E[\mathbf{v}_i \mathbf{w}_j^\top] = \mathbf{0}_i \qquad \text{for all } i, j, \qquad (9.5)$$

where E is the expectation operator. $\mathbf{Q}_i$ and $\mathbf{R}_i$ are referred to as the *process* and *measurement noise covariance* matrices, respectively. Both matrices are usually constant, in which case $\mathbf{Q}$ and $\mathbf{R}$ are used for the notation.

The KF consists of an *innovation process* and a *state prediction process*, which are performed sequentially. During the innovation process, the model output is given by

$$\hat{\mathbf{y}}_k = \mathcal{C} \hat{\mathbf{z}}_k + \mathcal{D} \hat{\mathbf{u}}_k \qquad (9.6)$$

presuming that $\hat{\mathbf{z}}_k$ and $\hat{\mathbf{u}}_k$ are known, where $\hat{\ }$ denotes an estimate. $\hat{\mathbf{z}}_k$ is also referred to as the *a priori state estimate*.

The a priori state estimate $\hat{\mathbf{z}}_k$ is updated and the *a posteriori state estimate* $\tilde{\mathbf{z}}_k$ is calculated from the residual of the measured $\mathbf{y}_k$ and the estimated $\hat{\mathbf{y}}_k$ outputs, which is weighted by the KALMAN *gain* $\mathbf{K}_k$ as

$$\tilde{\mathbf{z}}_k = \hat{\mathbf{z}}_k + \mathbf{K}_k (\mathbf{y}_k - \hat{\mathbf{y}}_k). \qquad (9.7)$$

The difference between the outputs is referred to as the *innovation error*, which is defined as

$$\mathbf{e}_k = \mathbf{y}_k - \hat{\mathbf{y}}_k. \qquad (9.8)$$

The KALMAN gain is computed by minimizing the mean-square estimation error using the *state covariance* matrix $\hat{\mathbf{P}}_k$ and the measurement noise

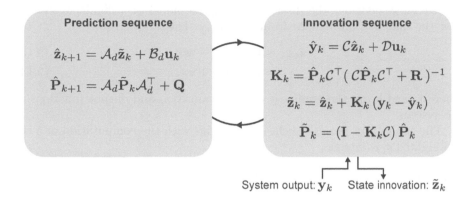

FIGURE 9.1
The recursive architecture of the KALMAN filter consisting of the innovation and prediction sequences.

covariance matrix $\mathbf{R}$ as well as the output matrix $\mathcal{C}$ as

$$\mathbf{K}_k = \hat{\mathbf{P}}_k \mathcal{C}^\top (\mathcal{C}\hat{\mathbf{P}}_k \mathcal{C}^\top + \mathbf{R})^{-1}. \tag{9.9}$$

During the innovation process, also the *estimation error covariance* matrix is updated as

$$\tilde{\mathbf{P}}_k = (\mathbf{I} - \mathbf{K}_k \mathcal{C})\hat{\mathbf{P}}_k. \tag{9.10}$$

where $\mathbf{I}$ is an identity matrix. At the state prediction process, the state of the system is then predicted as

$$\hat{\mathbf{z}}_{k+1} = \mathcal{A}_d \tilde{\mathbf{z}}_k + \mathcal{B}_d \mathbf{u}_k \tag{9.11}$$

and the state covariance is predicted from

$$\hat{\mathbf{P}}_{k+1} = \mathcal{A}_d \tilde{\mathbf{P}}_k \mathcal{A}_d^\top + \mathbf{Q}. \tag{9.12}$$

Figure 9.1 shows both the innovation and the prediction sequences of the KF and illustrates the recursive architecture of the filter.

9.4 Unscented KALMAN Filter

As previously mentioned, based on the general scheme of the KF, the UKF is able to observe the state of nonlinear systems without losing the computational efficiency of the KF. For the derivation of the governing equations of the UKF, we consider the discrete state-space model of a nonlinear system over the sampling interval Δt from t_k to t_{k+1}. The model reads

$$\mathbf{z}_{k+1} = \mathbf{f}_d[\mathbf{z}_k, \mathbf{u}_k] + \mathbf{w}_k, \tag{9.13}$$

$$\mathbf{y}_k = \mathbf{h}[\mathbf{z}_k, \mathbf{u}_k] + \mathbf{v}_k, \tag{9.14}$$

where $\mathbf{f}_d$ and $\mathbf{h}$ are the nonlinear discrete transition and output functions of the system.

The *innovation process* of the UKF begins with the computation of a set of weighted σ-*points* as

$$\hat{\sigma}_k = \left[\hat{\mathcal{Z}}_{0,k} \cdots \hat{\mathcal{Z}}_{i,k} \cdots \hat{\mathcal{Z}}_{i+n,k} \cdots \mid W_0^{(m)}\ W_0^{(c)} \cdots W_i^{(m)} \cdots W_{i+n}^{(m)} \cdots \right], \tag{9.15}$$

which approximate the GAUSSIAN distributed state at a defined time t_k. Here, $\hat{\mathcal{Z}}_{j,k}$ is the jth σ-point and W_j the corresponding weight factor with $j \in [0, 2n]\ \forall n \in \mathbb{N}$, where n is representing the dimension of the state. The first σ-point is located at the center of the distribution and corresponds to the mean value. The remaining $2n$ σ-points are distributed equidistantly along the eigenaxes of the covariance matrix of the state estimate $\hat{\mathbf{P}}_k$. Accordingly, the σ-points and the associated weights are calculated as

$$\hat{\mathcal{Z}}_{0,k} = \hat{\mathbf{z}}_k \qquad\qquad W_0^{(m)} = \frac{\lambda}{n+\lambda}, \tag{9.16}$$

$$W_0^{(c)} = \frac{\lambda}{n+\lambda} + 1 - \alpha^2 + \beta,$$

$$\hat{\mathcal{Z}}_{i,k} = \hat{\mathbf{z}}_k + \alpha[\sqrt{(n+\lambda)\hat{\mathbf{P}}_k}]_i \qquad W_i^{(m)} = W_i^{(c)} = \frac{1}{2(n+\lambda)}, \tag{9.17}$$

$$\hat{\mathcal{Z}}_{i+n,k} = \hat{\mathbf{z}}_k - \alpha[\sqrt{(n+\lambda)\hat{\mathbf{P}}_k}]_{i+n} \qquad W_{i+n}^{(m)} = W_{i+n}^{(c)} = \frac{1}{2(n+\lambda)}, \tag{9.18}$$

where $\hat{\mathbf{z}}_k$ is the *a priori state estimate*, which is assumed to be known at the first time step. The parameter λ reads

$$\lambda = \alpha^2(n+\kappa) - n, \tag{9.19}$$

where $10^{-4} \le \alpha \le 1$ scales the distance of the sigma points around $\hat{\mathbf{z}}_k$ and is set generally to a small positive value, cf. [324, 333]. κ is an adjustment parameter [310], which is optimally defined as

$$\kappa = n - 3. \tag{9.20}$$

In Equation 9.16, β is a further adjustment parameter, which is optimally $\beta = 2$ with the assumption of an ideal GAUSSIAN distribution [333].

The covariance matrix of the state estimate $\hat{\mathbf{P}}_k$ is always positive semi-definite. Accordingly, in Equations 9.17 and 9.18, the matrix square root can be computed by the CHOLESKY decomposition, as

$$\hat{\mathbf{P}}_k = \mathbf{L}_k \mathbf{L}_k^\top, \tag{9.21}$$

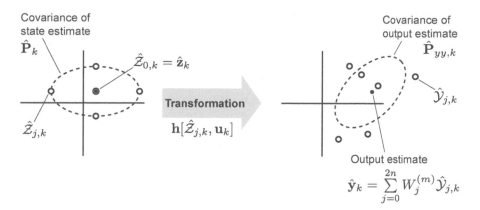

FIGURE 9.2
2D illustration of the unscented transformation of the state estimate to output
estimate using the σ-points.

where $\mathbf{L}_k$ and $\hat{\mathbf{P}}_k$ have the same eigenvectors and $\mathbf{L}_k$ corresponds to $\sqrt{\hat{\mathbf{P}}_k}$.
Consequently, in Equation 9.17, the ith column (respectively in Equation 9.18,
the $(i + n)^{\text{th}}$ column) of the expression within the brackets $[\cdot]$ will be scaled
by α and added to (respectively subtracted from) the state estimate $\hat{\mathbf{z}}_k$.

The next step of the innovation process of the UKF is the calculation of
the measurement covariance $\hat{\mathbf{P}}_{yy,k}$ and the cross-covariance $\hat{\mathbf{P}}_{xy,k}$ as

$$\hat{\mathbf{P}}_{yy,k} = \sum_{j=0}^{2n} W_j^{(c)} (\hat{\mathcal{Y}}_{j,k} - \hat{\mathbf{y}}_k)(\hat{\mathcal{Y}}_{j,k} - \hat{\mathbf{y}}_k)^\top + \mathbf{R}, \tag{9.22}$$

$$\hat{\mathbf{P}}_{xy,k} = \sum_{j=0}^{2n} W_j^{(c)} (\hat{\mathcal{Z}}_{j,k} - \hat{\mathbf{z}}_k)(\hat{\mathcal{Y}}_{j,k} - \hat{\mathbf{y}}_k)^\top, \tag{9.23}$$

where the σ-points of the output are calculated by the UT as

$$\hat{\mathcal{Y}}_{j,k} = \mathbf{h}[\hat{\mathcal{Z}}_{j,k}, \mathbf{u}_k] \tag{9.24}$$

and the output is estimated from the calculated σ-points as

$$\hat{\mathbf{y}}_k = \sum_{j=0}^{2n} W_j^{(m)} \hat{\mathcal{Y}}_{j,k}. \tag{9.25}$$

Figure 9.2 shows the UT process of the state estimate $\hat{\mathbf{z}}_k$ to output estimate
$\hat{\mathbf{y}}_k$ using the σ-points. First, the σ-points $(\hat{\mathcal{Z}}_{j,k})$ are determined according to
Equations 9.16–9.18. The σ-points are then propagated through the nonlinear
output function $\mathbf{h}$.

From both covariance matrices, the KALMAN *gain* of the UKF is computed as

$$\mathbf{K}_k = \hat{\mathbf{P}}_{zy,k}\hat{\mathbf{P}}_{yy,k}^{-1}, \tag{9.26}$$

with which the *a posteriori state estimate* is calculated from

$$\tilde{\mathbf{z}}_k = \hat{\mathbf{z}}_k + \mathbf{K}_k(\mathbf{y}_k - \hat{\mathbf{y}}_k) \tag{9.27}$$

considering the residual of the estimated and the measured outputs. Accordingly, the covariance matrix of the state estimate is updated as

$$\tilde{\mathbf{P}}_k = \hat{\mathbf{P}}_k - \mathbf{K}_k\hat{\mathbf{P}}_{yy,k}\mathbf{K}_k^\top. \tag{9.28}$$

The *state prediction process* of the UKF begins with the update of the σ-points considering the state innovation $\tilde{\mathbf{z}}_k$ and the updated covariance matrix $\tilde{\mathbf{P}}_k$ by

$$\tilde{\mathcal{Z}}_{0,k} = \tilde{\mathbf{z}}_k \qquad\qquad W_0^{(m)} = \frac{\lambda}{n+\lambda}, \tag{9.29}$$

$$W_0^{(c)} = \frac{\lambda}{n+\lambda} + 1 - \alpha^2 + \beta,$$

$$\tilde{\mathcal{Z}}_{i,k} = \tilde{\mathbf{z}}_k + \alpha[\sqrt{(n+\lambda)\tilde{\mathbf{P}}_k}]_i \qquad W_i^{(m)} = W_i^{(c)} = \frac{1}{2(n+\lambda)}, \tag{9.30}$$

$$\tilde{\mathcal{Z}}_{i+n,k} = \tilde{\mathbf{z}}_k - \alpha[\sqrt{(n+\lambda)\tilde{\mathbf{P}}_k}]_{i+n} \qquad W_{i+n}^{(m)} = W_{i+n}^{(c)} = \frac{1}{2(n+\lambda)}, \tag{9.31}$$

where $\tilde{\mathbf{z}}_k$ is the state innovation and all remaining terms, apart from the updated innovation matrix $\hat{\mathbf{P}}_k$, correspond to Equations 9.16–9.18.

As a next step, the updated σ-points are propagated through the nonlinear discrete transition function $\mathbf{f}_d$ and the state prediction is computed as

$$\hat{\mathcal{Z}}_{j,k+1} = \mathbf{f}_d[\tilde{\mathcal{Z}}_{j,k}, \mathbf{u}_k] \tag{9.32}$$

and the state is predicted from σ-points as

$$\hat{\mathbf{z}}_{k+1} = \sum_{j=0}^{2n} W_j^{(m)}\hat{\mathcal{Z}}_{j,k+1}. \tag{9.33}$$

As in the KF, the last step is the prediction of the state covariance matrix by

$$\hat{\mathbf{P}}_{k+1} = \sum_{j=0}^{2n} W_j^{(c)}(\hat{\mathcal{Z}}_{j,k+1} - \hat{\mathbf{z}}_{k+1})(\hat{\mathcal{Z}}_{j,k+1} - \hat{\mathbf{z}}_{k+1})^\top + \mathbf{Q}. \tag{9.34}$$

Figure 9.3 depicts the innovation and prediction sequences of the UKF.

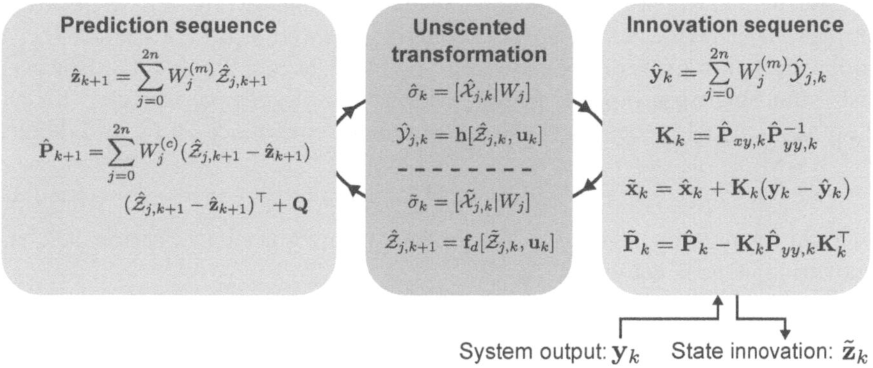

Prediction sequence

$$\hat{\mathbf{z}}_{k+1} = \sum_{j=0}^{2n} W_j^{(m)} \hat{\mathcal{Z}}_{j,k+1}$$

$$\hat{\mathbf{P}}_{k+1} = \sum_{j=0}^{2n} W_j^{(c)} (\hat{\mathcal{Z}}_{j,k+1} - \hat{\mathbf{z}}_{k+1})$$

$$(\hat{\mathcal{Z}}_{j,k+1} - \hat{\mathbf{z}}_{k+1})^\top + \mathbf{Q}$$

Unscented transformation

$$\hat{\sigma}_k = [\hat{\mathcal{X}}_{j,k} | W_j]$$

$$\hat{\mathcal{Y}}_{j,k} = \mathbf{h}[\hat{\mathcal{Z}}_{j,k}, \mathbf{u}_k]$$

$$- - - - - - - -$$

$$\tilde{\sigma}_k = [\tilde{\mathcal{X}}_{j,k} | W_j]$$

$$\hat{\mathcal{Z}}_{j,k+1} = \mathbf{f}_d[\tilde{\mathcal{Z}}_{j,k}, \mathbf{u}_k]$$

Innovation sequence

$$\hat{\mathbf{y}}_k = \sum_{j=0}^{2n} W_j^{(m)} \hat{\mathcal{Y}}_{j,k}$$

$$\mathbf{K}_k = \hat{\mathbf{P}}_{xy,k} \hat{\mathbf{P}}_{yy,k}^{-1}$$

$$\tilde{\mathbf{x}}_k = \hat{\mathbf{x}}_k + \mathbf{K}_k (\mathbf{y}_k - \hat{\mathbf{y}}_k)$$

$$\tilde{\mathbf{P}}_k = \hat{\mathbf{P}}_k - \mathbf{K}_k \hat{\mathbf{P}}_{yy,k} \mathbf{K}_k^\top$$

System output: $\mathbf{y}_k$ State innovation: $\tilde{\mathbf{z}}_k$

FIGURE 9.3
Similar to the KF, the UKF has an recursive architecture consisting of innovation and prediction sequences.

9.5 Adaptive Joint State-Parameter Observer

In this section, a system identification scheme is introduced for MDoF structures equipped with vibration control systems. The approach involves a joint state and parameter observer, which is based on the UKF and can consider abrupt changes, such as stiffness degradation of the structure. The goal of the scheme is to provide both the state and the relevant parameters of the structure, such as natural frequency and inherent damping, for the online tuning of control systems. The observer is designed to be integrated into an *open/closed-loop* scheme. Accordingly, both input and output signals of the system are used.

The UKF, as presented in Section 9.4, is able to estimate the state of nonlinear systems. However, for a joint state and parameter estimation, an *augmented state vector* is required. For a set of unknown parameters Θ_k with the dimension s, the augmented state vector at time t_k reads

$$\mathbf{z}_k^a = \begin{bmatrix} \mathbf{z}_k \\ \Theta_k \end{bmatrix} \tag{9.35}$$

where $\mathbf{z}_k^a \in \mathbb{R}^{(2n+s)\times 1}$ with the state vector $\mathbf{z}_k \in \mathbb{R}^{(2n)\times 1}$ as previously introduced. For the next time step t_{k+1}, it holds

$$\Theta_{k+1} = \Theta_k + \mathbf{w}_k^\Theta, \tag{9.36}$$

where $\mathbf{w}_k^\Theta$ is the process noise belonging to the unknown parameters. Accordingly, the state-space model must also be augmented, as thereafter shown by an example in Section 9.6.

For the identification of abrupt parameter changes, the UKF requires an adaptive framework, which is based on the updated state covariance $\tilde{\mathbf{P}}_k$. A similar framework was previously proposed by Bisht and Singh [334]. However, Bisht and Singh use in their framework a scalar trigger parameter β for the adaptation, which is based on the measurement covariance $\mathbf{P}_{yy,k}$ and defined as

$$\beta = \mathbf{e}_k^\top \mathbf{P}_{yy,k}^{-1} \mathbf{e}_k, \qquad (9.37)$$

with the innovation error $\mathbf{e}_k$. As previously introduced in Section 9.3, the innovation error is computed by

$$\mathbf{e}_k = \mathbf{y}_k - \hat{\mathbf{y}}_k \qquad (9.38)$$

and corresponds to the difference between the measured $\mathbf{y}_k$ and the estimated $\hat{\mathbf{y}}_k$ outputs of time t_k. The innovation error is expected to become smaller, when the UKF can estimate the system output accurately. In case of a system change, the accuracy of the UKF will drop. The same applies also for the measurement covariance $\mathbf{P}_{yy,k}$, which reads as

$$\mathbf{P}_{yy,k} = \sum_{j=0}^{2n} W_j^{(c)} (\mathcal{Y}_{j,k} - \hat{\mathbf{y}}_k)(\mathcal{Y}_{j,k} - \hat{\mathbf{y}}_k)^\top + \mathbf{R}. \qquad (9.39)$$

Compared to previous Equation 9.22 of $\hat{\mathbf{P}}_{yy,k}$, Equation 9.39 differs slightly, as the UKF scheme applied by Bisht and Singh begins first with the prediction process different than in Sections 9.3 and 9.4, where the filter method is introduced beginning with the innovation process.

To activate the adaptation of the state covariance $\tilde{\mathbf{P}}_k$, in their method, Bisht and Singh use a threshold β_0, which indicates a system change for $\beta > \beta_0$. The threshold β_0 is a scalar, which needs to be calculated based on the fact that β is Chi-square distributed [335] for a zero-mean GAUSSIAN innovation error with the number of DoFs corresponding to the number of measured outputs.

Accordingly, the method of Bisht and Singh requires information about the a posteriori output of the system, particularly for the computation of the measurement covariance with the sigma points $\mathcal{Y}$.

A similar method is also proposed by Rahimi et al. [336], where a time-dependent variable threshold is introduced to compensate the unknown a posteriori information.

To overcome the necessity of the a posteriori knowledge of the measurement covariance and the time dependent variable threshold, Schleiter and Altay [337] proposed the following trigger parameter formulation for the application of A-UKF to controlled MDoF structures:

$$\gamma = \mathbf{e}_k^\top \mathbf{R}^{-1} \mathbf{e}_k, \qquad (9.40)$$

where the innovation error $\mathbf{e}_k$ is normalized by the measurement noise covariance matrix $\mathbf{R}$. Accordingly, this formulation does not require a posteriori knowledge of the measurement covariance and, consequently, its calculation does not require UT.

Equation 9.40 simplifies itself for m sensors with identical properties, i.e., each sensor with the identical variance R_i, as

$$\gamma = \begin{bmatrix} e_1 \\ \vdots \\ e_m \end{bmatrix}^\top \begin{bmatrix} R_1 & \cdots & 0 \\ \vdots & \ddots & \vdots \\ 0 & \cdots & R_m \end{bmatrix}^{-1} \begin{bmatrix} e_1 \\ \vdots \\ e_m \end{bmatrix} = \frac{e_1^2}{R_1} + \cdots + \frac{e_m^2}{R_m} = \frac{e_1^2 + \cdots + e_m^2}{R_i}.$$

(9.41)

The innovation error corresponding to the output i is furthermore defined as the difference between measured and estimated outputs. Accordingly, it holds

$$e_i = y_i - \hat{y}_i = y_i - ([\mathcal{C}\hat{\mathbf{z}}]_i + [\mathcal{D}\mathbf{u}]_i),$$

(9.42)

where the expressions within the brackets $[\cdot]$ correspond to the ith element of the matrices $\mathcal{C}$ and $\mathcal{D}$, which are, as introduced previously, the output and feedthrough matrices, respectively.

The adaptation is initiated, when the trigger parameter γ of Equation 9.41 exceeds the threshold γ_0. In this formulation, the used threshold γ_0 is a time-independent scalar.

$$\gamma > \gamma_0 \ \rightarrow \ \text{adaptation}$$

(9.43)

As soon as the UKF converges and starts to predict the state accurately, the innovation error e_i corresponds only to the noises of the output y_i and the input u_i. This equilibrium stays valid until a system change occurs. Therefore, to define the threshold γ_0, we consider the equilibrium case before system change and add to the right side of Equation 9.42

$$e_i \approx n_y - n_u,$$

(9.44)

where n_y and n_u are the output and input measurement noises, respectively. On the other hand, if the system output is observed only by displacement and velocity sensors, the feedthrough matrix $\mathcal{D}$ in Equation 9.42 becomes zero and the innovation error corresponds only to the output measurement noise n_y.

For a system incorporating only accelerometers, if we assume that both noises n_y and n_u to be zero-mean GAUSSIAN, their superposition, the innovation error e_i, is to be a zero-mean GAUSSIAN as well with a cumulative variance of $2R_i$. For a system with only displacement and velocity sensors, the cumulative variance becomes R_i. We can now formulate the threshold γ_0 by rewriting Equations 9.41 and 9.43 as

$$\gamma = m\frac{e_i^2}{R_i} \approx m\delta z_{e,i}^2 > \gamma_0 \approx m\delta z_{e,0}^2 \ \rightarrow \ \text{adaptation},$$

(9.45)

where m is the number sensors and

$$\delta = \begin{cases} 2 & \text{for systems with accelerometers} \\ 1 & \text{for systems with displacement and velocity sensors.} \end{cases}$$

(9.46)

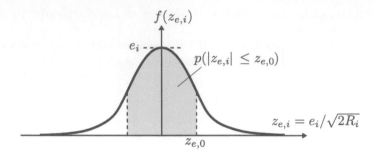

FIGURE 9.4
The zero-mean GAUSSIAN distribution of the innovation error e_i corresponding to the output i measured by an accelerometer.

For a mixed sensor system, δ can be defined individually in a similar manner.

As shown in Figure 9.4, $z_{e,i}$ is the standard normally distributed variable of the innovation error e_i, which reads

$$z_{e,i} = e_i/\sqrt{R}, \tag{9.47}$$

where $R = 2R_i$ for accelerometers and $R = R_i$ for displacement and velocity sensors.

Accordingly, on the right side of Equation 9.45, $z_{e,0}$ is a chosen variable, which defines the threshold γ_0. The effects of γ_0 on the accuracy of the observer will be investigated in Section 9.6 by parametric studies.

The updated state covariance matrix reads corresponding to the augmented state vector (Equation 9.35) as

$$\tilde{\mathbf{P}}_k = \begin{bmatrix} \tilde{\mathbf{P}}_{zz,k} & \tilde{\mathbf{P}}_{z\Theta,k} \\ \tilde{\mathbf{P}}_{z\Theta,k} & \tilde{\mathbf{P}}_{\Theta\Theta,k} \end{bmatrix}. \tag{9.48}$$

In case of an adaptation, in the updated state covariance matrix

$$\tilde{\mathbf{P}}_{\Theta\Theta,k} = \begin{bmatrix} \tilde{P}_{\Theta\Theta,11} & \cdots & \tilde{P}_{\Theta\Theta,1n} \\ \vdots & \tilde{P}_{\Theta\Theta,ii} & \vdots \\ \tilde{P}_{\Theta\Theta,n1} & \cdots & \tilde{P}_{\Theta\Theta,nn} \end{bmatrix}, \tag{9.49}$$

the entry $\tilde{P}_{\Theta\Theta,ii}$, which corresponds to the DoF i of the system parameter change, is replaced by a high constant covariance value P_a.

For the localization of the corresponding DoF i, the adaptation is applied sequentially for every DoFs with $i = \{1, \cdots, n\}$. Each adaptation is followed by an UKF and the corresponding trigger parameter γ_i is calculated according to Equation 9.40. The calculation with the DoF j, which yields the smallest

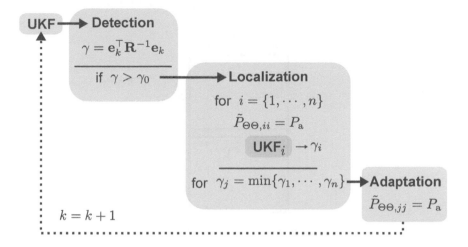

FIGURE 9.5
Adaptive scheme applied to the UKF for the joint state-parameter estimation of controlled structures exhibiting abrupt changes.

trigger parameter with $\gamma_j = \min\{\gamma_1, \cdots, \gamma_n\}$ indicates the correct adaptation and means that the corresponding DoF is localized. The matching $\tilde{P}_{\Theta\Theta,jj}$ is then updated with P_a. The processes of localization and adaptation are illustrated in Figure 9.5.

9.6 Parametric Studies

This section investigates the performance of the proposed adaptive joint state-parameter observer by means of numerical parametric studies. For this purpose, we consider a 2-DoF frame structure without and with a semi-active TMD attached to the top floor of the structure. The TMD can adapt its stiffness.

The 2DoF+TMD system is shown in Figure 9.6. Ground acceleration signals are applied and the structure exhibits abrupt stiffness change due to damages. Accelerometers measure both the motion of the 2DoF+TMD system as well as the ground acceleration.

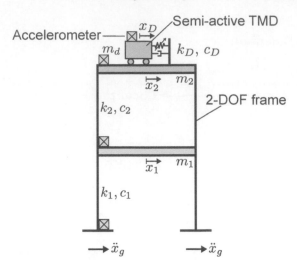

FIGURE 9.6
Numerically investigated 2-DoF frame structure with a semi-active TMD and sensors.

9.6.1   System Modeling

For the modeling of the 2DoF+TMD system, we consider the EoM

$$\mathbf{M}\ddot{\mathbf{x}}(t) + \mathbf{C}\dot{\mathbf{x}}(t) + \mathbf{K}(t)\mathbf{x}(t) = \mathbf{M}\ddot{x}_g(t)\boldsymbol{\nu}_u, \qquad (9.50)$$

where $\mathbf{M}$ and $\mathbf{C}$ are the time-invariant mass and damping matrices, respectively. $\mathbf{K}(t)$ is the stiffness matrix with time-variant stiffness values of the structure and time-variant stiffness value of the TMD. The matrices of the EoM read

$$\mathbf{M} = \begin{bmatrix} m_1 & 0 & 0 \\ 0 & m_2 & 0 \\ 0 & 0 & m_D \end{bmatrix}, \quad \mathbf{C} = \begin{bmatrix} c_1 + c_2 & -c_2 & 0 \\ -c_2 & c_2 + c_d & -c_D \\ 0 & -c_D & c_D \end{bmatrix},$$

$$\mathbf{K}(t) = \begin{bmatrix} k_1(t) + k_2(t) & -k_2(t) & 0 \\ -k_2(t) & k_2(t) + k_D(t) & -k_D(t) \\ 0 & -k_D(t) & k_D(t) \end{bmatrix}, \qquad (9.51)$$

where the parameters $m_{1,2}$, $c_{1,2}$ and $k_{1,2}$ are masses, damping coefficients and stiffness of the 2DoF-structure as shown in Figure 9.6. Correspondingly, m_D, c_D and k_D are the parameters associated with the semi-active TMD. In Equation 9.50, $\ddot{\mathbf{x}}$, $\dot{\mathbf{x}}$ and $\mathbf{x}$ are the acceleration, velocity and displacements vectors of the 2DoF+TMD system. Furthermore, in the EoM, $\ddot{x}_g$ is the ground acceleration and $\boldsymbol{\nu}_u$ the incidence vector distributing the ground acceleration to the DoFs of the system as

$$\boldsymbol{\nu}_u = \begin{bmatrix} 1 & \cdots & 1 \end{bmatrix}^\top. \qquad (9.52)$$

We derive the discrete state-space representation of the system from the following continuous transition and output equations assuming that the system behavior and its abrupt changes can be reflected by a stepwise linear representation as

$$\dot{\mathbf{z}}^a(t) = \mathcal{A}(t)\mathbf{z}^a(t) + \mathcal{B}\mathbf{u}(t) + \mathbf{w}(t), \tag{9.53}$$

$$\mathbf{y}(t) = \mathcal{C}(t)\mathbf{z}^a(t) + \mathcal{D}\mathbf{u}(t) + \mathbf{v}(t), \tag{9.54}$$

where the time-variant transition matrix $\mathcal{A}(t)$ and the input matrix $\mathcal{B}$ are defined as

$$\mathcal{A}(t) = \begin{bmatrix} \mathbf{0}_{3\times3} & \mathbf{I}_{3\times3} & \mathbf{0}_{3\times2} \\ -\mathbf{M}^{-1}\mathbf{K}(t)_{3\times3} & -\mathbf{M}^{-1}\mathbf{C}_{3\times3} & \mathbf{0}_{3\times2} \\ \mathbf{0}_{2\times3} & \mathbf{0}_{2\times3} & \mathbf{0}_{2\times2} \end{bmatrix}, \quad \mathcal{B} = \begin{bmatrix} \mathbf{0}_{3\times3} \\ \mathbf{I}_{3\times3} \\ \mathbf{0}_{2\times3} \end{bmatrix}. \tag{9.55}$$

In Equation 9.54, the time-variant output matrix $\mathcal{C}(t)$ and the feedthrough matrix $\mathcal{D}$ are written as

$$\mathcal{C}(t) = \begin{bmatrix} -\mathbf{M}^{-1}\mathbf{K}(t)_{3\times3} & -\mathbf{M}^{-1}\mathbf{C}_{3\times3} & \mathbf{0}_{3\times2} \end{bmatrix}, \quad \mathcal{D} = \begin{bmatrix} \mathbf{I}_{3\times3} \end{bmatrix}. \tag{9.56}$$

The augmented state, input and output vectors read

$$\mathbf{z}^a = \begin{bmatrix} x_1 & x_2 & x_D & \dot{x}_1 & \dot{x}_2 & \dot{x}_d & k_1 & k_2 \end{bmatrix}^\top, \tag{9.57}$$

$$\mathbf{y} = \begin{bmatrix} \ddot{x}_1 & \ddot{x}_2 & \ddot{x}_D \end{bmatrix}^\top, \tag{9.58}$$

$$\mathbf{u} = \begin{bmatrix} \ddot{x}_g & \ddot{x}_g & \ddot{x}_g \end{bmatrix}^\top, \tag{9.59}$$

corresponding to the DoFs of the system x_1, x_2 and x_D, the changing structural stiffnesses k_1 and k_2, the sensors monitoring the accelerations of the system with $\ddot{x}_1$, $\ddot{x}_2$ and $\ddot{x}_D$ as well as the ground acceleration $\ddot{x}_g$.

In Equations 9.53 and 9.54, the vectors $\mathbf{w}$ and $\mathbf{v}$ consider the system and measurement noises, respectively.

The discrete transition and input matrices from the continuous state-space representation are derived by employing the TAYLOR expansion of order p as

$$\mathcal{A}_d = e^{\mathcal{A}\Delta t} = \sum_{i=0}^{\infty} \frac{1}{i!}\mathcal{A}^i\Delta t^i \approx \mathbf{I} + \mathcal{A}\Delta t + \cdots + \frac{1}{p!}\mathcal{A}^p\Delta t^p, \tag{9.60}$$

$$\mathcal{B}_d = \int_0^{\Delta t} e^{\mathcal{A}\tau}\mathcal{B}\,\mathrm{d}\tau = \sum_{i=0}^{\infty} \frac{1}{(i+1)!}\mathcal{A}^i\mathcal{B}\Delta t^{i+1} \tag{9.61}$$

$$\approx \mathbf{0} + \mathcal{B}\Delta t + \cdots + \frac{1}{(p+1)!}\mathcal{A}^p\mathcal{B}\Delta t^{p+1},$$

where Δt is the sampling time. Compared with other methods, the explicit EULER method has the order of convergence $p = 1$ and the fourth-order RUNGE-KUTTA method corresponds to $p = 4$.

TABLE 9.1

Parameters of the 2DoF+TMD system.

Parameter		Value	Unit
Mass	m_1, m_2, m_D	1, 1, 0.1	t
Damping	c_1, c_2, c_D	0.1, 0.1, 0.051	kNs m^{-1}
Stiffness	$k_{1,0}, k_{2,0}, k_{D,0}$	12, 10, 0.36	kN m^{-1}

For the time step t_k, we formulate the discrete transition equation using $\mathcal{A}_d$ and $\mathcal{B}_d$ as

$$\mathbf{z}_{k+1}^a = \mathcal{A}_{d,k}\mathbf{z}_k^a + \mathcal{B}_d\mathbf{u}_k + \mathbf{w}_k. \tag{9.62}$$

The output equation (Equation 9.54) does not include any differential terms. Therefore, in the discrete version, the matrices $\mathcal{C}$ and $\mathcal{D}$ do not change and the system output reads

$$\mathbf{y}_k = \mathcal{C}\mathbf{z}_k^a + \mathcal{D}\mathbf{u}_k + \mathbf{v}_k. \tag{9.63}$$

9.6.2   Simulation Parameters and Load Cases

For the computation of the system, the discrete state-space representation of the 2DoF+TMD is implemented together with the proposed observer in Matlab. The solutions are performed in the time-domain. The convergence order is chosen as $p = 3$ with the time step size $\Delta t = 0.02$ s. The masses, damping coefficients and the initial stiffness values of the system are summarized in Table 9.1.

The initial natural frequencies of the frame structure without TMD are equal to $f_1 = 0.33$ Hz and $f_2 = 0.84$ Hz. The corresponding damping ratios are $D_1 = 0.92$ % and $D_2 = 2.49$ %. We assume that the initial stiffness values k_1 and k_2 of the structure drop by 10 % abruptly to $k_1 = 10.8$ kN m^{-1} and $k_2 = 9$ kN m^{-1} due to high interstory drift at corresponding time steps $t_{f,1}$ and $t_{f,2}$. This stiffness change corresponds to a cumulative reduction of the first natural frequency by 5 %, which matches with the real observation as introduced at the beginning of the section and reported by previous studies, such as by Naiem [332].

The parameters of the semi-active TMD are tuned to the first mode of the structure by the criteria of WARBURTON [55]. The mass of the TMD is chosen as 0.1 t. To focus on the identification procedure, during the calculations, the stiffness of the semi-active TMD kept constant as $k_D = k_{D,0}$. Accordingly, the mass ratio μ, the optimum natural frequency $f_{D,opt}$ and the optimum damping ratio $D_{D,opt}$ of the TMD yield

$$\mu = \frac{m_D}{\hat{m}_1} = 7.6\ \%, \quad f_{D,opt} = f_1 \frac{\sqrt{1-\frac{\mu}{2}}}{1+\mu} = 0.30\ \text{Hz},$$

$$D_{D,opt} = \sqrt{\frac{\mu(1-\frac{\mu}{4})}{4(1+\mu)(1-\frac{\mu}{2})}} = 13.42\ \%, \tag{9.64}$$

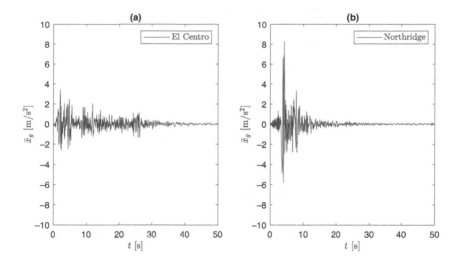

FIGURE 9.7
Time histories of the historic El Centro (a) and Northridge (b) earthquakes used in the numerical studies.

where $\hat{m}_1$ is the modal mass of the structure corresponding to the first mode. The damping c_D and initial stiffness $k_{D,0}$ parameters of the TMD (cf. Table 9.1) are calculated corresponding to m_D, $f_{D,opt}$ and $D_{D,opt}$.

As *seismic ground acceleration*, the records of the two historic California Earthquakes El Centro and Northridge are used. The time histories of the earthquakes are shown in Figure 9.7. The properties of the records are listed below [179].

- *El Centro*: N-S component recorded at the Imperial Valley Irrigation District substation in El Centro, during the Imperial Valley, California Earthquake of 1940. RMS = 0.05 g.

- *Northridge*: N-S component recorded at the County Hospital parking lot in Sylmar, during the Northridge, California Earthquake of 1994. RMS = 0.07 g.

For the input and output measurement noise, a white GAUSSIAN with an RMS value of 0.01 m s^{-2} is added to the output signals, which corresponds approximately to 2 % of the RMS value of the El Centro earthquake record.

During the localization stage of the observer, to facilitate the localization procedure, the corresponding stiffness parameter $\tilde{\Theta}_i$ of each DoF i is reduced manually before the subsequent UKF by 5 %. Further parameters of the observer are presented in Table 9.2. Here, $\hat{P}_0$ is the initial covariance matrix of the state estimate and P_a is the covariance value, which is used for the

TABLE 9.2

Parameters of the joint
state-parameter observer.

Observer parameter	Value
$\hat{\mathbf{P}}_0$	$10^{-6}\mathbf{I}_{8\times 8}$
P_a	1
$\mathbf{Q}$	$10^{-9}\mathbf{I}_{8\times 8}$
$\mathbf{R}$	$10^{-4}\mathbf{I}_{3\times 3}$
α	10^{-3}
β	2
κ	0
γ_0	108

adaptation of the $\tilde{P}_{\Theta\Theta,ii}$ value of the state covariance matrix as described in Section 9.5.

9.6.3 Study 1: Threshold Value

This study investigates numerically the performance of the observer for three different threshold values $\gamma_0 = \{16.2, 39.8, 108\}$ corresponding to $z_{e,0} = \{1.64, 2.58, 3\sqrt{2}\}$ with $\delta = 2$ (only accelerometers) and $m = 3$ (measured outputs: $\ddot{x}_1$, $\ddot{x}_2$ and $\ddot{x}_D$). The probability of exceedance of the threshold is $p = \{10, 1, 0.005\}$ %. All other parameters of both the system and the observer correspond to Tables 9.1 and 9.2.

The El Centro earthquake acceleration signal is applied to the structure. The time history of the computed trigger parameter γ is depicted and compared with different threshold values γ_0 in Figure 9.8 (a). During the ground excitation, the stiffness of the first floor of the structure drops. The stiffness of the second floor remains constant. Figure 9.8 (b) shows the corresponding time history of the real stiffness value k_1 and compares it with the predicted value $\hat{k}_1$, which is calculated by using $\gamma_0 = 108$ for the threshold. We assume that the initial estimate of the structural stiffness values $\hat{k}_1$ and $\hat{k}_2$ are known and correspond to $k_{1,0}$ and $k_{2,0}$ as given previously in Table 9.1.

We observe in Figure 9.8 (a) that during the stiffness drop at time step $t = 9$ s, the trigger parameter has a clear peak of $\gamma = 370$, which is detectable by the threshold $\gamma_0 = 108$. On the other hand, the first two threshold values $\gamma_0 = \{16.2, 39.8\}$ are too low to detect accurately the peak. As shown in Figure 9.8 (b), with the threshold value $\gamma_0 = 108$, the observer is able to track the stiffness drop accurately.

In Figure 9.9, the real displacement responses of the structure without and with the semi-active TMD are compared with the predicted displacement time histories, which are calculated by the observer using the threshold value $\gamma_0 = 108$ for the 2DoF+TMD system.

Figure 9.9 confirms the performance of the observer with the threshold

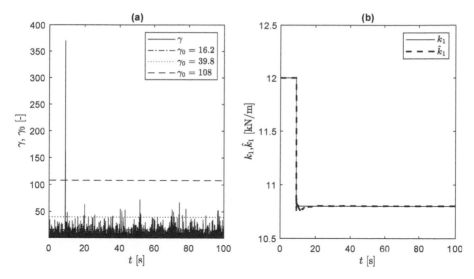

FIGURE 9.8
Time history of the trigger parameter γ computed for the El Centro earthquake compared with different threshold values γ_0 in logarithmic scale (a). Time histories of the real stiffness value of the first floor of the structure k_1 and its predicted value $\hat{k}_1$ (b), which is calculated by using $\gamma_0 = 108$.

$\gamma_0 = 108$. The observer tracks accurately the displacement responses for both without and with the TMD cases. For all further studies, the threshold value will remain as $\gamma = 108$.

The study shows that, when the threshold value γ_0 is chosen properly corresponding to the noise involved in the measurement system, the observer is able to track both the state and the parameters.

9.6.4 Study 2: Localization

This study investigates the accuracy of localization of the observer in case of a sequential stiffness reduction at both DoFs of the structure. In this case, the Northridge earthquake is applied to the structure, which a stronger event compared to the El Centro earthquake and is, therefore, expected to cause more significant damages on the structure. Accordingly, during the excitation, first, the stiffness value k_1 and, then, k_2 drops abruptly.

The time histories of the real stiffness values k_1 and k_2 of the structure are shown in Figure 9.10 and compared with their predicted time histories $\hat{k}_1$ and $\hat{k}_2$. In the same figure, also the displacement time histories are shown for the structure with the TMD at x_1 and x_2 with their corresponding predictions $\hat{x}_1$ and $\hat{x}_2$.

As we observe from Figure 9.10, this study shows that the observer is able

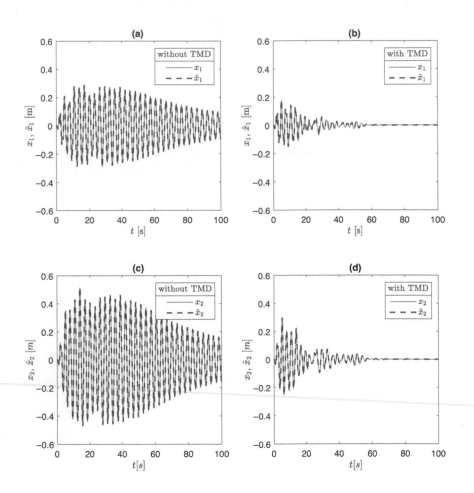

FIGURE 9.9

Time histories of the displacement of the structure without and with the TMD
at its DoFs x_1 (a,b) and x_2 (c,d) during the El Centro earthquake. The real
responses are compared with the predicted displacements $\hat{x}_1$ and $\hat{x}_2$.

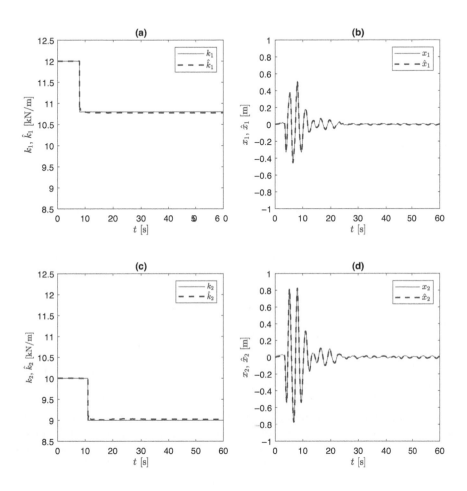

FIGURE 9.10
Time histories of the stiffness k_1 (a), k_2 (c) and displacement x_1 (b), x_2 (d) of the structure with the TMD during the Northridge earthquake compared with the predicted time histories $\hat{k}_1$, $\hat{k}_2$ and $\hat{x}_1$, $\hat{x}_2$.

to localize the stiffness changes and accurately track both the state and the stiffness of the structure accurately also in case of a sequential change in the structure.

9.6.5 Study 3: State Covariance

This study investigates the effects of the state covariance matrix on the accuracy of the observer. A high covariance value is expected to increase the sensitivity of the observer and reduce the required time for its reaction. This might be helpful particularly for the identification of MDoF structures with supplementary control devices. Due to the additional damping, the available time window for the identification of changes of such systems is expected to be shorter. On the other hand, very high state covariance values might cause the observer to react too sensitive and, accordingly, reduce its accuracy.

The observer is tested with different state covariance matrix values of $\hat{\mathbf{P}} = 1 \cdot \mathbf{I}$ to $\hat{\mathbf{P}} = 10^{-8} \cdot \mathbf{I}$. The adaptation capability of the observer is deactivated during this study. Accordingly, the state covariance matrix does not alter during the simulation time, i.e., $P_a = P_0$, where P_0 is the value corresponding to any of the diagonal elements of the state covariance matrix $\hat{\mathbf{P}}$.

The Northridge earthquake is applied to the structure without and with the TMD. The stiffness of the structure is assumed to be constant. However, this time, the initial estimation of the stiffness values are set as $\hat{k}_1 = 14.4$ kN m^{-1} and $\hat{k}_2 = 12$ kN m^{-1}, which are significantly higher than the real values $k_1 = 12$ kN m^{-1} and $k_2 = 10$ kN m^{-1}. We expect from the observer to converge to the true stiffness values. The time span required for the convergence will indicate the accuracy of the observer. All other system and observer parameters correspond to Tables 9.1 and 9.2.

Figure 9.11 shows the real stiffness value $k_1 = 12$ kN m^{-1} of the first floor of the structure without and with the TMD as well as the comparison with the predicted $\hat{k}_1$ values for state covariance values $P_0 = \{1, 10^{-4}, 10^{-8}\}$. We observe that with increasing P_0 the tracking accuracy also increases. With $P_0 = 10^{-8}$, the observer does not succeed to converge to the real k_1 in the shown time window. Particularly, with the TMD the observer seems to require more time confirming the effect of supplementary damping. With $P_0 = 1$, the observer catches the real k_1 value quickly. However, we observe also that due to the increased sensitivity, the tracking becomes unsteady showing some zigzags, which indicate that a further increase will cause the observer to significantly over or underestimate the stiffness of the structure. Therefore, we use this value for P_a, as also previously introduced in Table 9.2 and used in other studies, only after the activation of the adaptation of the UKF. The initial state covariance value remains $P_0 = 10^{-6}$.

Figure 9.12 shows the time histories of the real displacement x_1 of the structure without and with the TMD and compares them with their predictions $\hat{x}_1$ for the state covariance values $P_0 = 10^{-8}$ and $P_0 = 1$. The time histories confirm the previous findings that for $P_0 = 1$ the observer is very

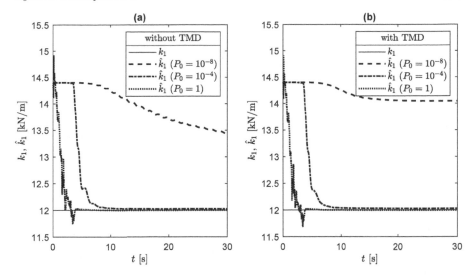

FIGURE 9.11
Time history of the stiffness k_1 of the structure without and with the TMD during the Northridge earthquake compared with the predicted time history $\hat{k}_1$ for different state covariance matrices $\hat{\mathbf{P}}$.

accurate and can track also the state of the structure. For $P_0 = 10^{-8}$, the observer struggles to predict the response of the structure.

The study confirms accordingly the initial expectations and shows a higher accuracy of the observer for higher state covariance values. On the other hand, the observer becomes unsteady at very high covariance values and starts to over- or underestimate the parameters.

9.6.6 Study 4: System Noise Covariance and Discretization Order

This study investigates the effects of modeling on the accuracy of the observer. For this purpose, we consider, as in the previous study, that the initial estimation of the stiffness values are set as $\hat{k}_1 = 14.4$ kN m^{-1} and $\hat{k}_2 = 12$ kN m^{-1}. The adaptation capability of the observer is deactivated. During the study, we alter both the system noise covariance value $\mathbf{Q}$ and the order of model discretization p. Here, the El Centro earthquake is applied to the structure without and with the TMD. All further system and observer parameters correspond to Tables 9.1 and 9.2.

Figure 9.13 shows real stiffness value of the first floor of the structure $k_1 = 12$ kN m^{-1} and compares it with the estimated time histories of $\hat{k}_1$ for different $\mathbf{Q} = \{10^{-14}\mathbf{I}_{8\times8}, 10^{-9}\mathbf{I}_{8\times8}\}$ and $p = \{1, 3, 4\}$. It is noteworthy that, as mentioned previously in Section 9.6.1, the explicit EULER method

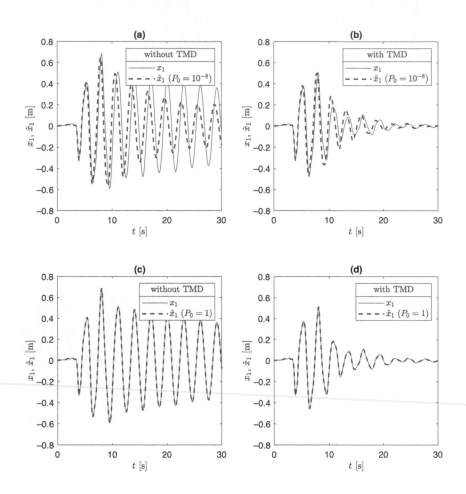

FIGURE 9.12

Time history of the displacement x_1 of the structure without and with the TMD during the Northridge earthquake compared with the predicted time history $\hat{x}_1$ for different state covariance matrices $\hat{\mathbf{P}}$.

corresponds to $p = 1$ and the fourth-order RUNGE-KUTTA method corresponds to $p = 4$. In previous studies, $p = 3$ is used.

In Figure 9.13 (a,b), the noise covariance is $\mathbf{Q} = 10^{-9}\mathbf{I}_{8\times8}$. At $p = 1$, the observer struggles to identify parameters of the structure both without and with the TMD. With the TMD the estimated stiffness converges to a wrong parameter, showing that, at least for the structure with the TMD, this discretization order is not sufficient. With increasing discretization order, the accuracy of the observer increases. At both $p = 3$ and 4, the observer tracks the real stiffness value both without and with the TMD very quickly.

In Figure 9.13 (c,d), the noise covariance is decreased to $\mathbf{Q} = 10^{-14}\mathbf{I}_{8\times8}$. We observe a clear deterioration in performance of the observer both without and with the TMD for $p = 1$. On the other hand, for higher orders both with $p = 3$ and $p = 4$, the performance of the observer is similar to the previous cases with $\mathbf{Q} = 10^{-14}\mathbf{I}_{8\times8}$.

Accordingly, the results of this study confirm that both the discretization order and the system noise covariance are relevant for the performance of the observer. With increasing order, also the accuracy of the observer increases. Furthermore, with increasing system noise covariance, the observer accuracy increases. However, the influence of the discretization order is higher. Consequently, for sufficiently high discretization order, the influence of system noise covariance becomes negligible.

9.7 Conclusion

This chapter is concerned with the methods used for the identification of system state and parameters. A general review of the latest developments is provided. The recursive KF method is described in detail with its extension for nonlinear systems—the UKF method. The governing equations of both methods are derived. For semi-active vibration control systems, a novel adaptive joint state-parameter observer is introduced. The approach is based on an adaptive formulation of the UKF and can identify both the response and the parameters of structures with supplementary damping devices in real-time. A highlight of the method is its capability to operate under abrupt changes, such as structural damage initiated stiffness degradation. The method is applied to a seismically excited MDoF structure with a semi-active TMD. Several numerical studies are performed concerning the estimation of the state and parameters and compared with true values showing the accuracy and performance of the observer.

General conclusions of the studies regarding the system identification are:

- The identification of structures incorporating semi-active vibration control systems imposes challenges due to the supplementary damping effects of the control devices.

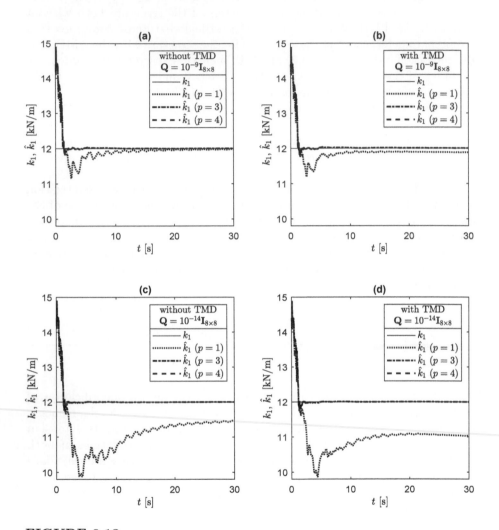

FIGURE 9.13
Time history of the stiffness k_1 of the structure without and with the TMD
during the El Centro earthquake compared with the predicted time history
$\hat{k}_1$ for different system noise covariance values $\mathbf{Q}$ and different order of model
discretization p.

- Abrupt changes in structures may occur due to damages, degradation, soil-structure interaction and temperature. To be able to control a structure, these changes must be identified in real-time.

- Recursive KF-based methods provide a robust identification scheme regarding the above listed challenges. Although, the KF expanded with UT can identify nonlinearities, abrupt changes require an adaptation of the state covariance.

- For this purpose, the structural change must be detected and localized by comparing the measured response of the structure, such as accelerations, with the estimated response and using the proposed formulations.

- System noise, discretization of the mathematical model and chosen filter parameter influence the performance of the observer as shown by the presented studies.

A

MATLAB Codes of Examples

A.1 Example 1.1: Responses of SDoF systems without damping

```
clc;clear;close all;

%% System parameters
omega_1 = 1.2;          % Natural frequency [rad/s]
k       = 1;            % Stiffness [kN/m]

x_0     = 1;            % Initial displacement [m]
dx_0    = 1;            % Initial velocity [m/s]
F_0     = .5;           % Load amplitude [kN]
omega   = 0.1*omega_1;  % Excitation frequency [rad/s]

%% Simulation parameters
t_end = 100;            % Simulation finish time [s]
dt    = 0.01;           % Time increment [s]
t     = 0:dt:t_end;     % Time [s]

eta   = omega/omega_1;  % Frequency ratio [-]

%% Calculation of response
% Transient displacement [m]
x_t = x_0*cos(omega_1*t) + ...
(dx_0/omega_1 - F_0/k*eta/(1-eta^2)) * sin(omega_1*t);

% Steady-state displacement [m]
x_s = F_0/k*1/(1-eta^2)*sin(omega*t);

% Total displacement [m]
x = x_t + x_s;

%% Plotting
plot(t,x_t)
```

```
hold on
plot(t,x_s)
plot(t,x)
xlabel('Time [s]');
ylabel('Displacement [m]');
legend({'Transient','Steady-state','Total'});
```

A.2 Example 1.3: Responses of SDoF systems with damping

```
clc;clear;close all;

%% System parameters
omega_1 = 1.2;          % Natural frequency [rad/s]
k       = 1;            % Stiffness [kN/m]
D       = 0.02;         % Damping ratio [-]

x_0     = 1;            % Initial displacement [m]
dx_0    = 1;            % Initial velocity [m/s]
F_0     = .5;           % Load amplitude [kN]
omega   = 0.1*omega_1;  % Excitation frequency [rad/s]

%% Simulation parameters
t_end = 100;            % Simulation finish time [s]
dt    = 0.01;           % Time increment [s]
t     = 0:dt:t_end;     % Time [s]

eta     = omega/omega_1;        % Frequency ratio [-]
omega_D = omega_1*(1-D^2)^.5;   % Damped frequency [rad/s]

R_d  = 1 / ((1-eta^2)^2+(2*D*eta)^2)^.5;
phi  = atan(2*D*eta/(1-eta^2));
C3   = F_0/k*R_d;
C1   = x_0 + C3*sin(phi);
C2   = 1/(1-D^2)^.5 * (dx_0/omega_1 + ...
C3*D*sin(phi) - C3*eta*cos(phi));

%% Calculation of response
% Transient displacement [m]
x_t = exp(-D*omega_1*t).*...
(C1*cos(omega_D*t)+C2*sin(omega_D*t));
```

```matlab
% Steady-state displacement [m]
x_s = C3*sin(omega*t-phi);

% Total displacement [m]
x = x_t + x_s;

%% Plotting
plot(t,x_t)
hold on
plot(t,x_s)
plot(t,x)
xlabel('Time [s]');
ylabel('Displacement [m]');
legend({'Transient','Steady-state','Total'});
```

A.3   Example 1.6: Newmark's integration method

```matlab
clc;clear;close all;

%% System paremeters
m = 1;                 % Mass [t]
k = 10;                % Stiffness [kN/m]
c = 0.316;             % Damping coefficient [kNs/m]

%% Simulation parameters
t_end = 10;            % Simulation finish time [s]
dt    = 0.01;          % Time increment [s]
t     = 0:dt:t_end;    % Time [s]
n     = length(t);     % Step [-]

%% Load function
F = zeros(1,n);
for i    = 1:round(0.3/dt)+1
    F(i) = t(i)/0.3;
end
for i    = round(0.3/dt)+2:round(0.6/dt)+1
    F(i) = 1-(t(i)-0.3)/0.3;
end

%% Initial conditions
x_0       = 0;         % Initial displacement [m]
x_dot_0   = 0;         % Initial velocity [m/s]
```

```
x(1)      = x_0;
x_dot(1) = x_dot_0;


%% Newmark Parameters
gamma = 0.5;
beta  = 0.25;        % Average acc. (Linear acc.: beta=1/6)

%% Initial calculations
x_ddot(1) = (F(1)-c*x_dot(1)-k*x(1)) / m;
k_hat     = k + gamma/beta/dt*c + 1/beta/(dt)^2*m;
a         = 1/beta/dt*m + gamma/beta*c;
b         = 1/2/beta*m+dt*(gamma/(2*beta)-1)*c;

%% Calculations for each time step
for i=1:n-1
    % Increments
    dF(i)      = F(i+1)-F(i);
    dF_hat(i)  = dF(i) + a*x_dot(i) + b*x_ddot(i);
    dx(i)      = dF_hat(i)/k_hat;
    dx_dot(i)  = gamma/beta/dt*dx(i) - gamma/beta*x_dot(i) ...
        + dt*(1-gamma/(2*beta))*x_ddot(i);
    dx_ddot(i) = 1/beta/(dt)^2*dx(i) - 1/beta/dt*x_dot(i) ...
        - 1/(2*beta)*x_ddot(i);

    % Response
    x(i+1)      = x(i)+dx(i);
    x_dot(i+1)  = x_dot(i)+dx_dot(i);
    x_ddot(i+1) = x_ddot(i)+dx_ddot(i);
end

%% Plotting
% Displacement time history
subplot(3,1,1)
plot(t,x)
xlabel('Time [s]');
ylabel('Displacement [m]');

% Velocity time history
subplot(3,1,2)
plot(t,x_dot)
xlabel('Time [s]');
ylabel('Velocity [m/s]');

% Acceleration time history
subplot(3,1,3)
```

```
plot(t,x_ddot)
xlabel('Time [s]');
ylabel('Acceleration [m/s^2]');
```

A.4 Example 1.7: Central difference method

```
clc;clear;close all;

%% System paremeters
m = 1;                   % Mass [t]
k = 10;                  % Stiffness [kN/m]
c = 0.316;               % Damping coefficient [kNs/m]

%% Simulation parameters
t_end = 10;             % Simulation finish time [s]
dt    = 0.01;           % Time increment [s]
t     = 0:dt:t_end;     % Time [s]
n     = length(t);      % Step [-]

%% Load function
F = zeros(1,n);
for i     = 1:round(0.3/dt)+1
    F(i) = t(i)/0.3;
end
for i     = round(0.3/dt)+2:round(0.6/dt)+1
    F(i) = 1-(t(i)-0.3)/0.3;
end

%% Initial conditions
x_0       = 0;         % Initial displacement [m]
x_dot_0   = 0;         % Initial velocity [m/s]
x(1)      = x_0;
x_dot(1)  = x_dot_0;

%% Initial calculations
x_ddot(1) = (F(1)-c*x_dot(1)-k*x(1)) / m;
x_1       = x(1) - dt*x_dot(1) + dt^2/2*x_ddot(1);   % i=-1
k_hat     = m/dt^2 + c/(2*dt);
a         = m/dt^2 - c/(2*dt);
b         = k - 2*m/dt^2;

%% Calculations for each time step
```

```
for i=1:n-1
    if i==1
        x_ = x_1;
    else
        x_ = x(i-1);
    end

    F_hat(i)  = F(i) - a*x_ - b*x(i);
    x(i+1)    = F_hat(i)/k_hat;
    x_dot(i)  = (x(i+1)-x_)/(2*dt);
    x_ddot(i) = (x(i+1)-2*x(i)+x_)/dt^2;
end

%% Plotting
% Displacement time history
subplot(3,1,1)
plot(t,x)
xlabel('Time [s]');
ylabel('Displacement [m]');

% Velocity time history
t(end)=[];
subplot(3,1,2)
plot(t,x_dot)
xlabel('Time [s]');
ylabel('Velocity [m/s]');

% Acceleration time history
subplot(3,1,3)
plot(t,x_ddot)
xlabel('Time [s]');
ylabel('Acceleration [m/s^2]');
```

A.5   Example 1.9: Modal analysis method

```
clc;clear;close all;

%% System paremeters
m = 1;          % Mass [t]
k = 5;          % Stiffness [kN/m]

K = [2*k -k     % Stiffness matrix
```

```
       -k k];

M = [m 0          % Mass matrix
     0 m];

F = [0            % Load vector amplitudes [kN]
     2];

w = 1;            % Excitation frequency [rad/s]

t = 0:0.1:100;   % Simulation time [s]

%% Solving eigenvalue problem
[Phi,Omega2] = eig(K,M);

Omega = sqrt(Omega2);
Freq  = Omega2^.5/2/pi;

% Simplification of the mode shapes
Phi_1 = Phi(1,1)./Phi(:,1);
Phi_2 = Phi(2,2)./Phi(:,2);


%% Generalized SDoF parameters
m_1 = Phi_1'*M*Phi_1;
m_2 = Phi_2'*M*Phi_2;

k_1 = Phi_1'*K*Phi_1;
k_2 = Phi_2'*K*Phi_2;

F_1 = Phi_1'*F;
F_2 = Phi_2'*F;

%% Steady state solution
X_0_1 = 1/k_1/(1-(w/Omega(1,1))^2);
q_1   = X_0_1* F_1*sin(w*t);

X_0_2 = 1/k_2/(1-(w/Omega(2,2))^2);
q_2   = X_0_2* F_2*sin(w*t);

x_1   = Phi_1.*q_1;
x_2   = Phi_2.*q_2;

x = x_1 +x_2;
```

```
%% Plotting
% Displacement time history
plot(t,x(1,:))
hold on
plot(t,x(2,:))
xlabel('Time [s]');
ylabel('Displacement [m]');
legend({'x1','x2'});
```

A.6   Example 1.11: Construction of the damping matrix

```
clc;clear;close all;

%% System paremeters
m = 1;                  % Mass [t]
k = 5;                  % Stiffness [kN/m]

K = [2*k -k             % Stiffness matrix
-k k];

M = [m 0                % Mass matrix
0 m];

omega_1 = 1.382;        % Natural frequencies
omega_2 = 3.618;        % [rad/s]

%% Measurement parameters
Omega_1 = 1;            % Excitation frequencies
Omega_2 = 5;            % [rad/s]

T_1 = (2*pi)/Omega_1  % Corresponding periods
T_2 = (2*pi)/Omega_2  % [s]

D_1 = 0.01;             % Measured damping ratios
D_2 = 0.03;             % [-]

alpha = 4*pi* (T_1*D_1 - T_2*D_2)...
/(T_1^2 - T_2^2);
beta  = 1/pi* (T_1^2*T_2*D_2-T_2^2*T_1*D_1)...
/(T_1^2 - T_2^2);

C = alpha*M + beta*K; % Damping matrix [kNs/m]
```

```
D_hat_1 = alpha/2 * 1/omega_1 ...      % Modal damping
+ beta/2*omega_1;                      % ratios
D_hat_2 = alpha/2 * 1/omega_2 ...      % [-]
+ beta/2*omega_2;

C_hat = [2*D_hat_1*omega_1*M(1,1) 0   % Modal damping
0 2*D_hat_2*omega_2*M(2,2)];% matrix
```

A.7 Example 1.12: Nonlinear vibrations

```
clc;clear;close all;

%% System parameters
m = 1;                  % Mass [t]
c = 0.316;              % Damping coefficient [kNm/s]

%% Simulation parameters
t_end = 10;             % Simulation finish time [s]
dt    = 0.1;            % Time increment [s]
t     = 0:dt:t_end;     % Time [s]
n     = length(t);      % Step [-]

%% Load function
F = zeros(1,n);
for i     = 1:round(0.3/dt)+1
    F(i) = t(i)/0.3;
end
for i     = round(0.3/dt)+2:round(0.6/dt)+1
    F(i) = 1-(t(i)-0.3)/0.3;
end

%% Initial conditions
x         = zeros(1,n); % Displacement vector [m]
x(1)      = 0;          % Initial displacement [m]
x_dot     = zeros(1,n); % Velocity vector [m/s]
x_dot(1) = 0;           % Initial velocity [m/s]

k_T      = zeros(1,n);  % Tangent stiffness vector [kN/m]
k_T(1) = 10;            % Initial stiffness value [kN/m]
fk       = zeros(1,n);  % Restoring force vector [kN]
fk(1)  = 0;             % Initial restoring force value [kN]
```

```
%% Initial calculations
gamma = 0.5;
beta  = 0.25;          % Average acc. (Linear acc.: beta=1/6)

x_ddot    = zeros(1,n);            % Acceleration vector
                                   % [m/s^2]
x_ddot(1) = (F(1) - c*x_dot(1) ... % Initial acceleration
             -fk(1))/m;            % [m/s^2]

a = m/beta/dt + c*gamma/beta;

b = m/beta/2 + c*dt*(gamma/beta/2-1);

epsilon = 0.001;       % Amount of tolerated error
                       % (Residual force)

%% Calculations for each time step
dF       = zeros(1,n);
dF_hat   = zeros(1,n);
k_T_hat  = zeros(1,n);
dR       = zeros(1,n);
k        = zeros(1,n);  % Number of iterations

for i=1:n-1
    dF(i)       = F(i+1) - F(i);
    dF_hat(i)   = dF(i) + a*x_dot(i) + b*x_ddot(i);
    k_T_hat(i)  = k_T(i) + c*gamma/beta/dt...
                  +m/beta/dt^2;

    x(i+1) = x(i);
    fk_    = fk(i);
    dR(i)  = dF_hat(i);
    dx_t=0;

    while abs(dR(i))>epsilon  % Check error condition
        dx     = dR(i)/k_T_hat(i);
        x(i+1) = x(i+1) + dx;

        if abs(x(i+1)) >= 0.05
            k_T(i+1)= 5;
            fk(i+1) = sign(x(i+1))*(0.5+k_T(i+1)...
                      *(abs(x(i+1))-0.05));
        else
```

```
                k_T(i+1)=10;
                fk(i+1) = k_T(i+1)*x(i+1);
            end

        df      = fk(i+1) - fk_ + a/dt*dx;
        fk_     = fk(i+1);
        dR(i)   = dR(i) - df;
        dx_t = dx_t+dx;  % Total displacement
                         % change of time step
        k(i) = k(i)+1;
    end

    dx_dot  = dx_t*gamma/beta/dt ...   % Velocity change
              -x_dot(i)*gamma/beta ...
              +x_ddot(i)*dt*(1-gamma/2/beta);

    dx_ddot = dx_t/beta/dt^2 ...       % Acceleration change
              -x_dot(i)/beta/dt ...
              -x_ddot(i)/2/beta;

    x_dot(i+1)  = x_dot(i)  + dx_dot;  % Velocity [m/s]
    x_ddot(i+1) = x_ddot(i) + dx_ddot; % Acceleration [m/s^2]
end
```

Bibliography

[1] F. Zhu, J.-T. Wang, F. Jin, and L.-Q. Lu. Real-Time Hybrid Simulation of Full-Scale Tuned Liquid Column Dampers to Control Multi-Order Modal Responses of Structures. *Engineering Structures*, 138:74–90, 2017.

[2] A. K. Chopra. *Dynamics of Structures: Theory and Applications to Earthquake Engineering*. Prentice Hall, Upper Saddle River, NJ, fifth edition, 2017.

[3] R. W. Clough and J. Penzien. *Dynamics of Structures*. Computers & Structures, Inc., Berkeley, CA, third edition, 2003.

[4] J. L. Humar. *Dynamics of Structures*. CRC Press/Taylor & Francis Croup, Boca Raton, FL, third edition, 2012.

[5] D. J. Inman and R. C. Singh. *Engineering Vibration*. Pearson International Edition. Pearson, Boston, MA, fourth edition, 2014.

[6] K. Meskouris. *Structural Dynamics: Models, Methods, Examples*. Ernst & Sohn, Berlin, 2000.

[7] C. W. de Silva. *Vibration – Fundamentals and Practice*. CRC Press, Abingdon, second edition, 2006.

[8] N. M. Newmark. A Method of Computation for Structural Dynamics. *Journal of the Engineering Mechanics Division*, 85(3):67–94, 1959.

[9] T. K. Caughey and M. E. J. O'Kelly. Classical Normal Modes in Damped Linear Dynamic Systems. *ASME Journal of Applied Mechanics*, 32:583–588, 1965.

[10] A. S. Veletsos and C. E. Ventura. Modal Analysis of Non-Classically Damped Linear Systems. *Earthquake Engineering and Structural Dynamics*, 14(2):217–243, 1986.

[11] A. Bayraktar, H. Keypour, and A. Naderzadeh. Application of Ancient Earthquake Resistant Method in Modern Construction Technology. In *Proceedings of the 15th World Conference on Earthquake Engineering*, 2012.

[12] M. M. Masoumi. Ancient Base Isolation System in Mausoleum of Cyrus the Great. *International Journal of Earthquake Engineering and Hazard Mitigation*, 4(1):13–18, 2016.

[13] A. Kareem, T. Kijewski, and Y. Tamura. Mitigation of Motions of Tall Buildings with Specific Examples of Recent Applications. *Wind and Structures*, 2(3):201–251, 1999.

[14] S. F. Masri. Effectiveness of Two-Particle Impact Dampers. *The Journal of the Acoustical Society of America*, 41(6):1553–1554, 1967.

[15] W. H. Reed III. Hanging-Chain Impact Dampers: A Simple Method for Damping Tall Flexible Structures. Technical report, NASA Langley Research Center, Hampton, VA, 1967.

[16] W. J. Nordell. Active Systems for Blast-Resistant Structures. Technical Report R-611, Naval Civil Engineering Laboratory, Port Hueneme, CA, 1969.

[17] J. T. P. Yao. Adaptive Systems for Seismic Structures. In *Proceedings of the NSF-UCEER Earthquake Engineering Research Conference*, pages 142–150, 1969.

[18] J. T. P. Yao. Concept of Structural Control. *ASCE Journal of Structural Division*, 98(7):1567–1574, 1972.

[19] G. W. Housner, L. A. Bergman, T. K. Caughey, A. G. Chassiakos, R. O. Claus, S. F. Masri, R. E. Skelton, T. T. Soong, B. F. Spencer, and J. T. P. Yao. Structural Control: Past, Present, and Future. *Journal of Engineering Mechanics*, 123(9):897–971, 1997.

[20] T. T. Soong and B. F. Spencer. Supplemental Energy Dissipation: State-of-the-Art and State-of-the-Practice. *Engineering Structures*, 24(3):243–259, 2002.

[21] B. F. Spencer and S. Nagarajaiah. State of the Art of Structural Control. *Journal of Structural Engineering*, 129(7):845–856, 2003.

[22] T. T. Soong and M. C. Constantinou, editors. *Passive and Active Structural Vibration Control in Civil Engineering*. Number 345 in Courses and Lectures. Springer, New York, NY, 1994.

[23] M. C. Constantinou, T. T. Soong, and G. F. Dargush. *Passive Energy Dissipation Systems for Structural Design and Retrofit*. Multidisciplinary Center for Earthquake Engineering Research, Buffalo, NY, 1998.

[24] H. Adeli and A. Saleh. *Control, Optimization, and Smart Structures: High-Performance Bridges and Buildings of the Future*. John Wiley & Sons, Inc., New York, NY, 1999.

[25] T. T. Soong and G. F. Dargush. *Passive Energy Dissipation Systems in Structural Engineering*. John Wiley & Sons, Inc., New York, NY, 1999.

[26] R. D. Hanson and T. T. Soong. *Seismic Design with Supplemental Energy Dissipation Devices*. Earthquake Engineering Research Institute, Oakland, CA, 2001.

[27] C. Petersen. *Schwingungsdämpfer im Ingenieurbau*. Maurer Söhne GmbH & Co. KG, Munich, 2001.

[28] J. J. Connor. *Introduction to Structural Motion Control*. Prentice Hall, Upper Saddle River, NJ, 2003.

[29] D. Hrovat, P. Barak, and M. Rabins. Semi-Active Versus Passive or Active Tuned Mass Dampers for Structural Control. *Journal of Engineering Mechanics*, 109(3):691–705, 1983.

[30] F. L. Lewis, D. L. Vrabie, and V. L. Syrmos. *Optimal Control*. John Wiley & Sons, Inc., Hoboken, NJ, third edition, 2012.

[31] L. A. Zadeh. Fuzzy Sets. *Information and Control*, 8(3):338–353, 1965.

[32] J.-S. R. Jang, C.-T. Sun, and E. Mizutani. *Neuro-Fuzzy and Soft Computing: A Computational Approach to Learning and Machine Intelligence*. Prentice Hall, Upper Saddle River, NJ, 1997.

[33] N. H. Siddique and H. Adeli. *Computational Intelligence: Synergies of Fuzzy Logic, Neural Networks, and Evolutionary Computing*. John Wiley & Sons, Inc., Chichester, 2013.

[34] E. H. Mamdani. Application of Fuzzy Algorithms for Control of Simple Dynamic Plant. *Proceedings of the Institution of Electrical Engineers*, 121(12):1585–1588, 1974.

[35] T. Takagi and M. Sugeno. Fuzzy Identification of Systems and Its Applications to Modeling and Control. *IEEE Transactions on Systems, Man, and Cybernetics*, SMC-15(1):116–132, 1985.

[36] Y. Tsukamoto. An Approach to Fuzzy Reasoning Method. In M. M. Gupta, R. K. Ragade, and R. R. Yager, editors, *Advances in Fuzzy Set Theory and Applications*, pages 137–149. North-Holland Publishing Company, Amsterdam, 1979.

[37] F. Naeim and J. M. Kelly. *Design of Seismic Isolated Structures*. John Wiley & Sons, Inc., New York, NY, 1999.

[38] H. Ozdemir. *Nonlinear Transient Dynamic Analysis of Yielding Structures*. Ph.D. Thesis, University of California Berkeley, 1976.

[39] M. A. Bhatti, K. S. Pister, and E. Polak. Optimal Design of an Earthquake Isolation System. Technical Report UCB/EERC-78/22, Earthquake Engineering Research Center, University of California, Berkeley, CA, 1978.

[40] J. M. Kelly, R. I. Skinner, and A. J. Heine. Mechanisms of Energy Absorption in Special Devices for Use in Earthquake Resistant Structures. *Bulletin of NZ Society for Earthquake Engineering*, 5(3):63–88, 1972.

[41] R. I. Skinner, J. M. Kelly, and A. J. Heine. Hysteretic Dampers for Earthquake-Resistant Structures. *Earthquake Engineering and Structural Dynamics*, 3(3):287–296, 1974.

[42] A. S. Pall and C. Marsh. Response of Friction Damped Braced Frames. *Journal of Structural Engineering*, 108(9):1313–1323, 1982.

[43] I. D. Aiken and J. M. Kelly. Earthquake Simulator Testing and Analytical Studies of Two Energy-Absorbing Systems for Multistory Structures. Technical Report UCB/EERC-90-03, University of California, Berkeley, CA, 1990.

[44] P. Mahmoodi. Structural Dampers. *Journal of the Structural Division*, 95(8):1661–1672, 1969.

[45] R.-H. Zhang, T. T. Soong, and P. Mahmoodi. Seismic Response of Steel Frame Structures with Added Viscoelastic Dampers. *Earthquake Engineering and Structural Dynamics*, 18(3):389–396, 1989.

[46] R.-H. Zhang and T. T. Soong. Seismic Design of Viscoelastic Dampers for Structural Applications. *Journal of Structural Engineering*, 118(5):1375–1392, 1992.

[47] M. C. Constantinou, M. D. Symans, P. Tsopelas, and D. P. Taylor. Fluid Viscous Dampers in Applications of Seismic Energy Dissipation and Seismic Isolation. In *Proceedings of the ATC-17-1 Seminar on Seismic Isolation, Passive Energy Dissipation, and Active Control*, volume 2, pages 581–592, 1993.

[48] N. Makris, M. C. Constantinou, and G. F. Dargush. Analytical Model of Viscoelastic Fluid Dampers. *Journal of Structural Engineering*, 119(11):3310–3325, 1993.

[49] D. P. Taylor and M. C. Constantinou. Fluid Dampers for Applications of Seismic Energy Dissipation and Seismic Isolation. In *Proceedings of the 11th World Conference on Earthquake Engineering*, 1996.

[50] A. M. Reinhorn, C. Li, and M. C. Constantinou. Experimental and Analytical Investigation of Seismic Retrofit of Structures with Supplemental Damping: Part 1 – Fluid Viscous Damping Devices. Technical Report 95-0028, State University of New York, Buffalo, NY, 1995.

[51] H. Frahm. Device for Damping Vibrations of Bodies – Patent US989958, 1911.

[52] J. P. Den Hartog. *Mechanical Vibrations*. McGraw-Hill, New York, NY, 1934.

[53] A. J. Clark. Multiple Passive Tuned Mass Dampers for Reducing Earthquake Induced Building Motion. In *Proceedings of the 9th World Conference on Earthquake Engineering*, volume V, 1988.

[54] G. B. Warburton and E. O. Ayorinde. Optimum Absorber Parameters for Simple Systems. *Earthquake Engineering and Structural Dynamics*, 8(3):197–217, 1980.

[55] G. B. Warburton. Optimum Absorber Parameters for Various Combinations of Response and Excitation Parameters. *Earthquake Engineering and Structural Dynamics*, 10(3):381–401, 1982.

[56] Q. S. Li, L.-H. Zhi, A. Y. Tuan, C.-S. Kao, S.-C. Su, and C.-F. Wu. Dynamic Behavior of Taipei 101 Tower: Field Measurement and Numerical Analysis. *Journal of Structural Engineering*, 137(1):143–155, 2011.

[57] T. Nagase. Earthquake Records Observed in Tall Buildings with Tuned Pendulum Mass Damper. In *Proceedings of the 12th World Conference on Earthquake Engineering*, Auckland, New Zealand, 2000.

[58] Y. Tamura, K. Fujii, T. Ohtsuki, T. Wakahara, and R. Kohsaka. Effectiveness of Tuned Liquid Dampers Under Wind Excitation. *Engineering Structures*, 17(9):609–621, 1995.

[59] H. F. Bauer. Oscillations of Immiscible Liquids in a Rectangular Container: A New Damper for Excited Structures. *Journal of Sound and Vibration*, 93(1):117–133, 1984.

[60] Y. Fujino, B. M. Pacheco, P. Chaiseri, and L. M. Sun. Parametric Studies on Tuned Liquid Damper (TLD) Using Circular Containers by Free-Oscillation Experiments. *Structural Engineering and Earthquake Engineering*, 1988(398):177–187, 1988.

[61] F. Welt and V. J. Modi. Vibration Damping Through Liquid Sloshing, Part I: A Nonlinear Analysis. *Journal of Vibration and Acoustics*, 114(1):10–16, 1992.

[62] L. M. Sun, Y. Fujino, B. M. Pacheco, and M. Isobe. Nonlinear Waves and Dynamic Pressures in Rectangular Tuned Liquid Damper (TLD). *Structural Engineering and Earthquake Engineering*, 1989(410):81–92, 1989.

[63] E. W. Graham and A. M. Rodriguez. The Characteristics of Fuel Motion Which Affect Airplane Dynamics. *Journal of Applied Mechanics*, 19:381–388, 1952.

[64] Raouf A Ibrahim. *Liquid Sloshing Dynamics: Theory and Applications*. Cambridge University Press, Cambridge, 2005.

[65] L. M. Sun, Y. Fujino, B. M. Pacheco, and P. Chaiseri. Modelling of Tuned Liquid Damper (TLD). *Journal of Wind Engineering and Industrial Aerodynamics*, 43(1):1883–1894, 1992.

[66] H. N. Abramson. The Dynamic Behavior of Liquids in Moving Containers – with Applications to Space Vehicle Technology. Technical Report SP-106, NASA, 1966.

[67] T. Konar and A. D. Ghosh. Flow Damping Devices in Tuned Liquid Damper for Structural Vibration Control: A Review. *Archives of Computational Methods in Engineering*, June 2020.

[68] H. Frahm. Means for Damping the Rolling Motion of Ships – Patent US970368, 1910.

[69] F. Sakai and S. Takaeda. Tuned Liquid Column Damper – New Type Device for Suppression of Building Vibrations. In *Proceedings of 1st International Conference on High-Rise Buildings*, pages 926–931, 1989.

[70] F. Sakai, T. T. Takaeda, and T. Tamaki. Damping Device for Tower-Like Structure – Patent US5070663, 1991.

[71] P. Irwin and B. Breukelman. Recent Applications of Damping Systems for Wind Response. In *Proceedings of the 6th World Congress of the Council on Tall Buildings and Urban Habitat*, 2001.

[72] A. Tamboli and C. Christoforou. Manhattans Mixed Construction Skyscrapers with Tuned Liquid and Mass Dampers. In *Proceedings of the 7th World Congress of the Council on Tall Buildings and Urban Habitat*, 2005.

[73] M. Gutierrez Soto and H. Adeli. Tuned Mass Dampers. *Archives of Computational Methods in Engineering*, 20(4):419–431, 2013.

[74] T. Balendra, C. M. Wang, and H. F. Cheong. Effectiveness of Tuned Liquid Column Dampers for Vibration Control of Towers. *Engineering Structures*, 17(9):668–675, 1995.

[75] T. Balendra, C. M. Wang, and G. Rakesh. Vibration Control of Various Types of Buildings Using TLCD. *Journal of Wind Engineering and Industrial Aerodynamics*, 83:197–208, 1999.

[76] C. C. Chang and C. T. Hsu. Control Performance of Liquid Column Vibration Absorbers. *Engineering Structures*, 20(7):580–586, 1998.

[77] H. Gao, K. C. S. Kwok, and B. Samali. Optimization of Tuned Liquid Column Dampers. *Engineering Structures*, 19(6):476–486, 1997.

[78] P. A. Hitchcock, K. C. K. Kwok, R. D. Watkins, and B. Samali. Characteristics of Liquid Column Vibration Absorbers (LCVA) – I. *Engineering Structures*, 19(2):126–134, 1997.

[79] P. A. Hitchcock, K. C. S. Kwok, R. D. Watkins, and B. Samali. Characteristics of Liquid Column Vibration Absorbers (LCVA) – II. *Engineering Structures*, 19(2):135–144, 1997.

[80] T. T. Soong. *Active Structural Control: Theory and Practice*. Longman Wiley, London, England, 1990.

[81] J. Rodellar, F. Ikhouane, F. Pozo, G. Pujol, L. Acho, and J. M. Rossell. The Art of Control Algorithms Design and Implementation. *Advances in Science and Technology*, 56:154–163, 2008.

[82] N. R. Fisco and H. Adeli. Smart Structures: Part I – Active and Semi-Active Control. *Scientia Iranica*, 18(3):275–284, 2011.

[83] F. Casciati, J. Rodellar, and U. Yildirim. Active and Semi-Active Control of Structures – Theory and Applications: A Review of Recent Advances. *Journal of Intelligent Material Systems and Structures*, 23(11):1181–1195, 2012.

[84] Y. Ikeda, K. Sasaki, M. Sakamoto, and T. Kobori. Active mass driver system as the first application of active structural control. *Earthquake Engineering and Structural Dynamics*, 30(11):1575–1595, 2001.

[85] M. Yamamoto, S. Aizawa, M. Higashino, and K. Toyama. Practical Applications of Active Mass Dampers with Hydraulic Actuator. *Earthquake Engineering and Structural Dynamics*, 30(11):1697–1717, 2001.

[86] A. Nishitani and Y. Inoue. Overview of the Application of Active/Semiactive Control to Building Structures in Japan. *Earthquake Engineering and Structural Dynamics*, 30(11):1565–1574, 2001.

[87] J. C. Wu and J. N. Yang. Active Control of Transmission Tower under Stochastic Wind. *Journal of Structural Engineering*, 124(11):1302–1312, 1998.

[88] H. Cao, A. M. Reinhorn, and T. T. Soong. Design of an Active Mass Damper for a Tall TV Tower in Nanjing, China. *Engineering Structures*, 20(3):134–143, 1998.

[89] F. Bossens and A. Preumont. Active Tendon Control of Cable-Stayed Bridges: A Large-Scale Demonstration. *Earthquake Engineering and Structural Dynamics*, 30(7):961–979, 2001.

[90] J. Rodellar, V. Mañosa, and C. Monroy. An Active Tendon Control Scheme for Cable-Stayed Bridges with Model Uncertainties and Seismic Excitation. *Journal of Structural Control*, 9(1):75–94, 2002.

[91] K. Tanida, M. Masao, Y. Koike, T. Murata, T. Kobori, K. Ishii, Y. Takenaka, and T. Arita. Development of V-Shaped Hybrid Mass Damper and Its Application to High-Rise Buildings. In *Proceedings of the 1st World Conference on Structural Control*, volume 3, pages 249–255, 1994.

[92] S. Yamazaki, N. Nagata, and H. Abiru. Tuned Active Dampers Installed in the Minato Mirai (MM) 21 Landmark Tower in Yokohama. *Journal of Wind Engineering and Industrial Aerodynamics*, 43(1):1937–1948, 1992.

[93] N. R. Fisco and H. Adeli. Smart Structures: Part II – Hybrid Control Systems and Control Strategies. *Scientia Iranica*, 18(3):285–295, 2011.

[94] M. Abe. Semi-Active Tuned Mass Dampers for Seismic Protection of Civil Structures. *Earthquake Engineering and Structural Dynamics*, 25(7):743–749, 1996.

[95] M. Setareh. Application of Semi-Active Tuned Mass Dampers to Base-Excited Systems. *Earthquake Engineering and Structural Dynamics*, 30(3):449–462, 2001.

[96] R. Yang, X. Zhou, and X. Liu. Seismic Structural Control Using Semi-Active Tuned Mass Dampers. *Earthquake Engineering and Engineering Vibration*, 1(1):111–118, 2002.

[97] J. Kang, H.-S. Kim, and D.-G. Lee. Mitigation of Wind Response of a Tall Building Using Semi-Active Tuned Mass Dampers. *The Structural Design of Tall and Special Buildings*, 20(5):552–565, 2011.

[98] N. Varadarajan and S. Nagarajaiah. Wind Response Control of Building with Variable Stiffness Tuned Mass Damper Using Empirical Mode Decomposition/Hilbert Transform. *Journal of Engineering Mechanics*, 130(4):451–458, 2004.

[99] S. N. Deshmukh and N. K. Chandiramani. LQR Control of Wind Excited Benchmark Building Using Variable Stiffness Tuned Mass Damper. *Shock and Vibration*, 2014, 2014.

[100] Z. Wang, H. Gao, H. Wang, and Z. Chen. Development of Stiffness-Adjustable Tuned Mass Dampers for Frequency Retuning. *Advances in Structural Engineering*, 22(2):473–485, 2019.

[101] S. Nagarajaiah. Adaptive Passive, Semiactive, Smart Tuned Mass Dampers: Identification and Control Using Empirical Mode Decomposition, Hilbert Transform, and Short-Term Fourier Transform. *Structural Control and Health Monitoring*, 16(7-8):800–841, 2009.

[102] C. Sun and S. Nagarajaiah. Study on Semi-Active Tuned Mass Damper with Variable Damping and Stiffness Under Seismic Excitations. *Structural Control and Health Monitoring*, 21(6):890–906, 2014.

[103] K. Karami, S. Manie, K. Ghafouri, and S. Nagarajaiah. Nonlinear Structural Control Using Integrated DDA/ISMP and Semi-Active Tuned Mass Damper. *Engineering Structures*, 181:589–604, 2019.

[104] S. K. Yalla and A. Kareem. Semiactive Tuned Liquid Column Dampers: Experimental Study. *Journal of Structural Engineering*, 129(7):960–971, 2003.

[105] V. D. La and C. Adam. General On-Off Damping Controller for Semi-Active Tuned Liquid Column Damper. *Journal of Vibration and Control*, 24(23):5487–5501, 2018.

[106] M. Matsuo. U-Shaped Tank Type Dynamic Vibration Absorbing Device – Patent JP09151986, 1997.

[107] T. Nomichi and H. Yoshida. Vibration-Proof Structure in Long-span Construction – Patent JP01239205, 1989.

[108] M. Yoshimura and K. Yamazaki. Vibration Damping Device for Vertical Vibration – Patent JP10220522, 1998.

[109] K. Kagawa and K. Fujita. Vibration Isolating Tank – Patent JP02278033, 1990.

[110] M. J. Hochrainer and F. Ziegler. Control of Tall Building Vibrations by Sealed Tuned Liquid Column Dampers. *Structural Control and Health Monitoring*, 13(6):980–1002, 2006.

[111] M. Reiterer and F. Ziegler. Control of Pedestrian-Induced Vibrations of Long-Span Bridges. *Structural Control and Health Monitoring*, 13(6):1003–1027, 2006.

[112] C. Fu. Application of Torsional Tuned Liquid Column Gas Damper for Plan-Asymmetric Buildings. *Structural Control and Health Monitoring*, 18(5):492–509, 2011.

[113] S. A. Mousavi, K. Bargi, and S. M. Zahrai. Optimum Parameters of Tuned Liquid Column–Gas Damper for Mitigation of Seismic-Induced Vibrations of Offshore Jacket Platforms. *Structural Control and Health Monitoring*, 20(3):422–444, 2013.

[114] M. Reiterer and A. Kluibenschedl. Liquid Damper for Reducing Vertical and/or Horizontal Vibrations in a Building or Machine Structure – Patent US20100200348, 2010.

[115] M. Yoshimura, K. Fujita, and A. Teramura. Hydrostatic Anti-Vibration System and Adjusting Method Therefor – Patent US5542220, 1996.

[116] A. Teramura and O. Yoshida. Development of Vibration Control System Using U-Shaped Water Tank. In *Proceedings of the 11th World Conference on Earthquake Engineering*, 1996.

[117] A. Ghosh and B. Basu. Seismic Vibration Control of Short Period Structures Using the Liquid Column Damper. *Engineering Structures*, 26(13):1905–1913, 2004.

[118] H. Kim and H. Adeli. Hybrid Control of Smart Structures Using a Novel Wavelet-Based Algorithm. *Computer-Aided Civil and Infrastructure Engineering*, 20(1):7–22, 2005.

[119] E. Sonmez, S. Nagarajaiah, C. Sun, and B. Basu. A Study on Semi-Active Tuned Liquid Column Dampers (STLCDs) for Structural Response Reduction Under Random Excitations. *Journal of Sound and Vibration*, 362:1–15, 2016.

[120] K.-W. Min, C.-S. Park, and J. Kim. Easy-to-Tune Reconfigurable Liquid Column Vibration Absorbers with Multiple Cells. *Smart Materials and Structures*, 24(6):065041, 2015.

[121] O. Altay. *Liquid Damper for the Control of Periodic and Random Vibrations of High-Rise Structures*. PhD thesis, RWTH Aachen University, 2013.

[122] O. Altay and S. Klinkel. A Semi-Active Tuned Liquid Column Damper for Lateral Vibration Control of High-Rise Structures: Theory and Experimental Verification. *Structural Control and Health Monitoring*, 25(12):e2270, 2018.

[123] M. Q. Feng. Application of Hybrid Sliding Isolation System to Buildings. *Journal of Engineering Mechanics*, 119(10):2090–2108, 1993.

[124] S. Narasimhan and S. Nagarajaiah. Smart Base Isolated Buildings with Variable Friction Systems: H∞ Controller and SAIVF Device. *Earthquake Engineering and Structural Dynamics*, 35(8):921–942, 2006.

[125] Y. L. Xu and C. L. Ng. Seismic Protection of a Building Complex Using Variable Friction Damper: Experimental Investigation. *Journal of Engineering Mechanics*, 134(8):637–649, 2008.

[126] Y. Peng and T. Huang. Sliding Implant-Magnetic Bearing for Adaptive Seismic Mitigation of Base-Isolated Structures. *Structural Control and Health Monitoring*, 26(10):e2431, 2019.

[127] E. Polak, G. Meeker, K. Yamada, and N. Kurata. Evaluation of an Active Variable-Damping Structure. *Earthquake Engineering and Structural Dynamics*, 23(11):1259–1274, 1994.

[128] J. N. Yang, J. C. Wu, K. Kawashima, and S. Unjoh. Hybrid Control of Seismic-Excited Bridge Structures. *Earthquake Engineering and Structural Dynamics*, 24(11):1437–1451, 1995.

[129] C.-H. Loh and M.-J. Ma. Control of Seismically Excited Building Structures Using Variable Damper Systems. *Engineering Structures*, 18(4):279–287, 1996.

[130] M. D. Symans and M. C. Constantinou. Seismic Testing of a Building Structure with a Semi-Active Fluid Damper Control System. *Earthquake Engineering and Structural Dynamics*, 26(7):759–777, 1997.

[131] M. D. Symans and M. C. Constantinou. Experimental Testing and Analytical Modeling of Semi-Active Fluid Dampers for Seismic Protection. *Journal of Intelligent Material Systems and Structures*, 8(8):644–657, 1997.

[132] S. J. Dyke, B. F. Spencer, M. K. Sain, and J. D. Carlson. Modeling and Control of Magnetorheological Dampers for Seismic Response Reduction. *Smart Materials and Structures*, 5(5):565, 1996.

[133] B. F. Spencer, S. J. Dyke, M. K. Sain, and J. D. Carlson. Phenomenological Model for Magnetorheological Dampers. *Journal of Engineering Mechanics*, 123(3):230–238, 1997.

[134] S. J. Dyke, B. F. Spencer, M. K. Sain, and J. D. Carlson. An Experimental Study of MR Dampers for Seismic Protection. *Smart Materials and Structures*, 7(5):693–703, 1998.

[135] L. M. Jansen and S. J. Dyke. Semiactive Control Strategies for MR Dampers: Comparative Study. *Journal of Engineering Mechanics*, 126(8):795–803, 2000.

[136] S.-B. Choi, S.-K. Lee, and Y.-P. Park. A Hysteresis Model for the Field-Dependent Damping Force of a Magnetorheological Damper. *Journal of Sound and Vibration*, 245(2):375–383, 2001.

[137] J. C. Ramallo, E. A. Johnson, and B. F. Spencer. "Smart" Base Isolation Systems. *Journal of Engineering Mechanics*, 128(10):1088–1100, 2002.

[138] G. Yang, B. F. Spencer, J. D. Carlson, and M. K. Sain. Large-Scale MR Fluid Dampers: Modeling and Dynamic Performance Considerations. *Engineering Structures*, 24(3):309–323, 2002.

[139] Y.-K. Wen. Method for Random Vibration of Hysteretic Systems. *Journal of the Engineering Mechanics Division*, 102(2):249–263, 1976.

[140] P. Y. Lin, L. L. Chung, and C. H. Loh. Semiactive Control of Building Structures with Semiactive Tuned Mass Damper. *Computer-Aided Civil and Infrastructure Engineering*, 20(1):35–51, 2005.

[141] M. Setareh, J. K. Ritchey, T. M. Murray, J.-H. Koo, and M. Ahmadian. Semiactive Tuned Mass Damper for Floor Vibration Control. *Journal of Structural Engineering*, 133(2):242–250, 2007.

[142] J. Y. Wang, Y. Q. Ni, J. M. Ko, and B. F. Spencer. Magneto-Rheological Tuned Liquid Column Dampers (MR-TLCDs) for Vibration Mitigation of Tall Buildings: Modelling and Analysis of Open-Loop Control. *Computers and Structures*, 83(25):2023–2034, 2005.

[143] S. Sarkar and A. Chakraborty. Optimal Design of Semiactive MR-TLCD for Along-Wind Vibration Control of Horizontal Axis Wind Turbine Tower. *Structural Control and Health Monitoring*, 25(2):e2083, 2018.

[144] T. Kobori, M. Takahashi, T. Nasu, N. Niwa, and K. Ogasawara. Seismic Response Controlled Structure with Active Variable Stiffness System. *Earthquake Engineering and Structural Dynamics*, 22(11):925–941, 1993.

[145] D. C. Nemir, Y. Lin, and R. A. Osegueda. Semiactive Motion Control Using Variable Stiffness. *Journal of Structural Engineering*, 120(4):1291–1306, 1994.

[146] K. Yamada and T. Kobori. Control Algorithm for Estimating Future Responses of Active Variable Stiffness Structure. *Earthquake Engineering and Structural Dynamics*, 24(8):1085–1099, 1995.

[147] W. N. Patten, C. Mo, J. Kuehn, and J. Lee. A Primer on Design of Semiactive Vibration Absorbers (SAVA). *Journal of Engineering Mechanics*, 124(1):61–68, 1998.

[148] N. Kurata, T. Kobori, M. Takahashi, N. Niwa, and H. Midorikawa. Actual Seismic Response Controlled Building with Semi-Active Damper System. *Earthquake Engineering and Structural Dynamics*, 28(11):1427–1447, 1999.

[149] F. Jabbari and J. E. Bobrow. Vibration Suppression with Resettable Device. *Journal of Engineering Mechanics*, 128(9):916–924, 2002.

[150] A. K. Agrawal, J. N. Yang, and W. L. He. Applications of Some Semi-active Control Systems to Benchmark Cable-Stayed Bridge. *Journal of Structural Engineering*, 129(7):884–894, 2003.

[151] H. Kurino, J. Tagami, K. Shimizu, and T. Kobori. Switching Oil Damper with Built-in Controller for Structural Control. *Journal of Structural Engineering*, 129(7):895–904, 2003.

[152] A. Nishitani, Y. Nitta, and Y. Ikeda. Semiactive Structural-Control Based on Variable Slip-Force Level Dampers. *Journal of Structural Engineering*, 129(7):933–940, 2003.

[153] A. Fukukita, T. Saito, and K. Shiba. Control Effect for 20-Story Benchmark Building Using Passive or Semiactive Device. *Journal of Engineering Mechanics*, 130(4):430–436, 2004.

[154] M. K. Bhardwaj and T. K. Datta. Semiactive Fuzzy Control of the Seismic Response of Building Frames. *Journal of Structural Engineering*, 132(5):791–799, 2006.

[155] J. N. Yang, J. Bobrow, F. Jabbari, J. Leavitt, C. P. Cheng, and P. Y. Lin. Full-Scale Experimental Verification of Resettable Semi-Active Stiffness Dampers. *Earthquake Engineering and Structural Dynamics*, 36(9):1255–1273, 2007.

[156] M. D. Symans and M. C. Constantinou. Semi-Active Control Systems for Seismic Protection of Structures: A State-of-the-Art Review. *Engineering Structures*, 21(6):469–487, 1999.

[157] F. Casciati, G. Magonette, and F. Marazzi. *Technology of Semiactive Devices and Applications in Vibration Mitigation*. John Wiley & Sons, Inc., Chichester, 2006.

[158] F. Y. Cheng, H. Jiang, and K. Lou. *Smart Structures: Innovative Systems for Seismic Response Control*. CRC Press/Taylor & Francis Group, Boca Raton, FL, 2008.

[159] Z.-D. Xu, X.-Q. Guo, J.-T. Zhu, and F.-H. Xu. *Intelligent Vibration Control in Civil Engineering Structures*. Intelligent Systems Series. Elsevier AP, London, 2016.

[160] Satish Nagarajaiah. Structural Vibration Damper with Continuously Variable Stiffness – Patent US6098969A, 2000.

[161] Chao Sun. Mitigation of Offshore Wind Turbine Responses Under Wind and Wave Loading: Considering Soil Effects and Damage. *Structural Control and Health Monitoring*, 25:e2117, 2017.

[162] C. Sun. Semi-Active Control of Monopile Offshore Wind Turbines Under Multi-Hazards. *Mechanical Systems and Signal Processing*, 99:285–305, January 2018.

[163] D. D. A. Csupor. Passive Stabilization Tanks – Patent US3678877, 1972.

[164] F. Ziegler. *Technische Mechanik der festen und flüssigen Körper: 101 Aufgaben mit Lösungen.* Springer, Vienna, third edition, 1998.

[165] S. D. Xue, J. M. Ko, and Y. L. Xu. Tuned Liquid Column Damper for Suppressing Pitching Motion of Structures. *Engineering Structures*, 22(11):1538–1551, 2000.

[166] Y. L. Xu, B. Samali, and K. C. S. Kwok. Control of Along-Wind Response of Structures by Mass and Liquid Dampers. *Journal of Engineering Mechanics*, 118(1):20–39, 1992.

[167] V. D. La and C. Adam. General On-Off Damping Controller for Semi-Active Tuned Liquid Column Damper. *Journal of Vibration and Control*, May 2016.

[168] X. T. Zhang, R. C. Zhang, and Y. L. Xu. Analysis on Control of Flow-Induced Vibration by Tuned Liquid Damper with Crossed Tube-Like Containers. *Journal of Wind Engineering and Industrial Aerodynamics*, 50:351–360, 1993.

[169] H. Ding, J.-T. Wang, L.-Q. Lu, and F. Zhu. A Toroidal Tuned Liquid Column Damper for Multidirectional Ground Motion-Induced Vibration Control. *Structural Control and Health Monitoring*, 27(8):e2558, 2020.

[170] L. Rozas, R. L. Boroschek, A. Tamburrino, and M. Rojas. A Bidirectional Tuned Liquid Column Damper for Reducing the Seismic Response of Buildings. *Structural Control and Health Monitoring*, 23(4):621–640, 2016.

[171] X. Tong, X. Zhao, and A. Karcanias. Passive Vibration Control of an Offshore Floating Hydrostatic Wind Turbine Model. *Wind Energy*, 21(9):697–714, 2018.

[172] C. Coudurier, O. Lepreux, and N. Petit. Modelling of a Tuned Liquid Multi-Column Damper. Application to Floating Wind Turbine for Improved Robustness Against Wave Incidence. *Ocean Engineering*, 165:277–292, 2018.

[173] B. Mehrkian and O. Altay. Mathematical Modeling and Optimization Scheme for Omnidirectional Tuned Liquid Column Dampers. *Journal of Sound and Vibration*, 484:115523, 2020.

[174] A. A. Taflanidis, D. C. Angelides, and G. C. Manos. Optimal Design and Performance of Liquid Column Mass Dampers for Rotational Vibration Control of Structures Under White Noise Excitation. *Engineering Structures*, 27(4):524–534, 2005.

[175] C. Coudurier, O. Lepreux, and N. Petit. Passive and Semi-Active Control of an Offshore Floating Wind Turbine Using a Tuned Liquid Column Damper. *IFAC-PapersOnLine*, 48(16):241–247, 2015.

[176] I. E. Idelchik. *Handbook of Hydraulic Resistance*. Begell House, Redding, CT, fourth edition, 2008.

[177] Y. A. Cengel and J. M. Cimbala. *Fluid Mechanics: Fundamentals and Applications*. McGraw-Hill Education, New York, NY, fourth edition, 2018.

[178] J. R. Dormand and P. J. Prince. A Family of Embedded Runge-Kutta Formulae. *Journal of Computational and Applied Mathematics*, 6(1):19–26, 1980.

[179] Y. Ohtori, R. E. Christenson, B. F. Spencer, and S. J. Dyke. Benchmark Control Problems for Seismically Excited Nonlinear Buildings. *Journal of Engineering Mechanics*, 130(4):366–385, 2004.

[180] I. D. Aiken, D. K. Nims, and J. M. Kelly. Comparative Study of Four Passive Energy Dissipation Systems. *Bulletin of the New Zealand Society for Earthquake Engineering*, 25(3):175–192, 1992.

[181] M. Dolce, D. Cardone, and R. Marnetto. Implementation and Testing of Passive Control Devices Based on Shape Memory Alloys. *Earthquake Engineering and Structural Dynamics*, 29(7):945–68, 2000.

[182] M. Indirli and M. G. Castellano. Shape Memory Alloy Devices for the Structural Improvement of Masonry Heritage Structures. *International Journal of Architectural Heritage*, 2(2):93–119, 2008.

[183] Y. Zhang and S. Zhu. A Shape Memory Alloy-Based Reusable Hysteretic Damper for Seismic Hazard Mitigation. *Smart Materials and Structures*, 16(5):1603–13, 2007.

[184] H. Li, C.-X. Mao, and J.-P. Ou. Experimental and Theoretical Study on Two Types of Shape Memory Alloy Devices. *Earthquake Engineering and Structural Dynamics*, 37(3):407–26, 2008.

[185] R. L. Boroschek, G. Farias, O. Moroni, and M. Sarrazin. Effect of SMA Braces in a Steel Frame Building. *Journal of Earthquake Engineering*, 11(3):326–42, 2007.

[186] O. E. Ozbulut, P. N. Roschke, P. Y. Lin, and C. H. Loh. GA-Based Optimum Design of a Shape Memory Alloy Device for Seismic Response Mitigation. *Smart Materials and Structures*, 19(6):065004, 2010.

[187] M. Dolce, D. Cardone, F. C. Ponzo, and C. Valente. Shaking Table Tests on Reinforced Concrete Frames without and with Passive Control Systems. *Earthquake Engineering and Structural Dynamics*, 34(14):1687–717, 2005.

[188] M. Dolce, D. Cardone, and F. C. Ponzo. Shaking-Table Tests on Reinforced Concrete Frames with Different Isolation Systems. *Earthquake Engineering and Structural Dynamics*, 36(5):573–96, 2007.

[189] R. Johnson, J. E. Padgett, M. Emmanuel Maragakis, R. DesRoches, and M. Saiid Saiidi. Large Scale Testing of Nitinol Shape Memory Alloy Devices for Retrofitting of Bridges. *Smart Materials and Structures*, 17(3):035018, 2008.

[190] H.-N. Li, Z. Huang, X. Fu, and G. Li. A Re-Centering Deformation-Amplified Shape Memory Alloy Damper for Mitigating Seismic Response of Building Structures. *Structural Control and Health Monitoring*, 25(9):e2233, 2018.

[191] C. Qiu and S. Zhu. Shake Table Test and Numerical Study of Self-Centering Steel Frame with SMA Braces. *Earthquake Engineering and Structural Dynamics*, 46(1):117–37, 2017.

[192] C. Fang, Y. Zheng, J. Chen, M. C. H. Yam, and W. Wang. Superelastic NiTi SMA Cables: Thermal-Mechanical Behavior, Hysteretic Modelling and Seismic Application. *Engineering Structures*, 183:533–49, 2019.

[193] A. S. Issa and M. S. Alam. Experimental and Numerical Study on the Seismic Performance of a Self-Centering Bracing System Using Closed-Loop Dynamic (CLD) Testing. *Engineering Structures*, 195:144–58, 2019.

[194] V. Torra, F. Martorell, F. C. Lovey, and M. L. Sade. Civil Engineering Applications: Specific Properties of NiTi Thick Wires and Their Damping Capabilities, a Review. *Shape Memory and Superelasticity*, 3(4):403–13, 2017.

[195] S. Casciati, V. Torra, and M. Vece. Local Effects Induced by Dynamic Load Self-Heating in NiTi Wires of Shape Memory Alloys. *Structural Control and Health Monitoring*, 25(4):e2134, 2018.

[196] O. E. Ozbulut, S. Hurlebaus, and R. Desroches. Seismic Response Control Using Shape Memory Alloys: A Review. *Journal of Intelligent Material Systems and Structures*, 22(14):1531–49, 2011.

[197] C. Fang and W. Wang. *Shape Memory Alloys for Seismic Resilience.* Springer, Singapore, 2020.

[198] F. Falk. Model Free Energy, Mechanics, and Thermodynamics of Shape Memory Alloys. *Acta Metallurgica*, 28(12):1773–1780, 1980.

[199] S. M. Foiles and M. S. Daw. Application of the Embedded Atom Method to Ni3Al. *Journal of Materials Research*, 2(1):5–15, 1987.

[200] L. Delaey, J. Ortin, and J. Van Humbeeck. Hysteresis Effects in Martensitic Non-Ferrous Alloys. In *Proceedings of the Phase Transformations'87*, pages 60–66, 1987.

[201] S. Govindjee and G. J. Hall. A Computational Model for Shape Memory Alloys. *International Journal of Solids and Structures*, 37(5):735–760, 2000.

[202] M. Brocca, L. C. Brinson, and Z. P. Bazant. Three-Dimensional Constitutive Model for Shape Memory Alloys Based on Microplane Model. *Journal of the Mechanics and Physics of Solids*, 50(5):1051–1077, 2002.

[203] K. Tanaka and S. Nagaki. A Thermomechanical Description of Materials with Internal Variables in the Process of Phase Transitions. *Ingenieur-Archiv*, 51(5):287–299, 1982.

[204] C. Liang and C. A. Rogers. One-Dimensional Thermomechanical Constitutive Relations for Shape Memory Materials. *Journal of Intelligent Material Systems and Structures*, 8(4):285–302, 1990.

[205] B. Raniecki, C. Lexcellent, and K. Tanaka. Thermodynamic Models of Pseudoelastic Behaviour of Shape Memory Alloys. *Archives of Mechanics-Archiwum Mechaniki Stosowanej*, 44(3):261–284, 1992.

[206] B. Raniecki and C. Lexcellent. RL-Models of Pseudoelasticity and Their Specification for Some Shape Memory Solids. *European Journal of Mechanics: A. Solids*, 13(1):21–50, 1994.

[207] B. Raniecki and C. Lexcellent. Thermodynamics of Isotropic Pseudoelasticity in Shape Memory Alloys. *European Journal of Mechanics: A. Solids*, 17(2):185–205, 1998.

[208] J. G. Boyd and D. C. Lagoudas. A Thermodynamical Constitutive Model for Shape Memory Materials. Part I. The Monolithic Shape Memory Alloy. *International Journal of Plasticity*, 12(6):805–42, 1996.

[209] S. Leclercq and C. Lexcellent. A General Macroscopic Description of the Thermomechanical Behavior of Shape Memory Alloys. *Journal of the Mechanics and Physics of Solids*, 44(6):953–980, 1996.

[210] F. Auricchio, R. L. Taylor, and J. Lubliner. Shape-Memory Alloys: Macromodelling and Numerical Simulations of the Superelastic Behavior. *Computer Methods in Applied Mechanics and Engineering*, 146(3-4):281–312, 1997.

[211] A. C. Souza, E. N. Mamiya, and N. Zouain. Three-Dimensional Model for Solids Undergoing Stress-Induced Phase Transformations. *European Journal of Mechanics: A. Solids*, 17(5):789–806, 1998.

[212] S. Reese and D. Christ. Finite Deformation Pseudo-Elasticity of Shape Memory Alloys – Constitutive Modelling and Finite Element Implementation. *International Journal of Plasticity*, 24(3):455–482, 2008.

[213] F. Auricchio and E. Bonetti. A New "Flexible" 3D Macroscopic Model for Shape Memory Alloys. *Discrete & Continuous Dynamical Systems – S*, 6(2):277, 2013.

[214] Benedikt Kohlhaas and Sven Klinkel. An FE^2 model for the analysis of shape memory alloy fiber-composites. *Computational Mechanics*, 55(2):421–437, 2015.

[215] M. Praster, M. Klassen, and S. Klinkel. An adaptive FE^2 approach for fiber–matrix composites. *Computational Mechanics*, 63(6):1333–1350, 2019.

[216] C. Cisse, W. Zaki, and T. Ben Zineb. A Review of Constitutive Models and Modeling Techniques for Shape Memory Alloys. *International Journal of Plasticity*, 76:244–84, 2016.

[217] E. J. Graesser and F. A. Cozzarelli. Shape-Memory Alloys as New Materials for Aseismic Isolation. *Journal of Engineering Mechanics*, 117(11):2590–2608, 1991.

[218] L. C. Brinson. One-Dimensional Constitutive Behavior of Shape Memory Alloys: Thermomechanical Derivation with Non-Constant Material Functions and Redefined Martensite Internal Variable. *Journal of Intelligent Material Systems and Structures*, 4(2):229–242, 1993.

[219] L. C. Brinson and R. Lammering. Finite Element Analysis of the Behavior of Shape Memory Alloys and Their Applications. *International Journal of Solids and Structures*, 30(23):3261–3280, 1993.

[220] K. Wilde, P. Gardoni, and Y. Fujino. Base Isolation System with Shape Memory Alloy Device for Elevated Highway Bridges. *Engineering Structures*, 22(3):222–229, 2000.

[221] W. Ren, H. Li, and G. Song. A One-Dimensional Strain-Rate-Dependent Constitutive Model for Superelastic Shape Memory Alloys. *Smart Materials and Structures*, 16(1):191–197, 2007.

[222] F. Auricchio, D. Fugazza, and R. Desroches. Rate-Dependent Thermo-Mechanical Modelling of Superelastic Shape-Memory Alloys for Seismic Applications. *Journal of Intelligent Material Systems and Structures*, 19(1):47–61, 2008.

[223] S. Zhu and Y. Zhang. A Thermomechanical Constitutive Model for Superelastic SMA Wire with Strain-Rate Dependence. *Smart Materials and Structures*, 16(5):1696–707, 2007.

[224] J. A. Shaw and S. Kyriakides. On the Nucleation and Propagation of Phase Transformation Fronts in a NiTi Alloy. *Acta Materialia*, 45(2):683–700, 1997.

[225] Y. Xiao, P. Zeng, L. Lei, and H. Du. Experimental Investigation on Rate Dependence of Thermomechanical Response in Superelastic NiTi Shape Memory Alloy. *Journal of Materials Engineering and Performance*, 24(10):3755–3760, 2015.

[226] C. Tatar and Z. Yildirim. Phase Transformation Kinetics and Microstructure of NiTi Shape Memory Alloy: Effect of Hydrostatic Pressure. *Bulletin of Materials Science*, 40(4):799–803, 2017.

[227] F. Auricchio and E. Sacco. Thermo-Mechanical Modelling of a Superelastic Shape-Memory Wire Under Cyclic Stretching–Bending Loadings. *International Journal of Solids and Structures*, 38(34-35):6123–45, 2001.

[228] J. Lubliner and F. Auricchio. Generalized Plasticity and Shape-Memory Alloys. *International Journal of Solids and Structures*, 33(7):991–1003, 1996.

[229] F. Auricchio and E. Sacco. A One-Dimensional Model for Superelastic Shape-Memory Alloys with Different Elastic Properties Between Austenite and Martensite. *International Journal of Non-Linear Mechanics*, 32(6):1101–14, 1997.

[230] T. Ikeda, F. A. Nae, H. Naito, and Y. Matsuzaki. Constitutive Model of Shape Memory Alloys for Unidirectional Loading Considering Inner Hysteresis Loops. *Smart Materials and Structures*, 13(4):916, 2004.

[231] H. Funakubo. *Shape Memory Alloys*. Precision Machinery and Robotics. Gordon and Breach Science Publishers, New York, NY, 1987.

[232] B. Raniecki and O. Bruhns. Thermodynamic Reference Model for Elastic-Plastic Solids Undergoing Phase Transformations. *Archives of Mechanics*, 43(2-3):343–376, 1991.

[233] J. Lemaitre and J.-L. Chaboche. *Mechanics of Solid Materials*. Cambridge University Press, Cambridge, 1994.

[234] J. Shaw. Thermomechanical Aspects of NiTi. *Journal of the Mechanics and Physics of Solids*, 43(8):1243–1281, 1995.

[235] A. Kaup, O. Altay, and S. Klinkel. Macroscopic Modeling of Strain-Rate Dependent Energy Dissipation of Superelastic SMA Dampers Considering Destabilization of Martensitic Lattice. *Smart Materials and Structures*, 29(2):025005, 2020.

[236] Y. Sato and K. Tanaka. Estimation of Energy Dissipation in Alloys Due to Stress-Induced Martensitic Transformation. *Res Mechanica*, 23(4):381–393, 1988.

[237] A. Sadjadpour and K. Bhattacharya. A Micromechanics Inspired Constitutive Model for Shape-Memory Alloys: The One-Dimensional Case. *Smart Materials and Structures*, 16(1):S51–S62, 2007.

[238] K. Tanaka. A Thermomechanical Sketch of Shape Memory Effect: One-Dimensional Tensile Behavior. *Res Mechanica*, 18:251–263, 1986.

[239] A. Kaup, H. Ding, J.-T. Wang, and O. Altay. Strain Rate Dependent Formulation of the Latent Heat Evolution of Superelastic Shape Memory Alloy Wires Incorporated in Multistory Frame Structures. *Journal of Intelligent Material Systems and Structures*, 2020.

[240] G. W. Housner. Analysis of the Taft Accelerogram of the Earthquake of 21 July 1952. Technical Report 26483, California Institute of Technology, Los Angeles, CA, 1953.

[241] Y. Gui, J.-T. Wang, F. Jin, C. Chen, and M.-X. Zhou. Development of a Family of Explicit Algorithms for Structural Dynamics with Unconditional Stability. *Nonlinear Dynamics*, 77(4):1157–1170, 2014.

[242] H. Mizuno, M. Sugimoto, T. Mori, M. Iiba, and T. Hirade. Dynamic Behaviour of Pile Foundation in Liquefaction Process – Shaking Table Tests Utilising Big Shear Box. In *Proceedings of the 12th World Conference on Earthquake Engineering*, 2000.

[243] M. Hakuno, M. Shidawara, and T. Hara. Dynamic Destructive Test of a Cantilever Beam, Controlled by an Analog-Computer. *Proceedings of the Japan Society of Civil Engineers*, 1969(171):1–9, 1969.

[244] K. Takanashi, K. Udagawa, M. Seki, T. Okada, and H. Tanaka. Non-Linear Earthquake Response Analysis of Structures by a Computer-Actuator on-Line System: Part 1 Detail of the System. *Transactions of the Architectural Institute of Japan*, 229, 1975.

[245] M. Nakashima, H. Kato, and E. Takaoka. Development of Real-Time Pseudo Dynamic Testing. *Earthquake Engineering and Structural Dynamics*, 21(1):79–92, 1992.

[246] M. Nakashima. Hybrid Simulation: An Early History. *Earthquake Engineering and Structural Dynamics*, 2020.

[247] M. Nakashima and N. Masaoka. Real-Time On-Line Test for MDOF Systems. *Earthquake Engineering and Structural Dynamics*, 28(4):393–420, 1999.

[248] T. Karavasilis, C.-Y. Seo, and J. Ricles. HybridFEM: A Program for Dynamic Time History Analysis of 2D Inelastic Framed Structures and Real-Time Hybrid Simulation. Technical Report 08-09, Lehigh University, 2008.

[249] C. Chen and J. M. Ricles. Large-Scale Real-Time Hybrid Simulation Involving Multiple Experimental Substructures and Adaptive Actuator Delay Compensation. *Earthquake Engineering and Structural Dynamics*, 41(3):549–569, 2012.

[250] Y. Chae, K. Kazemibidokhti, and J. M. Ricles. Adaptive Time Series Compensator for Delay Compensation of Servo-Hydraulic Actuator Systems for Real-Time Hybrid Simulation. *Earthquake Engineering and Structural Dynamics*, 42(11):1697–1715, 2013.

[251] M.-X. Zhou, J.-T. Wang, F. Jin, Y. Gui, and F. Zhu. Real-Time Dynamic Hybrid Testing Coupling Finite Element and Shaking Table. *Journal of Earthquake Engineering*, 18(4):637–653, 2014.

[252] F. Zhu, J.-T. Wang, F. Jin, M.-X. Zhou, and Y. Gui. Simulation of Large-Scale Numerical Substructure in Real-Time Dynamic Hybrid Testing. *Earthquake Engineering and Engineering Vibration*, 13(4):599–609, 2014.

[253] P. A. Bonnet, M. S. Williams, and A. Blakeborough. Evaluation of Numerical Time-Integration Schemes for Real-Time Hybrid Testing. *Earthquake Engineering and Structural Dynamics*, 37(13):1467–1490, 2008.

[254] A. Igarashi, H. Iemura, and T. Suwa. Development of Substructured Shaking Table Test Method. In *Proceedings of the 12th World Conference on Earthquake Engineering*, 2000.

[255] S.-Y. Chang. Explicit Pseudodynamic Algorithm with Unconditional Stability. *Journal of Engineering Mechanics*, 128(9):935–947, 2002.

[256] C. Chen and J. M. Ricles. Development of Direct Integration Algorithms for Structural Dynamics Using Discrete Control Theory. *Journal of Engineering Mechanics*, 134(8):676–683, 2008.

[257] F. Zhu, J.-T. Wang, F. Jin, and Y. Gui. Comparison of Explicit Integration Algorithms for Real-Time Hybrid Simulation. *Bulletin of Earthquake Engineering*, 14(1):89–114, 2016.

[258] P. A. Bonnet, M. S. Williams, and A. Blakeborough. Compensation of Actuator Dynamics in Real-Time Hybrid Tests. *Journal of Systems and Control Engineering*, 221(2):251–264, 2007.

[259] S. Mazzoni, F. McKenna, M. H. Scott, and G. L. Fenves. OpenSees Command Language Manual. Technical Report 264, Pacific Earthquake Engineering Research (PEER) Center, University of California, Berkeley, 2006.

[260] E. Kausel. Early History of Soil–Structure Interaction. *Soil Dynamics and Earthquake Engineering*, 30(9):822–832, 2010.

[261] M. Lou, H. Wang, X. Chen, and Y. Zhai. Structure–Soil–Structure Interaction: Literature Review. *Soil Dynamics and Earthquake Engineering*, 31(12):1724–1731, 2011.

[262] C. C. Mitropoulou, C. Kostopanagiotis, M. Kopanos, D. Ioakim, and N. D. Lagaros. Influence of Soil–Structure Interaction on Fragility Assessment of Building Structures. *Structures*, 6:85–98, 2016.

[263] V. Anand and S. R. Satish Kumar. Seismic Soil–Structure Interaction: A State-of-the-Art Review. *Structures*, 16:317–326, 2018.

[264] H. Ding, A. Kaup, J.-T. Wang, L.-Q. Lu, and O. Altay. Seismic Assessment of Shape Memory Alloy Dampers Considering Soil-Structure Interaction by RTHS. In *Proceedings of the 17th World Conference on Earthquake Engineering*, Sendai, Japan, 2020.

[265] J. Liu and B. Li. A Unified Viscous-Spring Artificial Boundary for 3-D Static and Dynamic Applications. *Science in China Series E Engineering & Materials Science*, 48(5):570–584, 2005.

[266] Q. Wang, J.-T. Wang, F. Jin, F.-D. Chi, and C.-H. Zhang. Real-Time Dynamic Hybrid Testing for Soil–Structure Interaction Analysis. *Soil Dynamics and Earthquake Engineering*, 31(12):1690–1702, 2011.

[267] L.-Q. Lu, J.-T. Wang, and F. Zhu. Improvement of Real-Time Hybrid Simulation Using Parallel Finite-Element Program. *Journal of Earthquake Engineering*, pages 1–19, 2018.

[268] M. I. Wallace, D. J. Wagg, and S. A. Neild. An Adaptive Polynomial Based Forward Prediction Algorithm for Multi-Actuator Real-Time Dynamic Substructuring. *Proceedings of the Royal Society A: Mathematical, Physical and Engineering Sciences*, 461(2064):3807–3826, 2005.

[269] T. Söderström and P. Stoica. *System Identification*. Prentice Hall International Series in Systems and Control Engineering. Prentice Hall, New York, NY, 1989.

[270] R. Brincker and C. E. Ventura. *Introduction to Operational Modal Analysis: Brincker/Introduction to Operational Modal Analysis.* John Wiley & Sons, Inc., Chichester, 2015.

[271] V. Zabel. *Operational Modal Analysis.* Habilitation, Bauhaus-Universität Weimar, 2019.

[272] S. Schleiter, O. Altay, and S: Klinkel. Experimental Incremental System Identification Method Using Separate Time Windows on Basis of Ambient Signals. In *Proceedings of the Experimental Vibration Analysis for Civil Structures*, volume 5, pages 694–704, San Diego, CA, 2018. Springer.

[273] A. W. Smyth, S. F. Masri, A. G. Chassiakos, and T. K. Caughey. On-Line Parametric Identification of MDOF Nonlinear Hysteretic Systems. *Journal of Engineering Mechanics*, 125(2):133–142, 1999.

[274] C.-H. Loh, C.-Y. Lin, and C.-C. Huang. Time Domain Identification of Frames Under Earthquake Loadings. *Journal of Engineering Mechanics*, 126(7):693–703, 2000.

[275] J.-W. Lin, R. Betti, A. W. Smyth, and R. W. Longman. On-Line Identification of Non-Linear Hysteretic Structural Systems Using a Variable Trace Approach. *Earthquake Engineering and Structural Dynamics*, 30(9):1279–1303, 2001.

[276] A. W. Smyth, S. F. Masri, E. B. Kosmatopoulos, A. G. Chassiakos, and T. K. Caughey. Development of Adaptive Modeling Techniques for Non-Linear Hysteretic Systems. *International Journal of Non-Linear Mechanics*, 37(8):1435–1451, 2002.

[277] J. N. Yang and S. Lin. On-Line Identification of Non-Linear Hysteretic Structures Using an Adaptive Tracking Technique. *International Journal of Non-Linear Mechanics*, 39(9):1481–1491, 2004.

[278] J. N. Yang and S. Lin. Identification of Parametric Variations of Structures Based on Least Squares Estimation and Adaptive Tracking Technique. *Journal of Engineering Mechanics*, 131(3):290–298, 2005.

[279] I. Yoshida. Damage Detection Using Monte Carlo Filter Based on Non-Gaussian Noises. In *Proceedings of the 8th International Conference on Structural Safety and Reliability*, 2001.

[280] M. S. Arulampalam, S. Maskell, N. Gordon, and T. Clapp. A Tutorial on Particle Filters for Online Nonlinear/Non-Gaussian Bayesian Tracking. *Proceedings of the IEEE Transactions on Signal Processing*, 50(2):174–188, 2002.

[281] I. Yoshida and T. Sato. Health Monitoring Algorithm by the Monte Carlo Filter Based on Non-Gaussian Noise. *Journal of Natural Disaster Science*, 24(2):101–107, 2002.

[282] O Maruyama and M Hoshiya. Nonlinear Filters Using Monte Carlo Integration for Conditional Random Fields. In *Proceedings 9th International Conference on Applications of Statistics and Probability in Civil Engineering*, 2003.

[283] S. J. Li, Y. Suzuki, and M. Noori. Identification of Hysteretic Systems with Slip Using Bootstrap Filter. *Mechanical Systems and Signal Processing*, 18(4):781–795, 2004.

[284] S. J. Li, Y. Suzuki, and M. Noori. Improvement of Parameter Estimation for Non-Linear Hysteretic Systems with Slip by a Fast Bayesian Bootstrap Filter. *International Journal of Non-Linear Mechanics*, 39(9):1435–1445, 2004.

[285] Y. Tanaka and T. Sato. Efficient System Identification Algorithm Using Monte Carlo Filter and Its Application. In *Proceedings SPIE, Health Monitoring and Smart Nondestructive Evaluation of Structural and Biological Systems III*, volume 5394, pages 464–474. International Society for Optics and Photonics, 2004.

[286] T. Chen, J. Morris, and E. Martin. Particle Filters for State and Parameter Estimation in Batch Processes. *Journal of Process Control*, 15(6):665–673, 2005.

[287] J. Ching, J. L. Beck, and K. A. Porter. Bayesian State and Parameter Estimation of Uncertain Dynamical Systems. *Probabilistic Engineering Mechanics*, 21(1):81–96, 2006.

[288] J. Ching, J. L. Beck, K. A. Porter, and R. Shaikhutdinov. Bayesian State Estimation Method for Nonlinear Systems and Its Application to Recorded Seismic Response. *Journal of Engineering Mechanics*, 132(4):396–410, 2006.

[289] E. N. Chatzi and A. W. Smyth. The Unscented Kalman Filter and Particle Filter Methods for Nonlinear Structural System Identification with Non-Collocated Heterogeneous Sensing. *Structural Control and Health Monitoring*, 16(1):99–123, 2009.

[290] R. E. Kalman. A New Approach to Linear Filtering and Prediction Problems. *Journal of Basic Engineering*, 82(1):35–45, 1960.

[291] C.-B. Yun and M. Shinozuka. Identification of Nonlinear Structural Dynamic Systems. *Journal of Structural Mechanics*, 8(2):187–203, 1980.

[292] M. Hoshiya and E. Saito. Structural Identification by Extended Kalman Filter. *Journal of Engineering Mechanics*, 110(12):1757–1770, 1984.

[293] C.-H. Loh and Y.-H. Tsaur. Time Domain Estimation of Structural Parameters. *Engineering Structures*, 10(2):95–105, 1988.

[294] M. Hoshiya and A. Sutoh. Extended Kalman Filter-Weighted Local Iteration Method for Dynamic Structural Identification. In *Proceedings of the 10th World Conference on Earthquake Engineering*, pages 3715–3720, 1992.

[295] C. G. Koh and L. M. See. Identification and Uncertainty Estimation of Structural Parameters. *Journal of Engineering Mechanics*, 120(6):1219–1236, 1994.

[296] L. Jeen-Shang and Z. Yigong. Nonlinear Structural Identification Using Extended Kalman Filter. *Computers and Structures*, 52(4):757–764, 1994.

[297] C.-H. Loh and I.-C. Tou. A System Identification Approach to the Detection of Changes in Both Linear and Non-Linear Structural Parameters. *Earthquake Engineering and Structural Dynamics*, 24(1):85–97, 1995.

[298] R. Ghanem and M. Shinozuka. Structural-System Identification I: Theory. *Journal of Engineering Mechanics*, 121(2):255–264, 1995.

[299] M. Shinozuka and R. Ghanem. Structural System Identification II: Experimental Verification. *Journal of Engineering Mechanics*, 121(2):265–273, 1995.

[300] D. Wang and A. Haldar. System Identification with Limited Observations and Without Input. *Journal of Engineering Mechanics*, 123(5):504–511, 1997.

[301] T. Sato, R. Honda, and T. Sakanoue. Application of Adaptive Kalman Filter to Identity a Five Story Frame Structure Using NCREE Experimental Data. In *Proceedings of the 8th International Conference on Structural Safety and Reliability*, 2001.

[302] H. Zhang, G. C. Foliente, Y. Yang, and F. Ma. Parameter Identification of Inelastic Structures Under Dynamic Loads. *Earthquake Engineering and Structural Dynamics*, 31(5):1113–1130, 2002.

[303] A. Corigliano and S. Mariani. Parameter Identification in Explicit Structural Dynamics: Performance of the Extended Kalman Filter. *Computer Methods in Applied Mechanics and Engineering*, 193(36–38):3807–3835, 2004.

[304] J. N. Yang, S. Lin, and L. Zhou. Identification of Parametric Changes for Civil Engineering Structures Using an Adaptive Kalman Filter. In *Proceedings Smart Structures and Materials 2004: Sensors and Smart Structures Technologies for Civil, Mechanical, and Aerospace Systems,*

volume 5391, pages 389–399. International Society for Optics and Photonics, 2004.

[305] S. Mariani and A. Corigliano. Impact Induced Composite Delamination: State and Parameter Identification Via Joint and Dual Extended Kalman Filters. *Computer Methods in Applied Mechanics and Engineering*, 194(50–52):5242–5272, 2005.

[306] J. N. Yang, S. Lin, H. Huang, and L. Zhou. An Adaptive Extended Kalman Filter for Structural Damage Identification. *Structural Control and Health Monitoring*, 13(4):849–867, 2006.

[307] O. Maruyama, M. Shinozuka, and M. K. Daigaku. Program EXKAL2 for Identification of Structural Dynamic Systems. Technical Report NCEER-89-0014, National Center for Earthquake Engineering Research, Buffalo, NY, 1989.

[308] S. Mariani and A. Ghisi. Unscented Kalman Filtering for Nonlinear Structural Dynamics. *Nonlinear Dynamics*, 49(1–2):131–150, 2007.

[309] M. Wu and A. W. Smyth. Application of the Unscented Kalman Filter for Real-Time Nonlinear Structural System Identification. *Structural Control and Health Monitoring*, 14(7):971–990, 2007.

[310] S. J. Julier, J. K. Uhlmann, and H. F. Durrant-Whyte. A New Approach for Filtering Nonlinear Systems. In *Proceedings of the American Control Conference*, volume 3, pages 1628–1632, Seattle, WA, 1995.

[311] S. J. Julier and J. K. Uhlmann. New Extension of the Kalman Filter to Nonlinear Systems. In *Proceedings of the SPIE, Signal Processing, Sensor Fusion, and Target Recognition VI*, volume 3068, pages 182–194, 1997.

[312] S. J. Julier, J. K. Uhlmann, and H. F. Durrant-Whyte. A New Method for the Nonlinear Transformation of Means and Covariances in Filters and Estimators. *IEEE Transactions on Automatic Control*, 45(3):477–482, 2000.

[313] S. J. Julier and J. K. Uhlmann. Unscented Filtering and Nonlinear Estimation. In *Proceedings of the IEEE*, volume 92, pages 401–422, 2004.

[314] E. A. Wan and R. Van Der Merwe. The Unscented Kalman Filter for Nonlinear Estimation. In *Proceedings of the IEEE 2000 Adaptive Systems for Signal Processing, Communications, and Control Symposium*, pages 153–158, 2000.

[315] J. R. Van Zandt. A More Robust Unscented Transform. In *Proceedings of the SPIE, Signal and Data Processing of Small Targets*, volume 4473, pages 371–380. International Society for Optics and Photonics, 2001.

[316] D. Tenne and T. Singh. The Higher Order Unscented Filter. In *Proceedings of the American Control Conference*, volume 3, pages 2441–2446. IEEE, 2003.

[317] A. Sitz, U. Schwarz, J. Kurths, and H. U. Voss. Estimation of Parameters and Unobserved Components for Nonlinear Systems from Noisy Time Series. *Physical Review*, E66(1):016210, 2002.

[318] C. Popescu and Y. Wong. The Unscented and Extended Kalman Filter for Systems with Polynomial Restoring Forces. In *Proceedings of the 44th AIAA/ASME/ASCE/AHS/ASC Structures, Structural Dynamics, and Materials Conference*, 2003.

[319] H. U. Voss, J. Timmer, and J. Kurths. Nonlinear Dynamical System Identification from Uncertain and Indirect Measurements. *International Journal of Bifurcation and Chaos*, 14(06):1905–1933, 2004.

[320] M. S. Miah, E. N. Chatzi, and F. Weber. Semi-Active Control for Vibration Mitigation of Structural Systems Incorporating Uncertainties. *Smart Materials and Structures*, 24(5):055016, 2015.

[321] M. S. Miah, E. N. Chatzi, V. K. Dertimanis, and F. Weber. Real-Time Experimental Validation of a Novel Semi-Active Control Scheme for Vibration Mitigation. *Structural Control and Health Monitoring*, 24(3), 2017.

[322] V. K. Dertimanis and E. N. Chatzi. LQR-UKF Active Comfort Control of Passenger Vehicles with Uncertain Dynamics. *IFAC-PapersOnLine*, 51(15):120–125, 2018.

[323] R. Van Der Merwe and E. A. Wan. Efficient Derivative-Free Kalman Filters for Online Learning. In *Proceedings of the European Symposium on Artificial Neural Networks*, pages 205–210. CiteSeerX, 2001.

[324] R. Van Der Merwe, E. Wan, and S. Julier. Sigma-Point Kalman Filters for Nonlinear Estimation and Sensor-Fusion: Applications to Integrated Navigation. In *Proceedings of the AIAA Guidance, Navigation, and Control Conference and Exhibit*, 2004.

[325] L. Chen, S. Seereeram, and R. K. Mehra. Unscented Kalman Filter for Multiple Spacecraft Formation Flying. In *Proceedings of the American Control Conference*, volume 2, pages 1752–1757, Denver, CO, 2003. IEEE.

[326] J. L. Crassidis and F. L. Markley. Unscented Filtering for Spacecraft Attitude Estimation. *Journal of Guidance, Control, and Dynamics*, 26(4):536–542, 2003.

[327] J. Choi, M. Bouchard, T. H. Yeap, and O. Kwon. A Derivative-Free Kalman Filter for Parameter Estimation of Recurrent Neural Networks and Its Applications to Nonlinear Channel Equalization. In *Proceedings of the 4th Int. ICSC Symposium on Engineering of Intelligent Systems*, pages 1–7. CiteSeerX, 2004.

[328] J. Choi, A. C. C. Lima, and S. Haykin. Kalman Filter-Trained Recurrent Neural Equalizers for Time-Varying Channels. *IEEE Transactions on Communications*, 53(3):472–480, 2005.

[329] D.-J. Lee and K. Alfriend. Adaptive Sigma Point Filtering for State and Parameter Estimation. In *Proceedings of the AIAA/AAS Astrodynamics Specialist Conference and Exhibit*, 2004.

[330] A. Romanenko and J. A. Castro. The Unscented Filter as an Alternative to the EKF for Nonlinear State Estimation: A Simulation Case Study. *Computers & Chemical Engineering*, 28(3):347–355, 2004.

[331] P. Li, T. Zhang, and B. Ma. Unscented Kalman Filter for Visual Curve Tracking. *Image and Vision Computing*, 22(2):157–164, 2004.

[332] F. Naeim. Performance of Extensively Instrumented Buildings During the January 17, 1994 Northridge Earthquake. Technical Report 97-7530.68, Mines and Geology, John A. Martin and Associates, Inc., Los Angeles, CA, 1997.

[333] S. S. Haykin. *Kalman Filtering and Neural Networks*. John Wiley & Sons, Inc., New York, NY, 2001.

[334] S. S. Bisht and M. P. Singh. An Adaptive Unscented Kalman Filter for Tracking Sudden Stiffness Changes. *Mechanical Systems and Signal Processing*, 49(1-2):181–195, 2014.

[335] Y. Bar-Shalom, X. R. Li, and T. Kirubarajan. *Estimation with Applications to Tracking and Navigation: Theory Algorithms and Software*. John Wiley & Sons, Inc., New York, NY, 2004.

[336] A. Rahimi, K. D. Kumar, and H. Alighanbari. Fault Estimation of Satellite Reaction Wheels Using Covariance Based Adaptive Unscented Kalman Filter. *Acta Astronautica*, 134:159–169, 2017.

[337] S. Schleiter and O. Altay. Identification of Abrupt Stiffness Changes of Structures with Tuned Mass Dampers under Sudden Events. *Structural Control and Health Monitoring*, 27(6):e2530, 2020.

Index